COLORING G
ANATOM
PHYSIOLOGY

COLORING GUIDE to ANATOMY & PHYSIOLOGY

JUDITH A. STONE
Suffolk County Community College

ROBERT J. STONE
Suffolk County Community College

Boston, Massachusetts   Burr Ridge, Illinios   Dubuque, Iowa
Madison, Wisconsin   New York, New York   San Francisco, California   St. Louis, Missouri

WCB/McGraw-Hill

A Division of The McGraw-Hill Companies

Book Team

Editor *Colin H. Wheatley*
Developmental Editor *Kristine Noel*
Production Editor *Julie L. Wilde*
Designer *K. Wayne Harms*
Visuals/Design Developmental Coordinator *Donna Slade*

Vice President and General Manager *Beverly Kolz*
Vice President, Publisher *Kevin Kane*
Vice President, Director of Sales and Marketing *Virginia S. Moffat*
Vice President, Director of Production *Colleen A. Yonda*
National Sales Manager *Douglas J. DiNardo*
Marketing Manager *Craig S. Marty*
Advertising Manager *Janelle Keeffer*
Production Editorial Manager *Renée Menne*
Publishing Services Manager *Karen J. Slaght*
Royalty/Permissions Manager *Connie Allendorf*

President and Chief Executive Officer *G. Franklin Lewis*
Senior Vice President, Operations *James H. Higby*
Corporate Senior Vice President, President of WCB Manufacturing *Roger Meyer*
Corporate Senior Vice President and Chief Financial Officer *Robert Chesterman*

Cover design by Pagecrafters Inc.

Interior design by David Lansdon

Figures 9.11a, 9.11b, 10.9b, and 11.7 drawn by Vincent Verdisco

Copyright © 1995 by Wm. C. Brown Communications, Inc. All rights reserved

ISBN-13: 978-0-697-17109-2
ISBN-10: 0-697-17109-4

No part of this publication may be reproduced, stored in a retrieval system, or transmitted, in any form or by any means, electronic, mechanical, photocopying, recording, or otherwise, without the prior written permission of the publisher.

Printed in the United States of America

To Karen, Andrew, Laura, and Vinny, who know where many small steps can lead.

Contents

Preface xv

1 Organization of the Body 1

    A. Introduction 1
    B. Body Cavities and Membranes 2
    C. Homeostasis 4

2 Chemistry 5

    A. Atomic Structure 5
    B. Isotopes 6
    C. Covalent Bonds 7
    D. Ionic Bonds 8
    E. Hydrogen Bonding 9
    F. Hydrophilic Attractions and Hydrophobic Interactions 10
    G. Solutions 11
    H. Chemical Reactions 12
    I. Reversible Reactions 13
    J. Ionization of Water 14
    K. pH 15
    L. Buffers 16
    M. Organic Chemistry 17
    N. Dehydration and Hydrolysis Reactions in Organic Compounds 18
    O. Protein Structure 20
    P. Nucleic Acids 22
    Q. ATP 23
    R. Enzymes 24
    S. Energy of Activation 25
    T. Effect of Inhibitor on Enzyme Action 26

3 Cells 27

    A. Cell Membrane 27
    B. Movement Across a Membrane 28
    C. Tonicity 30
    D. Cell Structure 32
    E. Chromosomes and DNA Replication 34
    F. Mitosis 36

    G. RNA Formation 38
    H. Protein Synthesis 39
    I. Genetic Code 40

4 Tissues 41

- A. Plasma Membrane Modifications 41
- B. Epithelial Tissue 42
- C. Connective Tissue 44
- D. Muscle Tissue 47
- E. Nervous Tissue 48

5 Integument 49

- A. General Anatomy 49
- B. Histology of Integument 50
- C. Hair 52
- D. Structure of the Nail 53
- E. Burns 54

6 Skeletal System 55

- A. Gross Anatomy of Bone 55
- B. Microscopic Anatomy of Bone 56
- C. Types of Bone Development 58
- D. Intramembranous Bone Formation 60
- E. Endochondral Bone Formation 62
- F. Growth and Remodeling 64
- G. Repair 65
- H. Whole Skeleton 66
- I. Skull 68
- J. Vertebrae and Ribs 70
- K. Pectoral Girdle and Upper Arm 72
- L. Pelvic Girdle and Lower Leg 74
- M. Articulations 76

7 Muscular System 79

- A. Gross Structure of Skeletal Muscle 79
- B. Skeletal Muscle Cell 80
- C. Structure of the Sarcomere 81
- D. Contraction of Skeletal Muscle 82
- E. Energetics 84
- F. The Neuromuscular Junction 86
- G. The Motor Unit 88
- H. Muscle Activity 89
- I. Skull 90
- J. Trunk 90
- K. Anterior Pectoral Girdle and Upper Arm 91

- L. Posterior Pectoral Girdle and Upper Arm 92
- M. Anterior Forearm 93
- N. Posterior Forearm 94
- O. Anterior Pelvic Girdle and Leg 95
- P. Posterior Pelvic Girdle and Leg 96
- Q. Anatomical Position 97
- R. Movement 98
- S. Levers 100

8 Nervous System 101

- A. Anatomical Organization 101
- B. Functional Organization 102
- C. Neurons and Neuroglia 103
- D. Neuroglia in Central Nervous System 104
- E. Membrane Potentials 106
- F. Action Potential 108
- G. Repolarization and Hyperpolarization 110
- H. Propagation 111
- I. Synapse 112
- J. Summation 114
- K. Generator Potential 115
- L. Sensory Receptors—Exteroceptors 116
- M. Sensory Receptors—Proprioceptors and Interoceptors 117
- N. Structure of the Brain 118
- O. Meninges and Cerebrospinal Fluid 119
- P. Structure of the Spinal Cord and Spinal Nerves 120
- Q. Spinal Reflex 121
- R. Functions of the Brain 122
- S. Cranial Nerves 124
- T. Autonomic Nervous System 126
- U. Functions of Hypothalamus 128
- V. Somatic Sensory System 130
- W. Somatic Motor System—Motor Cortex 131
- X. Somatic Motor System—Basal Ganglia 132
- Y. Function of the Cerebellum 133

9 Special Senses 135

- A. Gustatory and Olfactory Receptors 135
- B. General Anatomy of the Eye 136
- C. Chambers of the Eye 138
- D. Accommodation 139
- E. Rods and Cones 140
- F. Visual Pathway to the Brain 142
- G. General Anatomy of the Ear 143
- H. Hearing 144
- I. Sound Perception 146
- J. Dynamic Equilibrium 147
- K. Static Equilibrium 148

10 Endocrine System 151

A. Characteristics of Hormones 151
B. Intracellular Mechanism of Action 152
C. Plasma Membrane Mechanism of Action 153
D. Regulation of Hormones 154
E. Pituitary Gland 156
F. Pituitary Gland Regulation 158
G. Thyroid Gland 160
H. Parathyroid 161
I. Adrenal Glands (Suprarenal Glands) 162
J. Pancreas 164
K. Organs with Endocrine Function 166
L. Prostaglandins 168

11 Blood 169

A. Components of Blood 169
B. Red Blood Cells 170
C. White Blood Cells 172
D. Platelets 173
E. Clotting 174
F. ABO Blood Group 176
G. Rh Factor 178

12 Circulatory System 179

A. Overview of Circulation 179
B. Heart Anatomy 180
C. Cardiac Muscle Membranes 182
D. Regulation of Heartbeat 184
E. Cardiac Cycle 186
F. Blood Vessel Histology 187
G. Blood Pressure 188
H. Interstitial Fluid Balance 189
I. Vein Pump 190
J. Lymph Flow 192
K. Arteries of the Systemic System 193
L. Veins of the Systemic System 194
M. Blood Vessels of the Head and Neck 195
N. Abdominal Arteries 196
O. Hepatic Portal System 197
P. Fetal Circulation 198

13 Lymphatic System and Immunity 201

A. Lymphatic Organs 201
B. Secondary Lymphatic Tissue 202
C. Surface Barriers to Infection 204
D. Inflammation Defined 206

- E. Walling Off During Inflammation 208
- F. Cellular Response to Inflammation 209
- G. Phagocytosis and Fever 210
- H. Complement 211
- I. Immunological Surveillance 212
- J. Interferon 213
- K. Types of Specific Immunity 214
- L. B Cell Differentiation 215
- M. Antibody Structure 216
- N. Major Histocompatibility Complex Molecules 216
- O. Primary Differentiation of T Cells 218
- P. Secondary Differentiation of T Cells 220
- Q. T Cell Function 221

14 Respiratory System 223

- A. Anatomy of the Respiratory System—Airways 223
- B. Histology 224
- C. Thoracic Cavity 226
- D. Ventilation 227
- E. Lung Volumes 228
- F. Gas Pressures 230
- G. Oxygen Transport and Hemoglobin 232
- H. Carbon Dioxide Transport 234
- I. Respiratory Feedback 236
- J. Regulation of Respiratory Output 237

15 Digestive System 239

- A. Function of the Digestive System 239
- B. The Mouth 240
- C. Microscopic Anatomy of the Alimentary Canal 242
- D. Movements of the Digestive System 244
- E. Liver and Pancreas 246
- F. Carbohydrates 247
- G. Digestion of Carbohydrates 248
- H. Digestion of Proteins 250
- I. Digestion of Lipids 252
- J. Absorption by Carriers 254
- K. Water Balance in the Digestive System 255
- L. Nervous Control of the Digestive System 256
- M. Chemical Control 258

16 Metabolism 259

- A. Introduction to Metabolism 259
- B. Glucose Metabolism 260
- C. Cellular Energy Conversion 261
- D. Krebs Cycle and Oxidative Phosphorylation 262
- E. Synthesis of Triglycerides 264

F. Lipid Transport 265
  G. Proteins 266

17 Urinary System 267

  A. Organs of the Urinary System 267
  B. Abdominal Relationships 268
  C. Internal Anatomy of the Kidney 270
  D. Nephron Structure and Location 271
  E. Nephron Circulation 272
  F. Filtration, Reabsorption, Secretion—Defined 274
  G. Filtration 275
  H. Regulation of Glomerular Filtration Rate 276
  I. Reabsorption 278
  J. Mechanisms of Reabsorption 280
  K. Hormonal Regulation of Reabsorption 282
  L. Secretion 284
  M. Juxtaglomerular Apparatus 285

18 Fluids, Electrolytes, and Acid-Base Balance 287

  A. Fluid Compartments 287
  B. Electrolytes 288
  C. Fluid and Electrolyte Movement 289
  D. Water Balance 290
  E. Acid Regulation 292

19 Reproductive System 293

  A. Reproductive Function 293
  B. Development 294
  C. Male Anatomy 296
  D. Testes 297
  E. Male Hormonal Regulation 298
  F. Female Anatomy 300
  G. Ovarian Cycle 301
  H. Female Hormonal Regulation 302
  I. Uterine Cycle 304
  J. Menstrual Cycle Timing 305
  K. Pregnancy 306
  L. Mammary Glands 307
  M. Human Sexual Response 308

20 Development and Genetics 309

  A. Gametogenesis 309
  B. Meiosis 310
  C. Genes 312
  D. Autosomal Inheritance 313

E. Sex-Linked Inheritance 314
F. Genetic Variation 315
G. Fertilization 316
H. Cleavage and Implantation 317
I. Germ Layers, Extraembryonic Membranes, Placentation 318
J. Hormonal Regulation of Pregnancy 320

Preface

This book reviews the structure and function of the human body in a format that follows the usual organization of anatomy and physiology textbooks. Students taking a course in human anatomy and physiology can benefit from reinforcing and clarifying concepts presented in textbooks and in lectures. Other students may wish to review the material of a previous semester before beginning the next course, especially if time has passed since the last course. Those preparing for professional examinations can also benefit from this review.

This book is distinctive because each exercise begins with the coloring and labeling of a drawing. Labels and coloring directions are provided, and the questions that follow can usually be answered using the information in the drawing.

Students can use this book as a primary study guide by looking at the labeling key, or as a self test by filling in the blanks without consulting the key. The concepts are presented in a tutorial style in small increments. The answers are provided at the bottom of each page, thus providing immediate feedback to avoid the student building on misinformation. Review concepts are included and referenced back to their original presentation.

By labeling and coloring each drawing the student actively participates in the formation of each page. This visualization and practice constitutes important steps in the learning process and combats the student's tendency to mentally drift, which is a problem with passive study methods.

We would like to thank our colleagues at Suffolk Community College for their help and encouragement, especially Prof. Linda Sabatino for her information and advice on sexual development. Thanks to Mr. Vincent Verdisco for drawing human figures for several of the illustrations.

The reviewers of this manuscript were especially helpful with their positive criticism and valuable suggestions. The book is much stronger because of their work.

Sheila Cagle
Methodist Hospital of Indiana

Clementine A. deAngelis
Tarrant County Junior College

Diane E. Godin
Richland Community College

Gloria Hillert
Triton College

Stephen G. Lebsack
Linn-Benton Community College

Terry R. Martin
Kishwaukee College

Leslie J. Wiemerslage
Belleville Area College

We owe a special personal debt to Mr. Colin Wheatley of Wm. C. Brown and Ms. Jane DeShaw for not only standing by us but continuing to encourage us during a long delay caused by family illness.

How to Use This Book

Each chapter is divided into several topics. For each topic:

A. Follow the directions for the coloring and labeling of the drawing:
   1. Each structure to be colored or labeled is preceded by the symbol o. This circle should be colored in with the color to be used for the drawing. Some colors are suggested, such as red for arteries and blue for veins.
   2. When coloring over print use a light touch so that the print shows through.
   3. Some structures are too thin to be fully colored and should be traced by following their outline with a colored pencil.
   4. While coloring and labeling, observe the relationships between structures or concepts as indicated by lines, arrows, or similar colors.
   5. Try to use the same color for a particular organ or type of structure as you go through a chapter so that relationships stand out not only within a drawing but from page to page.

B. Use the information provided by the drawing to answer the questions. A labeled and colored drawing has most of the information needed to answer the questions in each exercise.

C. Answer the questions in order. Information revealed in a question may lead to the answer to following questions.

D. Be as neat as possible when writing the labels to keep the drawing from becoming too cluttered.

Chapter 1 Introduction

A. Levels of Organization

Color (trace) and label:
1. ependymal cell
   a. ◯ inorganic chemical
   b. ◯ organic chemical
   c. ◯ organelle
2. ◯ cuboidal epithelium
3. ◯ brain
4. ◯ nervous system

Label:
a. cell
b. tissue
c. organ
d. organ system

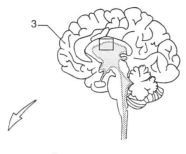
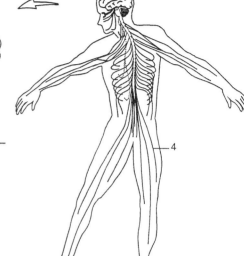
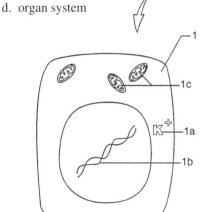

Figure 1.1. Organization of nervous system.

Exercise 1.1:

1. The structural and functional unit of living material is the cell.
   a. It contains functional subunits called _____ .
   b. Cell function also depends upon the presence of specific _____ .
2. Groups of similar cells that perform a particular function are _____ .
3. The structures that contain different tissues organized to perform a general function are _____ .
4. A group of organs that work together to carry out a general function is a(n) _____ .

Answers to Exercise 1.1: 1a. organelles; 1b. chemicals; 2. tissues; 3. organs; 4. organ system.

B. Body Cavities and Membranes

Color (trace) and label:

1. ○ cranial cavity (brain)
2. ○ vertebral canal (spinal cord)
3. ○ pleural cavity (lungs)
4. ○ pericardial cavity (heart)
5. ○ abdominopelvic cavity
6. ○ visceral membrane
    a. visceral pleura
    b. visceral pericardium
    c. visceral peritoneum
7. ○ parietal membrane
    a. parietal pleura
    b. parietal pericardium
    c. parietal peritoneum
8. ○ organ
9. ○ serous fluid
10. ○ body wall

Exercise 1.2:

_____ 1. The brain is contained in the _____ cavity.

_____ 2. The spinal cord is contained in the _____ cavity.

_____ 3. The lungs, heart, and most abdominal viscera are covered with the _____ part of serous membranes.

_____ 4. The cavities containing these organs are lined with the _____ part of serous membranes.

_____ 5. The visceral and parietal membranes allow these organs to slide freely within their cavities because of the _____ between the membranes.

Answers to Exercise 1.2: 1. cranial; 2. vertebral; 3. visceral; 4. parietal; 5. serous fluid.

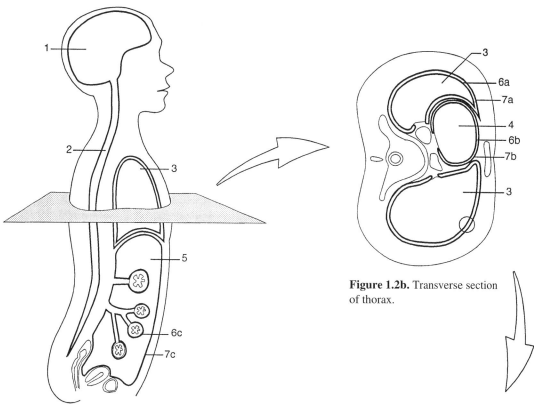

Figure 1.2b. Transverse section of thorax.

Figure 1.2a. Midsagittal view of head and trunk.

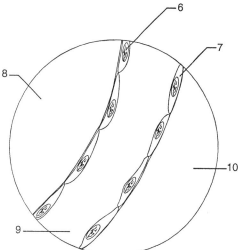

Figure 1.2c. Serous membrane.

C. Homeostasis

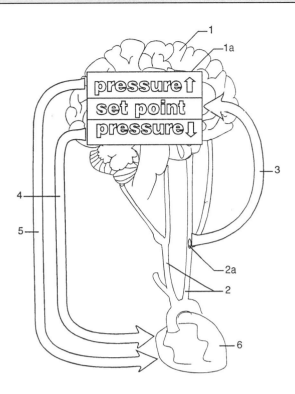

Color and label:
1. ◯ brain
   a. cardiovascular center
2. ◯ blood vessels
   a. pressure receptors
3. ◯ sensory input (nerves)
4. ◯ excitatory motor output (nerves)
5. ◯ inhibitory motor output (nerves)
6. ◯ heart

Figure 1.3. Negative feedback control of blood pressure.

Exercise 1.3:

_____ 1. Blood is pumped to the brain by the _____.

_____ 2. Blood pressure receptors are found in the _____.

_____ 3. Information about blood pressure is sent to the brain stem by _____ nerves.

_____ 4. In the brain, the blood pressure is compared to a set point in the _____ center.

_____ a. If the blood pressure is higher than the set point, then motor neurons _____ (slow down, speed up) the heart.

_____ b. If the blood pressure is lower than the set point, then motor neurons _____ (slow down, speed up) the heart.

_____ c. This type of control mechanism is called _____.

_____ 5. If blood pressure falls, then blood flow to the brain _____ (increases, decreases).

_____ a. Blood pressure receptors then send this information to the _____ in the brain.

_____ b. This causes the heart rate to _____ (increase, decrease), which causes the blood flow to the brain to _____ (increase, decrease).

_____ 6. If the blood pressure increases, this feedback loop causes the heart rate to _____ (increase, decrease).

_____ 7. The process whereby the body remains within narrow limits for its various physiological conditions is called _____.

Answers to Exercise 1.3: 1. heart; 2. blood vessels; 3. sensory; 4. cardiovascular; a. slow down; b. speed up; c. negative feedback; 5. decreases; a. cardiovascular center; b. increase, increase; 6. decrease; 7. homeostasis.

Chapter 2: Chemistry

A. Atomic Structure

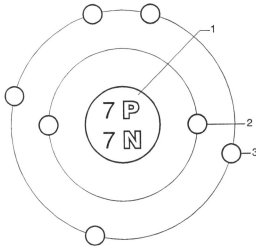

Color and label:
- ○ P (protons)
- ○ N (neutrons)
1. ○ nucleus
2. ○ electrons in first shell
3. ○ electrons in second shell

Figure 2.1. Atomic structure of nitrogen.

Exercise 2.1:

1. The nucleus contains _____ and _____ .
    a. Nitrogen's atomic number (number of protons) is _____ .
    b. Nitrogen's atomic mass (number of protons + number of neutrons) is _____ .
2. How many electrons does nitrogen have in the first shell? second shell? total?
3. Since protons are positive and electrons are negative, an *atom* (as it is shown here) has _____ (a positive charge, negative charge, no charge).
4. The first shell can hold a maximum of 2 electrons, the second a maximum of 8. Does nitrogen have a complete first shell? complete second shell?
5. The number of electrons in the outermost shell determines the chemical properties of an atom. Therefore, nitrogen would behave most like _____ (carbon with 6 electrons, oxygen with 8 electrons, phosphorus with 15 electrons).
6. Atoms are not stable unless their shells are filled with electrons.
    a. Would you expect nitrogen to be stable in the form that it is shown here?
    b. How many electrons would be needed to complete nitrogen's second shell?
    c. If we add electrons to complete nitrogen's second shell, would its charge change?
7. Groups of atoms with the same number of protons are called _____ (elements, compounds).

Answers to Exercise 2.1: 1. protons, neutrons; 1a. 7; 1b. 7 + 7 = 14; 2. 2, 5, 7; 3. no charge; 4. yes, no; 5. phosphorus (because it has 5 electrons in its outermost shell); 6a. no; 6b. 3; 6c. yes (−3); 7. elements.

B. Isotopes

Color:
- ○ P (protons) and their number
- ○ N (neutrons) and their number
1. ○ nucleus
2. ○ electrons

Label:
a. carbon-12
b. carbon-13
c. carbon-14

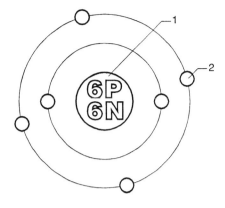

a ─────────────

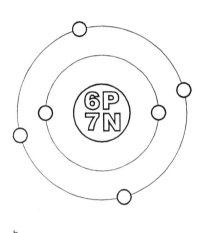

b ─────────────

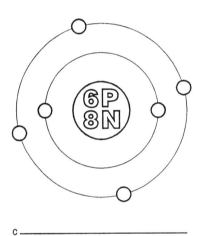

c ─────────────

Figure 2.2. Isotopes of carbon.

Exercise 2.2:

1. The atoms of an element may exist in different forms called isotopes.
    a. Which vary in isotopes—the number of protons, neutrons, or electrons?
    b. Is the atomic number the same or different for each isotope?

2. What is the atomic mass for each carbon isotope shown in figure 2.2?

3. Do you expect the chemical properties of these isotopes to be similar or different?

4. Carbon and some other lighter elements have radioactive isotopes, which have unstable nuclei when the mass is the heaviest for that element. Which carbon isotope would you expect to be radioactive?

Answers to Exercise 2.2: 1a. neutrons; 1b. same; 2. 12 for carbon-12, 13 for carbon-13, 14 for carbon-14; 3. similar (because they have the same number of outermost electrons); 4. 14.

C. Covalent Bonds

Color and label:
○ + electropositive
○ – electronegative

Label each atom and color its electrons:
1. ○ hydrogen
2. ○ oxygen

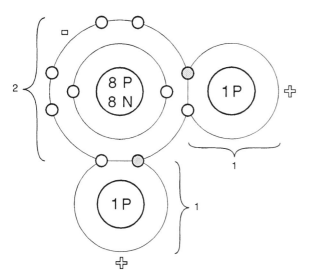

Figure 2.3a. Water (H_2O).

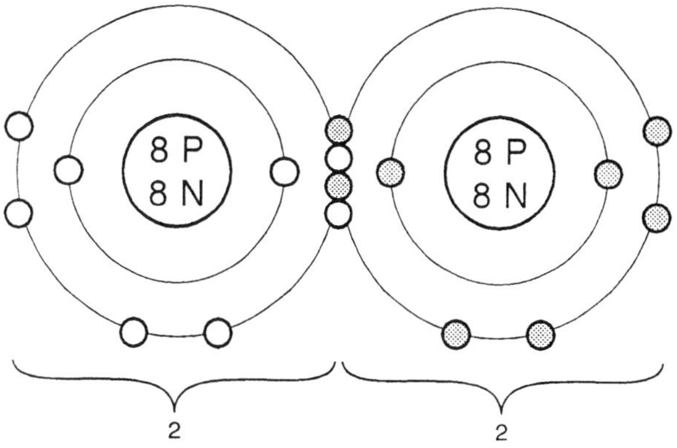

Figure 2.3b. Molecular oxygen (O_2).

Exercise 2.3:

_____ 1. Covalent bonding involves the sharing of electrons. Which molecules show covalent bonds?

_____   a. Single covalent bonds (sharing of two electrons) occur in _____ (H_2O, O_2).

_____   b. Double bonds (sharing of four electrons) occur in _____ (H_2O, O_2).

_____ 2. In the figures shown, the outer shell of oxygen has _____ (number) electrons and hydrogen has _____ (number) electrons.

_____ 3. Does covalent bonding create complete outer shells?

_____ 4. For covalent bonds between small atoms, shared electrons are drawn to the atom with the greater number of protons, creating a polar bond.

_____   a. Which molecule shows polar covalent bonding?

_____   b. For this molecule, the electrons are drawn toward _____ (oxygen or hydrogen).

_____   c. The electronegative end, the electron-rich end, is labeled _____ (+, –).

_____   d. The electropositive end, the electron-poor end, is labeled _____ (+, –).

_____ 5. Molecular oxygen has a _____ (polar, nonpolar) covalent bond. Explain.

_____ 6. Does the water molecule have a positive charge, negative charge, or no charge? Does molecular oxygen have a positive charge, negative charge, or no charge?

_____ 7. Name two reasons for chemical stability in covalent bonding.

Answers to Exercise 2.3: 1. H_2O, O_2; 1a. H_2O; 1b. O_2; 2. 8, 2 (shared electrons are counted for each atom they encircle); 3. yes; 4a. H_2O; 4b. oxygen; 4c. –; 4d. +; 5. nonpolar; protons in each oxygen atom have equal pull on the shared electrons; 6. no charge, no charge; 7. complete outer shell, no charge on molecule.

D. Ionic Bonds

Label each atom and color its electrons:
1. ○ sodium
2. ○ chlorine
3. ○ calcium

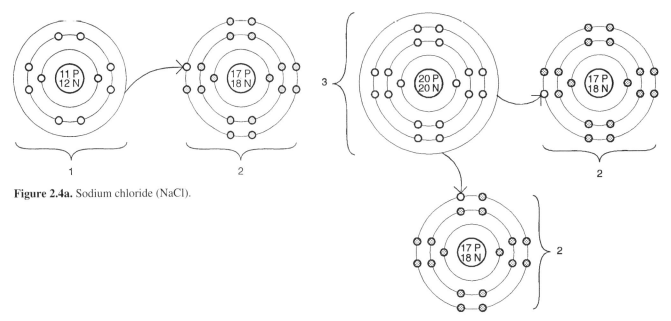

Figure 2.4a. Sodium chloride (NaCl).

Figure 2.4b. Calcium chloride (CaCl₂).

Exercise 2.4:

_____ 1. Ionic bonding involves the transfer of electrons from one atom to another, forming ions. Which molecules show ionic bonding?

_____   a. For NaCl, _____ loses electrons nd _____ gains electrons.

_____   b. For NaCl, what are the charges on the newly formed *ions?*

_____ 2. Ionic bonding is the attraction of positive and negative ions. Which ions are attracted to each other in NaCl?

_____ 3. For CaCl₂, _____ loses electrons and _____ gain electrons.

_____   a. What is the charge on the calcium ion? chloride ion?

    b. Why is only one chlorine needed to accept sodium's electrons, while two chlorines are needed to accept calcium's electrons?

_____ 4. Does the sodium chloride molecule have a positive charge, negative charge, or no charge? calcium chloride molecule?

5. Notice that covalent and ionic bonding result in complete outer shells.

_____   a. When an atom has a half filled outer shell, it is more likely to form _____ (covalent, ionic) bonds.

_____   b. When an atom has a nearly empty or nearly full outer shell, it is more likely to form _____ (covalent, ionic) bonds.

6. Name two reasons for chemical stability in ionic bonding.

7. The molecular weight equals the sum of the atomic masses of each atom in the molecule. Calculate the molecular weight for each of the following. (see atomic mass, p. 5)

   a. H_2O d. $CaCl_2$

   b. O_2 e. $C_6H_{12}O_6$ (glucose)

   c. NaCl

8. A mole (mol) is the mass in grams that numerically equals the molecular weight. Calculate the grams present in each of the following.

   a. 1 mol O_2 e. 2 mol O_2

   b. 1 mol NaCl   f. 0.15 mol NaCl

   c. 1 mol $CaCl_2$ g. 4 mol $CaCl_2$

   d. 1 mol $C_6H_{12}O_6$ h. 0.3 mol $C_6H_{12}O_6$

Answers to Exercise 2.4: 1. NaCl, $CaCl_2$; 1a. sodium, chlorine; 1b. Na^{+1}, Cl^{-1}; 2. Na^{+1}, Cl^{-1}; 3. calcium, chlorine; 3a. Ca^{+2}, Cl^{-1}; 3b. sodium loses only one electron, but calcium loses two; 4. no charge for both (the number of positive charges for Na^+ or Ca^{+2} equals the number of negative charges for Cl^-); 5a. covalent; 5b. ionic; 6. complete outer shell, no charge on molecule; 7a. 2(1) + 16 = 18; 7b. 2(16) = 32; 7c. 23 + 35 = 58; 7d. 40 + 2(35) = 110; 7e. 6(12) + 12(1) + 6(16) = 180; 8a. 32 grams; 8b. 58 grams; 8c. 110 grams; 8d. 180 grams; 8e. 2(32) = 64; 8f. 0.15(58) = 8.7 grams; 8g. 4(110) = 440 grams; 8h. 0.3(180) = 54 grams.

E. Hydrogen Bonding

Label each atom and color its electrons:
1. ◯ O (oxygen)
2. ◯ H (hydrogen)

Label and trace:
3. hydrogen bond

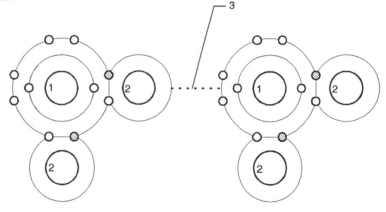

Figure 2.5. Water molecules (H_2O).

Exercise 2.5:

1. Which ends of the water molecule are electropositive (+) and electronegative (−)? Label accordingly.

2. Hydrogen bonds are attractions between an electropositive hydrogen and an electronegative atom. Is there a transfer of electrons? a sharing of electrons?

3. In this example, the hydrogen bond is _____ (within a molecule, between molecules).

Answers to Exercise 2.5: 1. hydrogen is electropositive and oxygen is electronegative. 2. no, no; 3. between molecules (although hydrogen bonding can also occur within molecules).

F. Hydrophilic Attractions and Hydrophobic Interactions

Label and color:
1. ○ Na⁺
2. ○ Cl⁻
3. ○ fats (nonpolar)

Color atoms in water:
4. ○ hydrogen
5. ○ oxygen

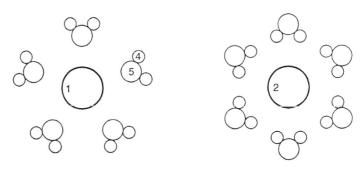

Figure 2.6a. Sphere of hydration for sodium chloride.

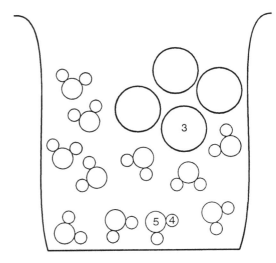

Figure 2.6b. Water and fat molecules.

Exercise 2.6:

_____ 1. The electropositive (+) end of the water molecule is _____ and the electronegative (−) end is _____ . Label accordingly. (see covalent bond, p. 7)

_____ a. A positively charged ion is attracted to the _____ end of the water molecule.

_____ b. A negatively charged ion is attracted to the _____ end of the water molecule.

_____ 2. The effective size of charged particles in water _____ (increases, remains the same, decreases).

_____ 3. Are charged ions hydrophilic (attracted to water) or hydrophobic (have no affinity to water)?

_____ 4. Are water molecules attracted to each other? (see figure 2.6b)

_____ 5. If a nonpolar substance is mixed with water, will spheres of hydration form around the substance?

_____ 6. What can you say about the chemical nature of hydrophobic substances?

Answers to Exercise 2.6: 1. hydrogen, oxygen; 1a. electronegative (oxygen); 1b. electropositive (hydrogen); 2. increases; 3. hydrophilic; 4. yes; 5. no (the nonpolar substance will not attract the polar water molecules, while water molecules will attract each other excluding the nonpolar material); 6. they are nonpolar.

G. Solutions

Color and label:
1. ◯ solute
2. ◯ solvent

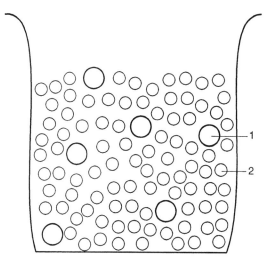

Figure 2.7. Components of a solution.

Exercise 2.7:

_____ 1. The substance present in greater amounts is the _____ .

_____ 2. The substance present in lesser amounts is the _____ .

_____ 3. Which would dissolve in a *polar* solvent—polar or nonpolar solutes?

_____ 4. Which would dissolve in a *nonpolar* solvent—polar or nonpolar solutes?

_____ 5. The most common solvent in the body is _____ .

_____ 6. Which will dissolve in water—polar or nonpolar solutes?

_____ 7. The cell membrane, a nonpolar fluid, can act as a solvent. Will polar or nonpolar solutes dissolve in the membrane (and, therefore, pass through it)?

8. The concentration of a solution can be expressed as a percentage equal to the grams of solute per 100 ml solution. What are the concentrations of the following solutions:

_____ a. 9.5 grams NaCl in 1 liter (1000 ml) solution

_____ b. 3 grams $C_6H_{12}O_6$ in 200 ml solution

9. The concentration of a solution can also be defined by molarity. Molarity equals the number of moles of solute per liter of solution. What are the molarities of the following solutions? (see ionic bonds, #8, p. 9)

_____ a. 29 grams NaCl in 1 liter solution

_____ b. 220 grams $CaCl_2$ in 1 liter solution

_____ c. 270 grams $C_6H_{12}O_6$ in 2 liters solution

Answers to Exercise 2.7: 1. solvent; 2. solute; 3. polar; 4. nonpolar; 5. water; 6. polar; 7. nonpolar; 8a. 0.95% NaCl; 8b. 1.5% $C_6H_{12}O_6$; 9a. 0.5M NaCl; 9b. 2M $CaCl_2$; 9c. 0.75M $C_6H_{12}O_6$.

H. Chemical Reactions

Color:
- reactants (chemicals going into reaction)
- products (end result of reaction)

Label:
a. synthesis reaction
b. decomposition reaction
c. exchange reaction

a. _____

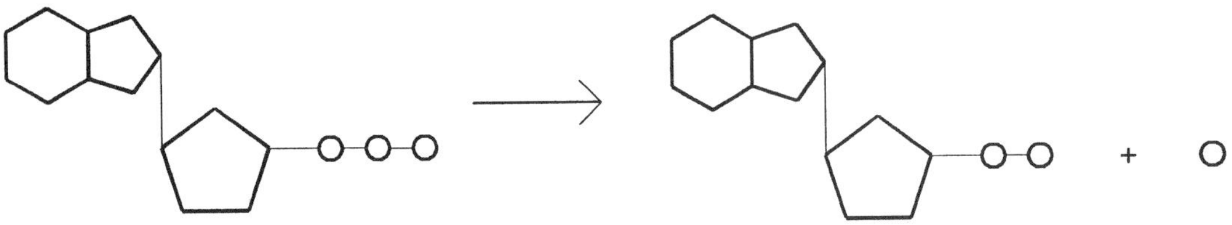

b. _____

c. _____

Figure 2.8 Types of chemical reactions.

Exercise 2.8:

1. Bond formations that produce larger molecules are anabolic reactions. An example of an anabolic reaction is _____ .

2. Breakage of bonds that results in the formation of smaller molecules are catabolic reactions. An example of a catabolic reaction is _____ .

3. Endergonic reactions require an outside source of energy, which is then stored as chemical bond energy. Which are endergonic—anabolic or catabolic reactions?

4. Breakage of chemical bonds releases energy during exergonic reactions. Which are exergonic—anabolic or catabolic reactions?

5. If an endergonic reaction is to occur, with what kind of reaction must it be coupled? Why?

6. Which reaction is a combination of decomposition and synthesis reactions?

7. Are exchange reactions exergonic or endergonic?

8. If propane gas from a kitchen stove or gas grill is mixed with oxygen from the atmosphere, will they react (burn) spontaneously?

9. What do you have to do for this reaction to occur?

10. When energy is needed to start a chemical reaction, it is called _____ .

Answers to Exercise 2.8: 1. synthesis reaction; 2. decomposition reaction; 3. anabolic reactions; 4. catabolic reactions; 5. exergonic reactions (the exergonic reaction supplies the energy needed to run the endergonic reaction); 6. exchange reaction; 7. either, depending upon the reaction; 8. no; 9. light a match; 10. energy of activation.

I. Reversible Reactions

Color:
1. ○ A
2. ○ B
3. ○ C
4. ○ D

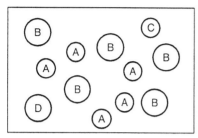
Solution 1

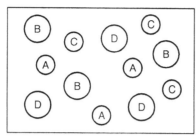
Solution 2

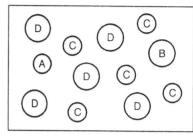
Solution 3

Figure 2.9. $A + B \rightleftharpoons C + D$.

Exercise 2.9:

_____ 1. Molecules are always in motion. In order for molecules to react, they must collide. Molecules are more likely to collide when they are in _____ (low, high) concentration.

_____ 2. Reversible reactions are reactions that can go in either direction. When the concentrations of A and B are high, _____ and _____ form.

_____ a. Which solution shows the *highest* concentrations of A and B?

_____ b. For this solution, which is more likely to occur: $A + B \rightarrow C + D$ or $C + D \rightarrow A + B$?

_____ 3. When the concentrations of C and D are high, _____ and _____ form.

_____ a. Which solution shows the *highest* concentration of C and D?

_____ b. For this solution, which is more likely to occur: $A + B \rightarrow C + D$ or $C + D \rightarrow A + B$?

_____ 4. For solution 2, can A and B collide and react? Can C and D collide and react? Therefore, must all the A and B convert to C and D before the reverse reaction occurs?

_____ 5. At chemical equilibrium, the rate of the reaction $A + B \rightarrow C + D$ is the same as the rate of the reaction $C + D \rightarrow A + B$. Do all reactions stop at chemical equilibrium?

_____ 6. The rate of the reaction in one direction may not be the same as the rate of the reaction in the opposite direction. Therefore, are the concentrations of A, B, C, and D necessarily the same at chemical equilibrium?

_____ 7. If a mixture of A, B, C, and D are at chemical equilibrium and we add more A, is the mixture still at chemical equilibrium? What reaction will increase?

_____ 8. If a mixture of A, B, C, and D are at chemical equilibrium and we add more D, is the mixture still at chemical equilibrium? What reaction will increase?

Answers to Exercise 2.9: 1. high; 2. C, D; 2a. solution 1; 2b. $A + B \rightarrow C + D$; 3. A, B; 3a. solution 3; 3b. $C + D \rightarrow A + B$; 4. yes, yes, no (when the concentration of C and D is high enough for them to react, the reverse reaction increases in frequency); 5. no (the reactions continue at equal rates, but in opposite directions); 6. no; 7. no, $A + B \rightarrow C + D$; 8. no, $C + D \rightarrow A + B$.

J. Ionization of Water

Color:
1. O electrons from oxygen
2. O electrons from hydrogen

Label:
3. hydroxide ion
4. hydrogen ion

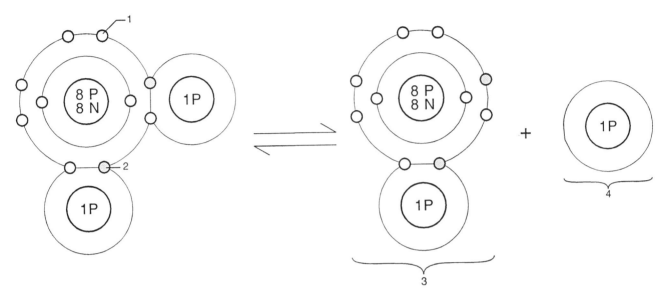

Figure 2.10. Dissociation of water.

Exercise 2.10:

_____ 1. Water molecules are extremely stable, but small numbers of them dissociate (and rejoin). When water dissociates, _____ and _____ ions form.

_____ 2. The atoms that make up the hydroxide ion are _____ and _____ . The charge on this ion is _____ .

_____ 3. The subatomic particle that makes up the hydrogen ion is the _____ . The charge on this ion is _____ .

_____ 4. Since there are equal numbers of H⁺ and OH⁻ ions in pure water, water has _____ (a positive, a negative, no) charge.

_____ 5. Is the dissociation of water a reversible reaction?

_____ 6. A solution with more protons than OH⁻ ions is a(n) _____ .

_____ 7. A solution with more OH⁻ ions than protons is a(n) _____ .

Answers to Exercise 2.10: 1. hydroxide, hydrogen; 2. O, H, –1; 3. proton, +1; 4. no; 5. yes; 6. acid; 7. base.

K. pH

Color:
1. ○ H⁺ ion concentration
2. ○ OH⁻ ion concentration

List the pH for each hydrogen ion concentration.

Label:
a. strong acid
b. weak acid
c. neutral
d. weak base
e. strong base

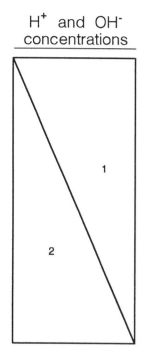

H⁺ and OH⁻ concentrations	H⁺ ion concentration (molar)		pH scale
	1.0	= 10^0	___ a _____
	0.1	= 10^{-1}	___
	0.01	= 10^{-2}	___
	0.001	= 10^{-3}	___
	0.0001	= 10^{-4}	___
	0.00001	= 10^{-5}	___
	0.000001	= 10^{-6}	___ b _____
	0.0000001	= $10^{-⑦}$ → **7** c _____
	0.00000001	= 10^{-8}	___ d _____
	0.000000001	= 10^{-9}	___
	0.0000000001	= 10^{-10}	___
	0.00000000001	= 10^{-11}	___
	0.000000000001	= 10^{-12}	___
	0.0000000000001	= 10^{-13}	___
	0.00000000000001	= 10^{-14}	___ e _____

Figure 2.11. pH scale.

Exercise 2.11:

_____ 1. An acid increases the concentration of _____ ions. An acid has a _____ (high, low) pH.

_____ 2. A base increases the concentration of _____ ions. A base has a _____ (high, low) pH.

_____ 3. An acid is a proton _____ (donor, acceptor).

_____ 4. A base is a proton _____ (donor, acceptor).

_____ 5. A one point change in pH represents what kind of change in concentration of H⁺ ions?

_____ 6. When an acid and a base are mixed, _____ and _____ form. What happens to the pH?

7. Determine whether each of the following fluids is a strong acid, weak acid, weak base, or strong base.

_____ a. urine—pH 5.0–6.9

_____ b. gastric juices—pH 1.0–3.0

_____ c. saliva—pH 6.3–6.9

_____ d. blood—pH 7.40

_____ e. bile—pH 8.0–8.3

Answers to Exercise 2.11: 1. H⁺, low; 2. OH⁻, high; 3. donor (H⁺ ions are protons); 4. acceptor (OH⁻ ions combine with protons); 5. tenfold; 6. salt, water, pH approaches neutral; 7a. weak acid; 7b. strong acid; 7c. weak acid; 7d. weak base; 7e. weak base.

L. Buffers

Color:

1. ○ a. OH⁻ added to the buffer
   b. H₂O formed
   c. corresponding arrow
2. ○ a. H⁺ added to the buffer
   b. corresponding arrow

Label:

a. buffer + base
b. buffer + acid

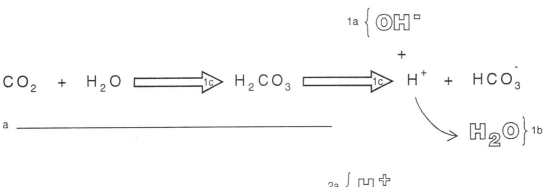

a _____

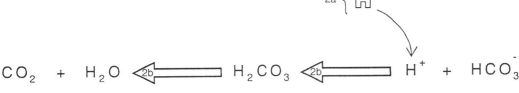

b _____

Figure 2.12. Bicarbonate buffer.

Exercise 2.12:

_____ 1. When CO₂ and H₂O combine, they form _____ , which dissociates to _____ and _____ .

_____ 2. When H⁺ and HCO₃⁻ combine, they form _____ , which breaks down to _____ and _____ .

_____ 3. When a base is added to the buffer, the reaction goes to the _____ (right, left).

_____   a. OH⁻ ions combine with _____ to form _____ .

_____   b. As H⁺ ions are removed, more form by the breakdown of _____ .

_____   c. The pH _____ (increases, decreases, remains the same).

_____ 4. When an acid is added to the buffer, the reaction goes to the _____ (right, left).

_____   a. H⁺ ions combine with _____ to make _____ ,

_____   b. which forms _____ and _____ .

_____   c. The pH _____ (increases, decreases, remains the same).

Answers to Exercise 2.11: 1. H₂CO₃ , H⁺, HCO₃⁻; 2. H₂CO₃, CO₂, H₂O; 3. right; 3a. H⁺, H₂O; 3b. H₂CO₃; 3c. remains the same; 4. left; 4a. HCO₃⁻, H₂CO₃; 4b. CO₂, H₂O; 4c. remains the same.

M. Organic Chemistry

Color the symbols for each atom:
- ○ C (carbon)
- ○ H (hydrogen)
- ○ O (oxygen)
- ○ N (nitrogen)

Circle:
- ○ –COOH (carboxyl group)
- ○ –NH₃ (amino group)

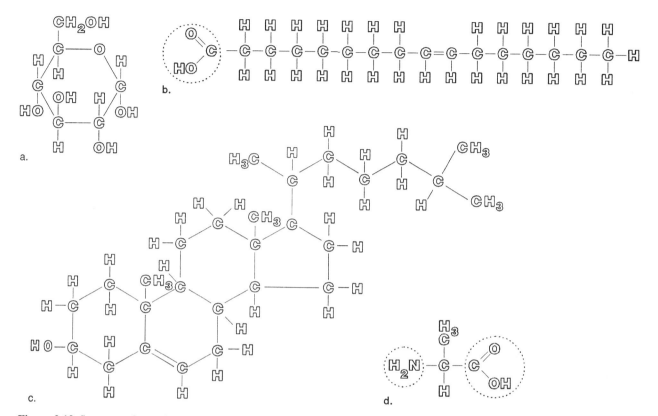

Figure 2.13. Structure of organic compounds.

Exercise 2.13:

_____ 1. A carbohydrate has a C:H:O ratio of 1:2:1. The carbohydrate is _____ (a, b, c, d).

_____ 2. Lipids have a high proportion of carbons and hydrogens and a low proportion of oxygens. Fatty acids are lipids containing a carboxyl group. The fatty acid is _____ (a, b, c, d).

_____ 3. Saturated fatty acids have only single bonds between carbons (thus, a maximum number of hydrogens per carbon), while unsaturated fatty acids have at least one double or triple carbon bond (and fewer hydrogens per carbon). This is a(n) _____ (saturated, unsaturated) fatty acid.

_____ 4. Steroids are lipids that have a carbon ring structure. The steroid is _____ (a, b, c, d).

_____ 5. Amino acids have a central carbon with four attachments: hydrogen, carboxyl group, amino group, and a radical (any of a variety of groups). The amino acid is _____ (a, b, c, d). The radical for the amino acid shown is _____ .

_____ 6. The organic compounds that contain nitrogen are _____ .

Answers to Exercise 2.13: 1. a; 2. b; 3. unsaturated; 4. c; 5. d, –CH₃; 6. amino acids.

N. Dehydration and Hydrolysis Reactions in Organic Compounds

Color:
- ○ –OH (block letters)
- ○ –H (block letters)

Trace:
1. ○ peptide bond

Label:
a. monosaccharide
b. disaccharide
c. glycerol
d. saturated fatty acids
e. unsaturated fatty acid
f. triacylglycerol*
g. amino acid
h. dipeptide
i. water
j. dehydration reaction
k. hydrolysis reaction

*also called triglyceride or neutral fat

Exercise 2.14:

_____ 1. When two monsaccharides combine, a _____ and _____ form.

_____ 2. When glycerol and three fatty acids combine, a _____ and _____ form.

_____ 3. When two amino acids combine, a _____ and _____ form.

_____ 4. These reactions are _____ (dehydration, hydrolysis) reactions.

_____   a. The product that forms in each case is _____ .

_____   b. These reactions are _____ (anabolic, catabolic) reactions. (see chemical reactions, p. 12)

_____   c. Is dehydration synthesis exergonic or endergonic?

_____ 5. The reactions that involve breakdown of disaccharides, triacylglycerol, and dipeptides are _____ (dehydration, hydrolysis) reactions.

_____   a. The molecule that is added is _____ .

_____   b. Hydrolysis reactions are _____ (anabolic, catabolic) reactions.

_____   c. Is hydrolysis exergonic or endergonic?

_____ 6. Peptide bonds form between the _____ group of one amino acid and the _____ group of the second amino acid. Peptide bonds are _____ (covalent, ionic, hydrogen) bonds.

_____ 7. A protein is a chain of many amino acids. Which reaction repeated many times would form a protein?

_____ 8. A polysaccharide is a combination of many monosaccharides. Which reaction repeated many times would build a polysaccharide?

_____ 9. The breakdown of polysaccharides and proteins occurs by _____ (dehydration, hydrolysis) reactions.

Answers to Exercise 2.14: 1. disaccharide, water; 2. triacylglycerol, water; 3. dipeptide, water; 4. dehydration; 4a. water; 4b. anabolic; 4c. endergonic; 5. hydrolysis; 5a. water; 5b. catabolic; 5c. exergonic; 6. carboxyl, amino, covalent; 7. dehydration reaction; 8. dehydration reaction; 9. hydrolysis.

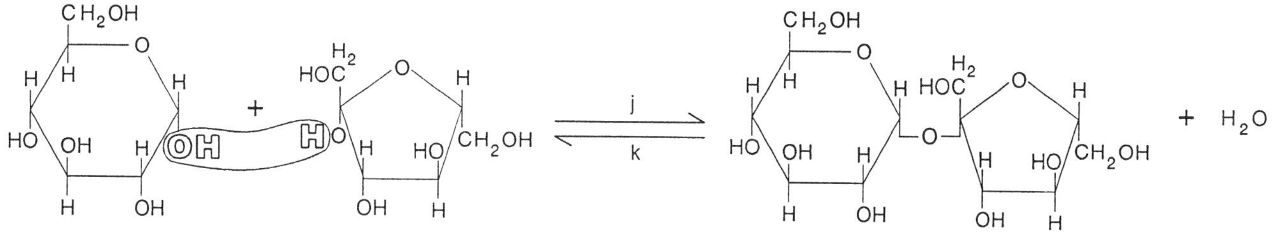

a _____   a _____   b _____   i _____

Figure 2.14a. Carbohydrates.

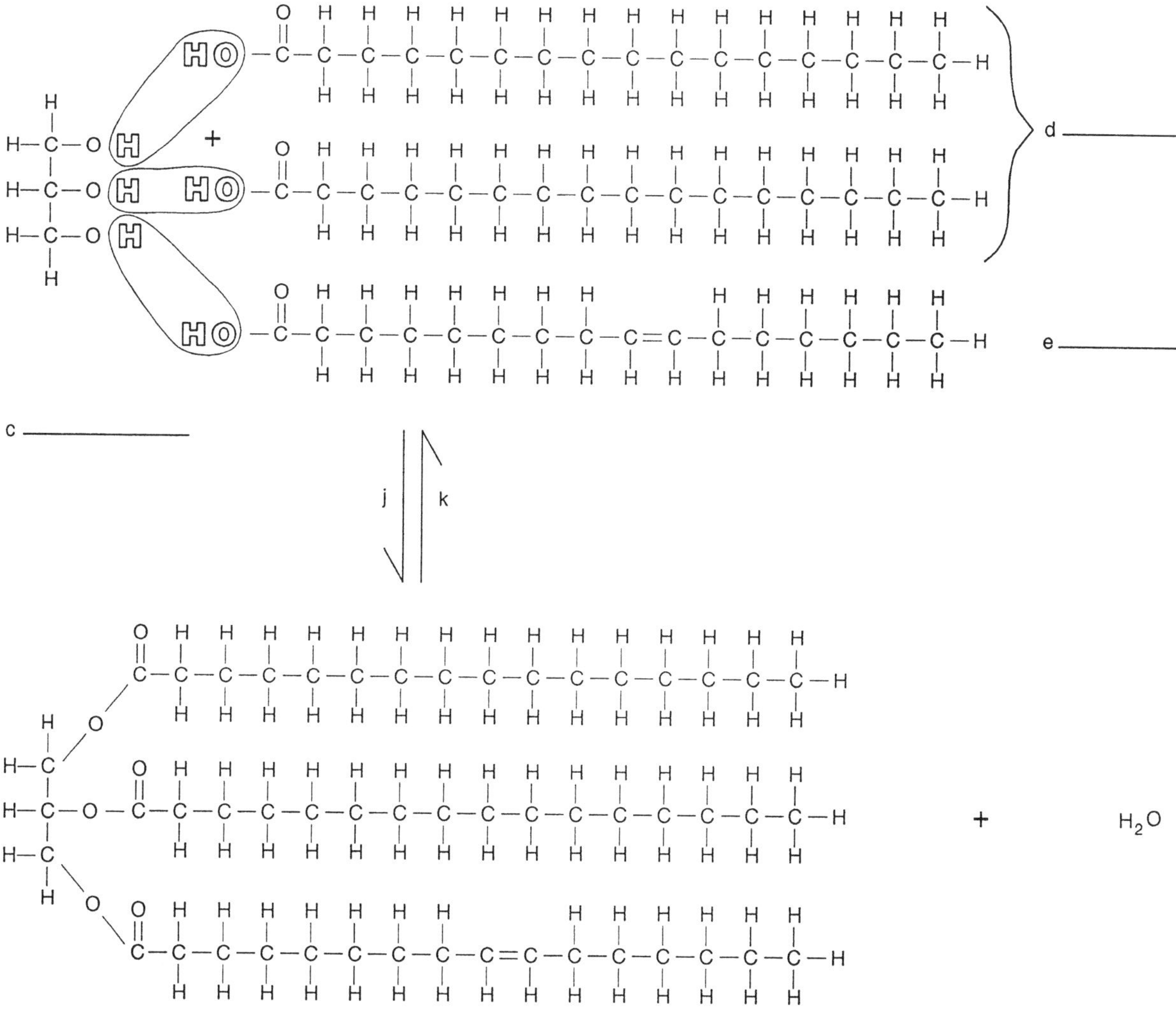

d _____

e _____

c _____

f _____   i _____

Figure 2.14b. Lipids.

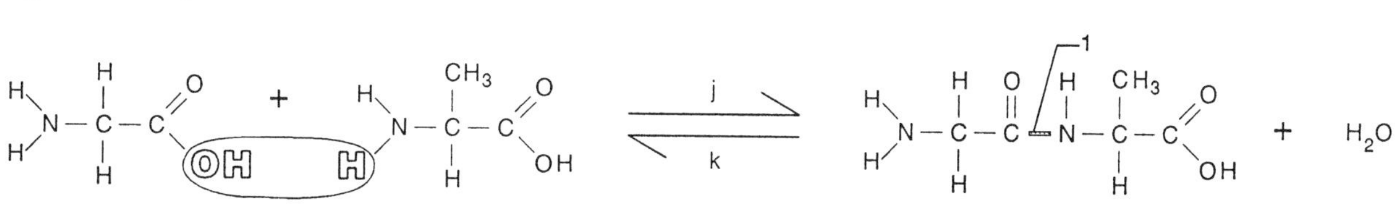

g _____   g _____   h _____   i _____

Figure 2.14c. Proteins.

Dehydration and Hydrolysis Reactions in Organic Compounds

O. Protein Structure

Color:
- ○ H (hydrogen)
- ○ O (oxygen)
- ○ R (radicals)
1. ○ backbone of amino acid chain
2. ○ peptide bonds (figure 2.15a)
3. ○ hydrogen bonds (figure 2.15b)

Color each polypeptide chain (figure 2.15c and d):
4. ○
5. ○
6. ○
7. ○
8. ○

Label:
a. primary structure
b. secondary structure
c. tertiary structure
d. quaternary structure

Exercise 2.15:

1. Compare the two amino acids in figure 2.14c.
   a. What is the same in each?
   b. What is different? What symbol is given in figure 2.15a?

2. The peptide bond between two amino acids is a(n) _____ (covalent, ionic, hydrogen) bond.

3. The protein backbone is made of _____ and _____ .

4. The atoms or functional groups that stick out of the backbone are _____ , _____ , and _____ .

5. If hydrogen is in the vicinity of an electronegative atom, what type of bond will form? (see bonding, p. 7) Is oxygen electronegative or electropositive?

6. The amino acid chain coils or is pleated at the _____ structure.

7. What bonds help to stabilize this secondary structure?

8. The R groups can vary in charge, polarity, and size. How do these characteristics alter the position of neighboring R groups?

9. Sulfur containing radicals near each other can form covalent bonds between the sulfurs. Does this help to stabilize the structure?

10. When the amino acid chain is folded on itself in this manner, it is called the _____ structure.

11. How many polypeptide chains make up the tertiary structure of a protein? Quaternary structure of a protein?

12. To summarize, the secondary, tertiary, and quaternary structures may be stabilized by what kinds of interactions?

13. An increase in temperature increases the movement of atoms. Can this break hydrogen bonds and other types of stabilizing attractions?

Answers to Exercise 2.15: 1a. central C, –NH₂ group, –COOH group, hydrogen; 1b. one amino acid has an additional hydrogen, while the other has a –CH₃ group; the symbol R is used; 2. covalent bond; 3. nitrogen, carbon; 4. hydrogen, oxygen, radical; 5. hydrogen bond, electronegative; 6. secondary; 7. hydrogen bonds; 8. radicals can be attracted to each other or repelled by each other, thus distorting the secondary structure. 9. yes; 10. tertiary; 11. one, more than one; 12. hydrogen bonds, attraction of opposite charges, hydrophobic interactions, sulfur to sulfur covalent bonds; 13. yes (the protein may denature and no longer function properly).

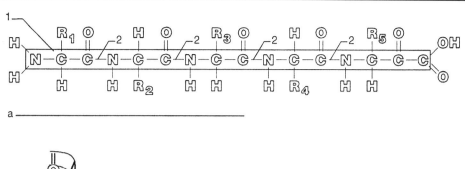
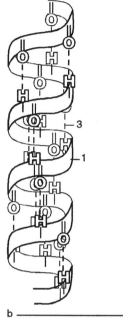
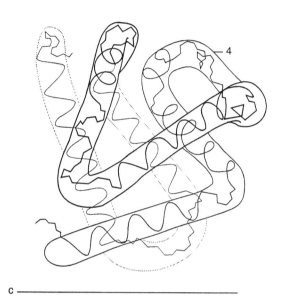
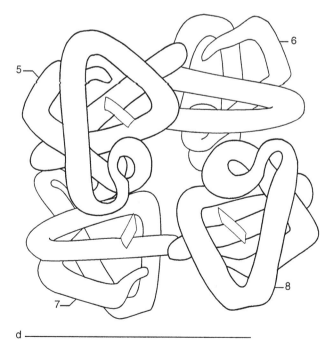

Figure 2.15. Structural levels of protein.

Protein Structure

P. Nucleic Acids

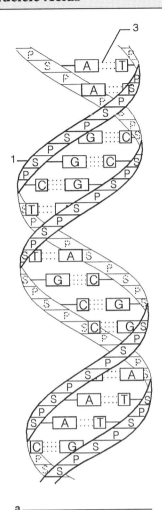

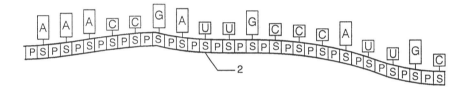

Color and label:
- ○ A (adenine)
- ○ G (guanine)
- ○ T (thymine)
- ○ C (cytosine)
- ○ U (uracil)
- ○ P (phosphates)
1. ○ deoxyribose
2. ○ ribose

Trace and label:
3. hydrogen bonds

Label:
a. DNA
b. RNA

Figure 2.16. Types of nucleic acids.

Exercise 2.16:

_____ 1. DNA is _____ (single, double) stranded.

_____ 2. The backbone of DNA contains _____ and _____.

_____ 3. The central runs of the DNA "ladder" are the nitrogenous bases _____, _____, _____, and _____.

_____ a. These bases are attached to _____ in the DNA backbone.

_____ b. The larger bases (purines) are _____ and _____.

_____ c. The smaller bases (pyrimidines) are _____ and _____.

_____ d. Adenine always pairs with _____.

_____ e. Guanine always pairs with _____.

_____ f. What kind of bond holds the bases together?

_____ 4. RNA is _____ (single, double) stranded.

_____ 5. What components are present in DNA, but not in RNA?

_____ 6. What components are present in RNA, but not in DNA?

_____ 7. Name three kinds of RNA.

Answers to Exercise 2.16: 1. double; 2. phosphates, deoxyribose; 3. adenine, guanine, thymine, cytosine; 3a. deoxyribose; 3b. adenine, guanine; 3c. thymine, cytosine; 3d. thymine; 3e. cytosine; 3f. hydrogen bond; 4. single; 5. deoxyribose, thymine; 6. ribose, uracil; 7. messenger RNA, transfer RNA, ribosomal RNA.

Q. ATP

Color and label:
1. ○ adenine
2. ○ ribose
3. ○ phosphate

Label:
a. ATP
b. ADP

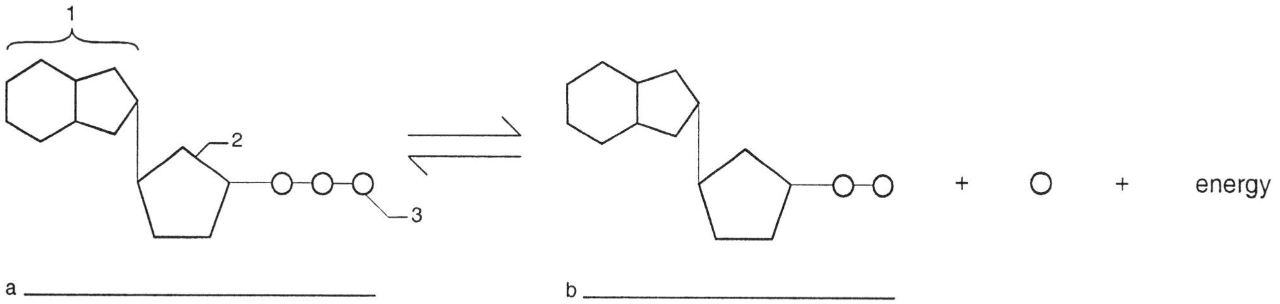

Figure 2.17. ATP formation and breakdown.

Exercise 2.17:

_____ 1. ATP contains _____ (number) phosphates.

_____ 2. ADP contains _____ (number) phosphates.

_____ 3. The breakdown of ATP is an _____ (endergonic, exergonic) reaction. (see types of reactions, p. 12)

_____ 4. ATP formation is an _____ (endergonic, exergonic) reaction.

_____ 5. Our bodies get the energy that is used to make ATP from _____ .

Answers to Exercise 2.17: 1. 3; 2. 2; 3. exergonic; 4. endergonic; 5. the breakdown of organic molecules.

R. Enzymes

Color and label:
1. ◯ substrate
2. ◯ enzyme
3. ◯ active site
4. ◯ product

Label:
5. enzyme-substrate complex

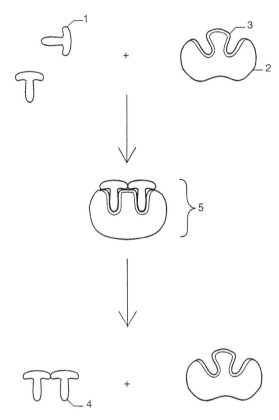

Figure 2.18. Enzyme action.

Exercise 2.18:

_____ 1. The molecule the enzyme reacts with is the _____ .

_____   a. Do the substrates and active site have similar shapes?

_____   b. Would you expect the substrates and active site to be chemically compatible?

_____   c. Since physical and chemical fit are necessary for enzyme substrate interaction, would you expect enzymes to be specific or nonspecific in their reactions?

_____ 2. Does the shape of the enzyme shown change when it reacts with its substrates?

_____ 3. How does the enzyme help the substrate molecules react?

_____ 4. Is the enzyme used up in the reaction?

_____ 5. Since enzymes are proteins, what would happen to the shape of an enzyme as it is heated?

_____ 6. How will heating the system change the ability of the enzyme to function?

_____ 7. If the system cools, the rate of the reaction _____ (increases, remains the same, decreases).

_____ 8. Can a change in pH alter enzyme activity?

_____ 9. What would you expect the optimum pH to be for enzymes in the body?

Answers to Exercise 2.18: 1. substrate; 1a. yes; 1b. yes; 1c. specific; 2. yes (this enzyme shows induced fit, although not all enzymes do); 3. correctly orients substrate molecules; 4. no; 5. secondary and tertiary structures might change with high temperatures (hydrogen bonds and attractive forces between opposite charges break); 6. increases rate of reaction (more collisions), high temperature decreases rate of reaction (shape of active site may change); 7. decreases; 8. yes (change in charge on active site changes chemical fit of substrate); 9. same as pH of fluid where enzyme is found.

S. Energy of Activation

Color energy of activation curves and their arrows. Label:
1. ○ without enzyme
2. ○ with enzyme

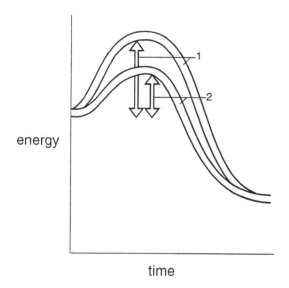

Figure 2.19. Effect of enzyme on energy of activation.

Exercise 2.19:

_____ 1. The presence of an enzyme _____ (increases, decreases) the energy of activation.

_____ 2. How does the enzyme do this?

Answers to Exercise 2.19: 1. decreases; 2. by aligning the reactants (it increases the probability that the reaction will occur).

T. Effect of Inhibitor on Enzyme Action

Color and label:
1. ○ substrate
2. ○ enzyme
3. ○ active site
4. ○ competitive inhibitor
5. ○ noncompetitive inhibitor

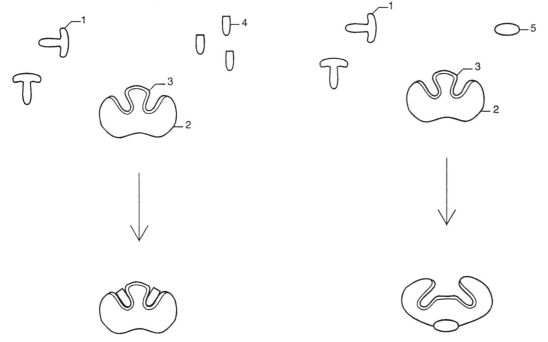

Figure 2.20a. Competitive inhibition.　　　　**Figure 2.20b.** Noncompetitive inhibition.

Exercise 2.20:

_____ 1. The competitive inhibitor reacts with the part of the enzyme called the _____ .

_____ 2. The purpose of the competitive inhibitor on the active site is to _____ .

_____ 3. The noncompetitive inhibitor reacts with the part of the enzyme called the _____ .

_____ 4. The purpose of the noncompetitive inhibitor on the active site is to _____ .

Answers to Exercise 2.20: 1. active site; 2. block it; 3. part other than active site; 4. indirectly make the active site inaccessible to the substrate.

Chapter 3 Cells

A. Cell Membrane

Color (or trace) and label:
1. phospholipids
   - 1a. ○ phosphate
   - 1b. ○ fatty acid
2. ○ cholesterol
3a. ○ integral protein
3b. ○ peripheral protein
4. ○ carbohydrate

Label:
5. glycocalyx
5a. glycoprotein
5b. glycolipid
6. channel
7. enzyme
8. carrier
9. marker
10. receptor

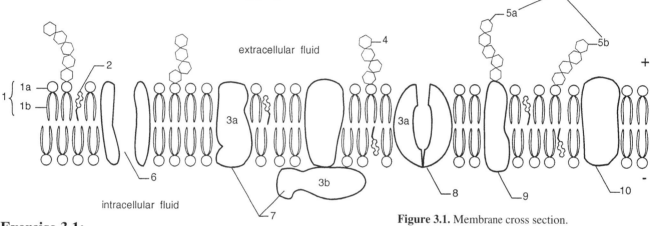

Figure 3.1. Membrane cross section.

Exercise 3.1:

1. The primary membrane components are ____ and ____ .
   a. Phospholipids contain hydrophilic ____ and hydrophobic ____ .
   b. Integral (intrinsic) proteins are primarily ____ (polar, nonpolar).
   c. Peripheral (extrinsic) proteins are primarily ____ (polar, nonpolar).
2. Carbohydrates chemically bound to membrane proteins are called ____ and carbohydrates chemically bound to membrane lipids are called ____ .
   a. Glycoproteins and glycolipids are parts of the ____ .
   b. The glycocalyx is on the ____ (inside, outside) surface of the membrane.
3. The outside of the membrane is usually ____ (positive (+), negative (−)).
4. What membrane components serve the following functions?
   a. stabilize the membrane
   b. transport water soluble (polar) substances
   c. allow movement of nonpolar substances across membrane (see solutions, p. 11)
   d. act as receptors
   e. aid in cell recognition
5. Where are the membrane enzymes located?

Answers to Exercise 3.1: 1. phospholipids, proteins; 1a. phosphates, fatty acids; 1b. nonpolar; 1c. polar; 2. glycoproteins, glycolipids; 2a. glycocalyx; 2b. outside surface; 3. positive; 4a. cholesterol; 4b. channels, carriers; 4c. phospholipids; 4d. proteins; 4e. glycocalyx, protein markers; 5. within membrane and on surface.

B. Movement Across a Membrane

Color:
1. ○ solute
2. ○ solvent (not shown in figure 3.2e–h)
3. ○ membrane
4. ○ carrier
5. ○ cellular debris (or bacterium)
6. ○ receptor

Label:
a. simple diffusion
b. osmosis
c. facilitated diffusion
d. active transport
e. phagocytosis
f. pinocytosis
g. receptor-mediated endocytosis
h. exocytosis

Exercise 3.2:

1. For figure 3.2a–d,

   a. the higher concentration of solute is in the area ____ (above, below) the membrane.

   b. the higher concentration of solvent is in the area ____ (above, below) the membrane.

   c. The solvent is ____ . (see solutions, p. 11)

2. Which transport mechanisms shown in figure 3.2a–d are described below?

   a. movement against a concentration gradient

   b. movement along a concentration gradient, does not require a membrane

   c. carrier mediated movement down a concentration gradient

   d. movement of water across a membrane

   e. requires ATP

   f. can move materials *into* or *out of* the cell

3. Can water move against a concentration gradient?

4. The rate of diffusion decreases as the size of the particle increases. How does sphere of hydration affect the diffusion rate of charged particles? (see sphere of hydration, p. 10)

5. Which transport mechanisms shown in figure 3.2e–h are described below?

   a. movement of large molecules across the membrane

   b. movement of large molecules out of the cell (secretion)

   c. outpocketing (evagination) of membrane to engulf bacteria or cellular debris

   d. infolding (invagination) of membrane to entrap small amounts of water and solutes

   e. specific reaction of substances with membrane receptors and invagination of the membrane

   f. requires ATP

6. Endocytosis is the process of transporting large molecules into the cell by enclosing them in membrane. Which of the transport mechanism(s) shown are examples of endocytosis?

Answers to Exercise 3.2: 1a. above; 1b. below; 1c. water; 2a. active transport; 2b. diffusion; 2c. facilitated diffusion; 2d. osmosis (also, diffusion); 2e. active transport; 2f. simple diffusion, osmosis, facilitated diffusion, active transport; 3. no (water can only move by diffusion); 4. decreases it; 5a. exocytosis and all types of endocytosis; 5b. exocytosis; 5c. phagocytosis; 5d. pinocytosis; 5e. receptor-mediated endocytosis; 5f. phagocytosis, pinocytosis, receptor-mediated endocytosis, and exocytosis; 6. phagocytosis, pinocytosis, and receptor-mediated endocytosis.

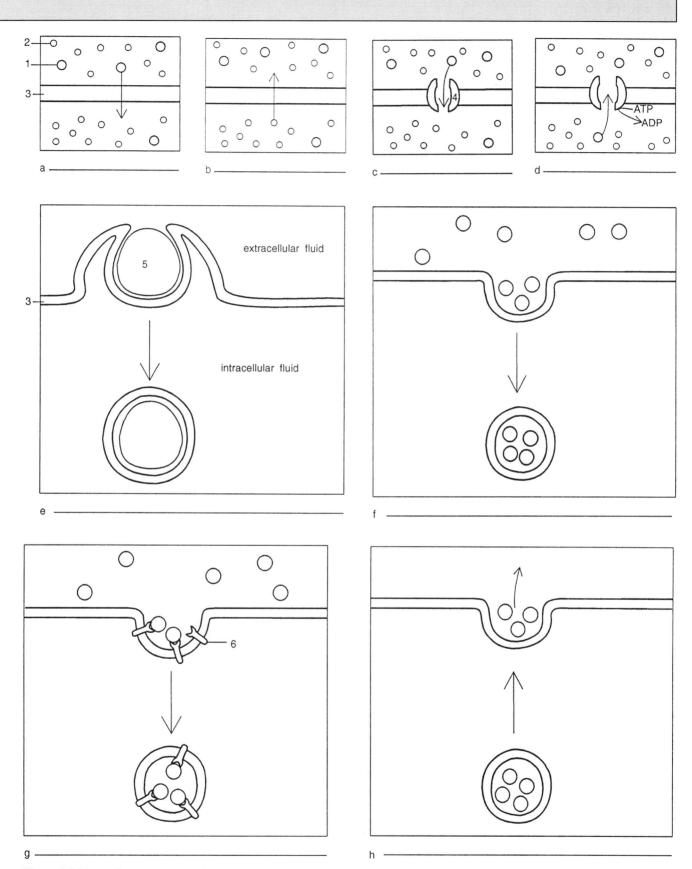

Figure 3.2. Types of transport across the membrane.

C. Tonicity

Color and label:
1. ○ solute
2. ○ solvent (water)

Label:
3. cell

Exercise 3.3:

1. The membrane shown in figure 3.3 is selectively permeable, i.e., allows only specific materials through.

    a. As shown, does solute cross the membrane?

    b. Does water cross the membrane?

    c. This movement of water across the membrane is called _____ .

2. The percentages shown represent the concentration of _____ (solute, water). (see solutions, p. 11)

3. The cell in figure 3.3a has a _____ (high, low) concentration of solute when compared with the surrounding hypotonic solution.

    a. Therefore, the cell's water concentration is _____ (high, low) when compared with the surrounding solution.

    b. Does water enter or leave the cell?

    c. What happens to the cell size?

    d. This process is called _____ .

4. The cell in figure 3.3c has a _____ (high, low) concentration of solute when compared with the surrounding hypertonic solution.

    a. Therefore, the cell's water concentration is _____ (high, low) when compared with the surrounding solution.

    b. Does water enter or leave the cell?

    c. What happens to the cell size?

    d. This process is called _____ .

5. Is the concentration gradient for the solute in the same or the opposite direction as the concentration gradient for the water?

6. Does a solute or water concentration gradient exist for cells in an isotonic solution?

7. For cells in an isotonic solution, does the movement of water across the membrane stop?

8. The pressure required to prevent the movement of water across a selectively permeable membrane is called _____ .

9. As the solute concentration on one side of a selectively permeable membrane increases, the rate of diffusion of water also increases. Does osmotic pressure increase, remain the same, or decrease?

Answers to Exercise 3.3: 1a. no; 1b. yes; 1c. osmosis; 2. solute; 3. high; 3a. low; 3b. enter; 3c. increase (cell may burst); 3d. hemolysis; 4. low; 4a. high; 4b. leave; 4c. decrease; 4d. crenation; 5. opposite; 6. no; 7. no (diffusion of water into and out of the cell is the same); 8. osmotic pressure; 9. increase.

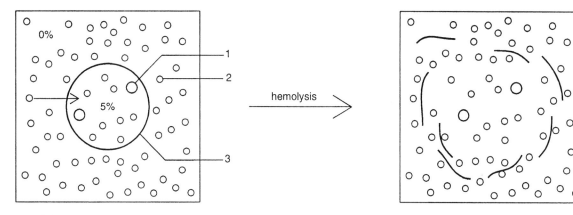

a. cell in hypotonic solution

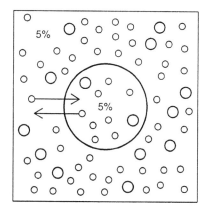

b. cell in isotonic solution

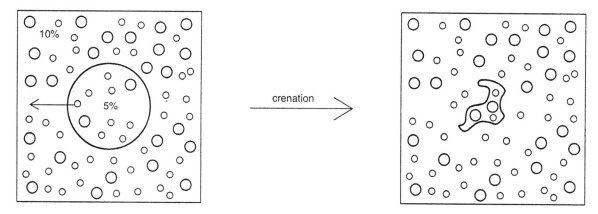

c. cell in hypertonic solution

Figure 3.3. Osmosis.

Tonicity

D. Cell Structure

Label:
1. plasma membrane
2. cytoplasm
3. nucleus
4. chromatin
5. nuclear pores

Color and label:
6. ○ nuclear membrane
7. ○ nucleolus
8. ○ ribosomes
   a. free
   b. attached
9. ○ endoplasmic reticulum
   a. rough endoplasmic reticulum
   b. smooth endoplasmic reticulum
10. ○ Golgi
11. ○ mitochondrion
12. ○ lysosome
13. ○ peroxisome
14. ○ inclusions
15. ○ centrioles
   a. centrosome
   b. basal body
16. ○ a. cilia
   b. flagellum
17. ○ microvilli

Label:
18. microfilaments
19. intermediate filaments
20. microtubules

Exercise 3.4:

1. Name the part of the cell that matches the description in the following chart:
   a. defines the cell surface
   b. matrix outside nucleus, contains organelles
   c. intracellular transport of materials, site of some chemical reactions (lipid metabolism, protein synthesis)
   d. site of protein synthesis, may be attached to endoplasmic reticulum or free
   e. process, sort, package, and deliver proteins to various cell parts and for possible secretion to outside
   f. ATP production from breakdown of glucose and other organic compounds
   g. contain potent digestive enzymes that can break down cells and/or cell parts
   h. contain enzymes that metabolize hydrogen peroxide
   i. site of spindle fiber attachment during cell division
   j. causes movement of materials across cell surface, contains microtubules
   k. elongated projection for movement of the cell, contains microtubules
   l. folds of plasma membrane that increase its surface area
   m. contains genetic material, encased by double membrane
   n. found within nucleus, site of ribosomal RNA production
   o. DNA content of cell
   p. large aggregates of specific chemicals, such as melanin, glycogen, lipids
   q. movement in muscle cells, hold shape of cell
   r. make up centrioles, cilia, flagella, spindle fibers
   s. structural support of cell
   t. anchor for cilia and flagella

2. What structures can be seen using the light microscope?
3. What structures can only be seen with electron microscopy?
4. Are all structures found in every cell?
5. Are some structures found in every cell?

Answers to Exercise 3.4: 1a. plasma membrane; 1b. cytoplasm; 1c. endoplasmic reticulum; 1d. ribosomes; 1e. Golgi; 1f. mitochondria; 1g. lysosome; 1h. peroxisome; 1i. centrosome; 1j. cilia; 1k. flagella; 1l. microvilli; 1m. nucleus; 1n. nucleolus; 1o. chromatin; 1p. inclusions; 1q. microfilaments; 1r. microtubules; 1s. intermediate filaments; 1t. basal bodies; 2. cytoplasm, nucleus, nucleolus, plasma membrane, nuclear membrane, chromatin, flagellum, cilia; 3. all other structures; 4. no (the number and kind of organelles present depend upon the function of the cell); 5. yes (mitochondria are needed in every cell; centrioles are in every dividing cell; nuclei are in most cells, but not mature red blood cells).

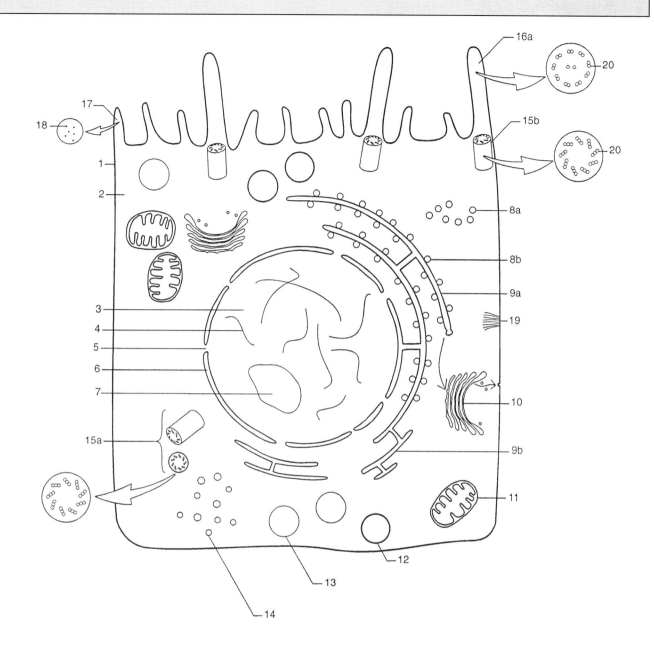

Figure 3.4a. Generalized cell.

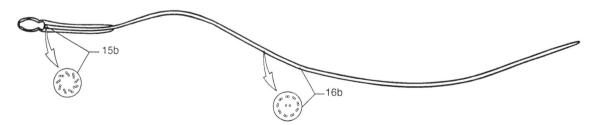

Figure 3.4b. Spermatocyte.

E. Chromosomes and DNA Replication

Color (or trace):

1. ○ chromosome
1a. chromatid
2a. ○ DNA backbone (original strand)
2b. ○ DNA backbone (new strand)

Label:

3. centromere
4. nucleotide

Exercise 3.5:

_____ 1. The genetic material in our cells is _____ and is found in _____ (cell structure).

_____ a. When a chromosome replicates, it forms two _____ that are held together at their _____ .

_____ b. When the centromeres separate, the two chromatids become two _____ .

_____ 2. Are both strands of DNA needed to make new DNA?

_____ 3. In order for DNA replication to occur, what bonds must break? (see DNA, p. 22)

_____ 4. How are the bases in the new DNA strands determined?

_____ 5. Does DNA replication occur along the entire length of DNA?

_____ 6. What is the distribution of original and new DNA in each chromatid that forms?

_____ 7. What is the distribution of old and new DNA in each new chromosome that forms?

_____ 8. Is the genetic material in each new chromosome identical to the genetic material in the original chromosome?

_____ 9. What is the fate of the two new chromosomes?

Answers to Exercise 3.5: 1. DNA, chromosomes; 1a. chromatids, centromeres; 1b. chromosomes; 2. yes (each strand serves as a template); 3. hydrogen bonds; 4. adenine always pairs with thymine, cytosine always pairs with guanine; 5. yes; 6. one original and one new strand of DNA per chromatid; 7. one original and one new strand of DNA per chromosome; 8. yes; 9. one to each new cell.

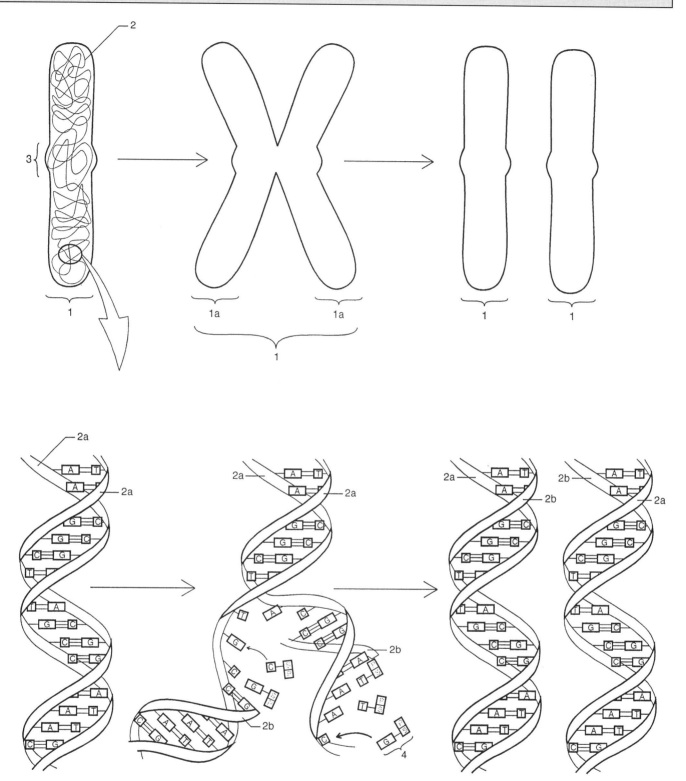

Figure 3.5. Chromosomes and DNA replication.

F. Mitosis

Color (trace) and label:
1. ○ nuclear membrane
2. ○ nucleolus
3. ○ long chromosome
4. ○ short chromosome
5. ○ centromere
6. ○ spindle fibers
7. ○ centrioles

Label:
8. chromatin
9. chromatid
10. cleavage furrow
    a. interphase
    b. prophase
    c. metaphase
    d. anaphase
    e. telophase
    f. daughter cells

Exercise 3.6:

_____

1. During interphase, DNA is uncoiled and has a mottled appearance in the nucleus. DNA in this form is called _____ .

_____

2. During cell division, DNA condenses into shortened rodlike structures called _____ , each containing two _____ .

    a. The cells shown in figure 3.6b contain _____ (number) pairs of chromosomes. How many pairs are present in most human cells?

    b. These chromosomes are visible throughout the four stages of mitosis: _____ , _____ , _____ , and _____ .

_____

    c. Since chromosomes appear in chromatid form, DNA replication must have occurred during _____ .

_____

3. Chromosomes line up in the center of the cell in _____ (a single row, pairs).

    a. When the centromeres separate, each chromatid becomes a new _____ .

_____

    b. Since one chromatid from each original chromosome moves to each pole, the cells that form will be _____ (identical, different).

_____

    c. Are the daughter cells that form genetically the same as the original cell? (see DNA replication, p. 35)

_____

4. Movement of chromosomes to the center of the cell and later to their respective poles depends upon lengthening and shortening of _____ .

_____

5. The spindle fibers are attached to the _____ and the _____ .

_____

6. Are the daughter cells the same size as the original cell?

7. Name the stage in cell division when each of the following occurs:

_____

    a. spindle fibers appear

_____

    b. centrioles move to poles

_____

    c. centromeres divide

_____

    d. chromosomes become visible

_____

    e. cytokinesis (division of cytoplasm)

_____

    f. nuclear membrane disappears

_____

    g. nuclear membrane reappears

_____

    h. chromosomes line up in center of cell

_____

    i. chromosomes move to poles

_____

    j. spindle fibers disappear

_____

    k. nucleolus disappears

_____

    l. nucleolus reappears

Answers to Exercise 3.6: 1. chromatin; 2. chromosomes, chromatids; 2a. 2, 23; 2b. prophase, metaphase, anaphase, telophase; 2c. interphase; 3. a single row; 3a. chromosome; 3b. identical; 3c. yes; 4. spindle fibers; 5. centrioles, centromeres; 6. no; 7a. prophase; 7b. prophase; 7c. anaphase; 7d. prophase; 7e. telophase; 7f. prophase; 7g. telophase; 7h. metaphase; 7i. anaphase; 7j. telophase; 7k. prophase; 7l. telophase.

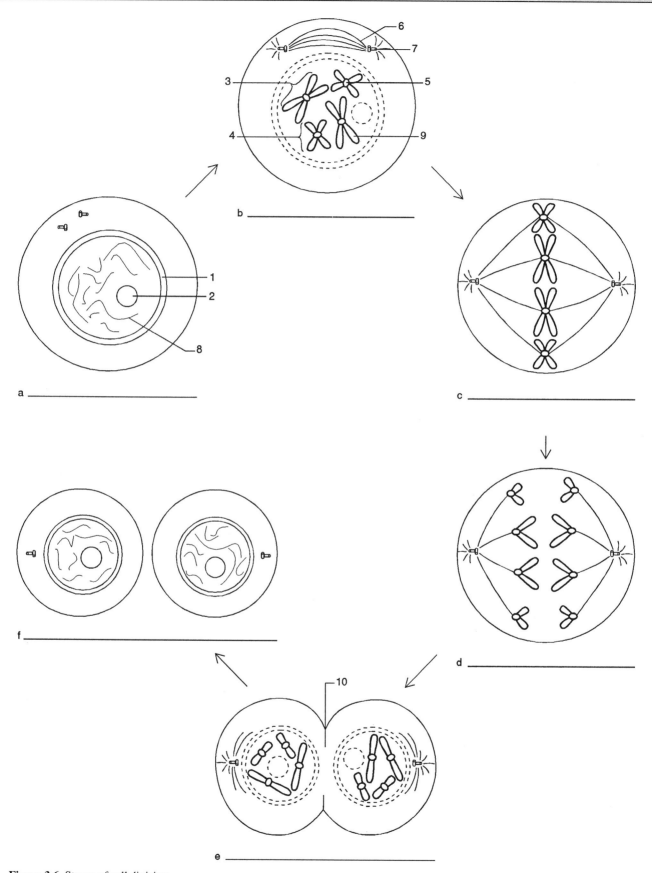

Figure 3.6. Stages of cell division.

G. RNA Formation

Color:

A ○ adenine
T ○ thymine
C ○ cytosine
G ○ guanine
U ○ uracil

Color the sugar-phosphate backbone and label:

1. DNA
   a. ○ DNA (sense strand)
   b. ○ DNA (complementary strand)

○ **Color the sugar-phosphate backbones the same. Label:**

2. messenger RNA
3. RNA nucleotides

Label:

4. gene
5. chromosome

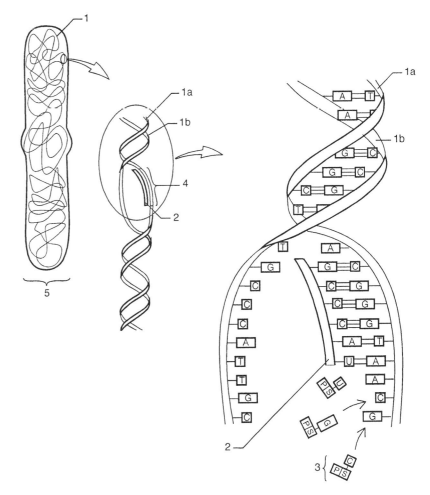

Figure 3.7. RNA formation.

Exercise 3.7:

_____ 1. Are both strands of DNA needed to make RNA?

_____ 2. Does RNA formation occur along the entire length of the DNA (chromosome)?

_____ 3. The section of DNA that is copied during RNA formation is called a _____.

_____ 4. How are the RNA nucleotides different from DNA nucleotides? (see nucleic acids, p. 22)

_____ 5. Adenine in DNA always pairs with _____ in RNA nucleotides.

_____ 6. What kinds of RNA are made by this complementary copying of DNA?

Answers to Exercise 3.7: 1. no; 2. no; 3. gene; 4. ribose and uracil present, but no thymine; 5. uracil; 6. messenger RNA, transfer RNA, ribosomal RNA.

H. Protein Synthesis

Color and label:
1. nuclear membrane
2. ○ DNA
3. ○ mRNA
4. ribosome
   a. ○ large subunit
   b. ○ small subunit
5. ○ transfer RNA
6. ○ amino acids

Label:
7. nucleus
8. nuclear pore
9. cytoplasm

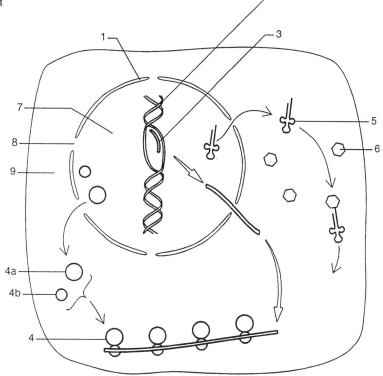

Figure 3.8. Transcription and translation.

Exercise 3.8:

1. Which RNA performs each of the following functions?
   a. transports amino acids to assembly site
   b. contains blueprint for protein synthesis
   c. site of protein assembly
2. Messenger RNA formation occurs in the _____ (cell structure). The name of this process is _____ .
3. Messenger RNA leaves the nucleus through _____ .
4. Messenger RNA is read at the _____ (cell structure). The name of this process is _____ .
5. Amino acids are transported to the ribosomes by _____ .
6. Each transfer RNA carries a specific amino acid. Since there are 20 different amino acids, the minimum number of transfer RNAs needed for protein synthesis is _____ .

Answers to Exercise 3.8: 1a. tRNA; 1b. mRNA; 1c. ribosomes; 2. nucleus, transcription; 3. nuclear pores; 4. ribosomes, translation; 5. tRNA; 6. 20.

I. Genetic Code

Color:
A ○ adenine
U ○ uracil
C ○ cytosine
G ○ guanine

Color (trace) and label:
1. ○ ○ ○ ○ amino acids
2. ○ peptide bond
3. ○ ribosome
4. ○ transfer RNA
5. ○ mRNA (backbone)

Label:
6. codon
7. anticodon

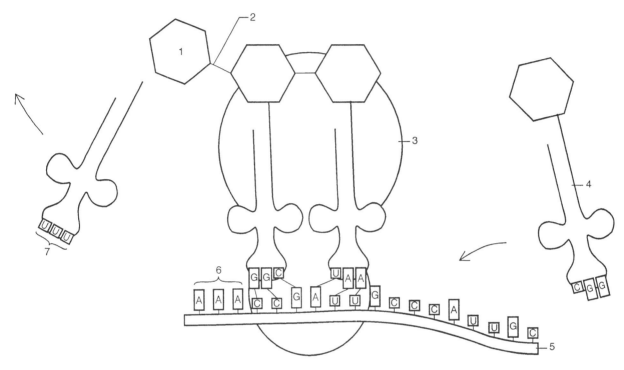

Figure 3.9. Protein assembly.

Exercise 3.9:

1. How many mRNA bases code for one amino acid?
2. The mRNA codon is complementary to the _____ on transfer RNA.
3. How does transfer RNA correctly read (translate) messenger RNA?
4. For a protein containing 100 amino acids, what is the minimum number of mRNA bases needed?
5. The sequence of bases in mRNA determines the sequence of amino acids in a protein. If one base is different (because of a DNA mutation), could this change the corresponding amino acid?
6. If some amino acids are not present in the cell when protein synthesis is occurring, will the protein be completed?

Answers to Exercise 3.9: 1. 3; 2. anticodon; 3. anticodon (tRNA) and codon (mRNA) complement; 4. 3 × 100 = 300; 5. yes; 6. no.

Chapter 4 Tissues

A. Plasma Membrane Modifications

Color and label:
1. ○ epithelial cells
2. ○ underlying connective tissue
3. ○ microvilli
   a. microfilaments
4. ○ cilia
   a. microtubules

Color (or trace) and label:
5. ○ membrane
6. ○ glycoprotein
7. ○ tonofilaments
8. ○ connexions

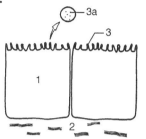
Figure 4.1a. Microvilli.

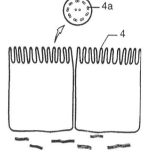
Figure 4.1b. Cilia.

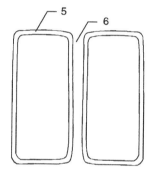
Figure 4.1c. No membrane modification.

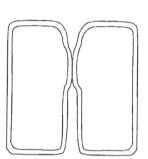
Figure 4.1d. Tight junction.

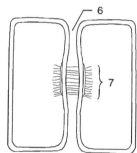
Figure 4.1e. Adhering junction.

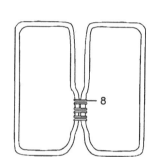
Figure 4.1f. Gap junction.

Exercise 4.1:

_____ 1. Which of the drawings show surface modifications?

_____   a. Are these surface modifications found in epithelial, connective, muscle, or nervous tissue?

_____   b. What surface modification contains microtubules? What is its function?

_____   c. What surface modification contains microfilaments? What is its function?

_____ 2. Which of the drawings show cell junctions?

3. Match each junction type with its description:

_____   a. prevents movement of materials between cells, found between epithelial cells

_____   b. holds cells together, found in skin and cardiac muscle

_____   c. transmits impulse and small ions from cell to cell, found in heart, smooth muscle, some nerve cells

_____ 4. Adhering junctions that exist in a band around cells are called _____, while those that exist as spots are called _____.

Answers to Exercise 4.1: 1. a and b; 1a. epithelial tissue; 1b. cilia, move materials along outer surface of cells; 1c. microvilli, increase surface area for membrane transport; 2. d, e, f; 3a. tight junctions; 3b. adhering junctions; 3c. gap junctions; 4. intermediate junctions, desmosomes.

B. Epithelial Tissue

Label:
a. simple squamous
b. simple cuboidal
c. simple columnar
d. pseudostratified and grandular
e. stratified squamous
f. stratified cuboidal
g. stratified columnar
h. transitional

Color (trace). Label one drawing.
1. ○ free surface (of epithelial cells)
2. ○ basement membrane

Color and label:
3. ○ epithelial cells
4. ○ underlying connective tissue
5. ○ blood vessels (red)
6. ○ cilia
7. ○ goblet cell (unicellular gland)

Label:
8. duct (of multicellular gland)
9. secretory portion (of multicellular gland)

Exercise 4.2:

1. Which epithelial tissues are one layer thick?
2. Which epithelial tissues have more than one layer?
3. Are all the epithelial cells in the same tissue the same shape?
4. What is the name of the layer between the epithelium and connective tissue that contains noncellular materials as well as collagenous and reticular fibers?
5. Are there any blood vessels in the epithelium itself? How does epithelium obtain nutrients and oxygen and get rid of wastes?
6. Which are more likely to serve as absorptive or secretory surfaces—simple or stratified tissues?
7. Which are more resistant to wear and tear—simple or stratified tissues?
8. Which are in line—the nuclei in cells of simple columnar or pseudostratified epithelium?
9. Which vary in appearance—simple columnar or pseudostratified nuclei?
10. Match the epithelial tissue with its description.
    a. single layer, all cells touch basement membrane but not all cells reach free surface
    b. cells change shape when stretched versus contracted
    c. more than one layer thick, surface cells are flattened
    d. more than one layer thick, surface cells are irregular in shape
11. Are the glands that secrete their contents to the surface called exocrine or endocrine glands?
12. Are the mucus-secreting goblet cells unicellular or multicellular glands?
13. What type of epithelium lines the duct of the multicellular gland shown in figure 4.2d?
14. Holocrine glands release their secretions when cells burst. What is this process called?
15. Merocrine glands release their secretions when vesicles containing secretions bind with the plasma membrane. What is this process called?
16. Are glands without ducts called exocrine or endocrine glands?
17. How are the secretions of endocrine glands transported to their target cells?

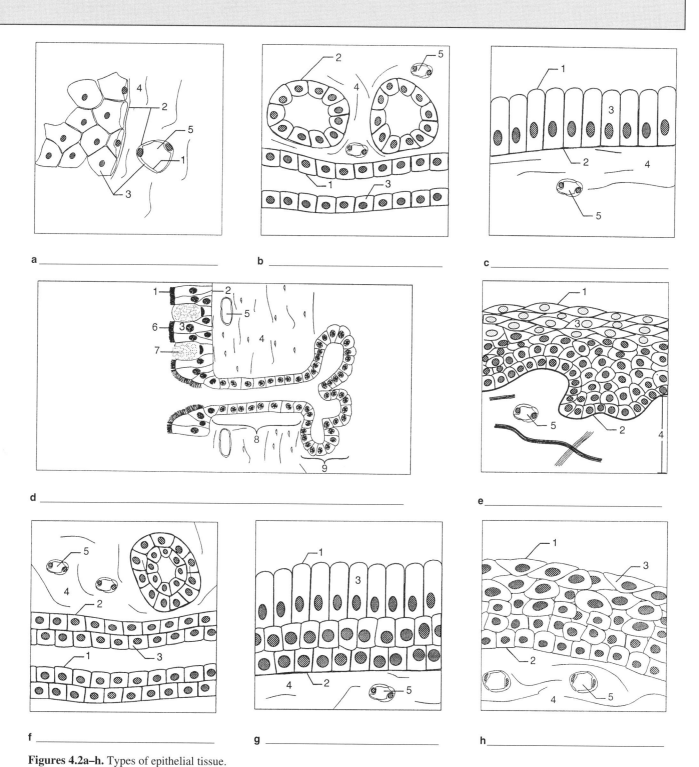

Figures 4.2a–h. Types of epithelial tissue.

Answers to Exercise 4.2: 1. simple squamous, simple cuboidal, simple columnar, pseudostratified; 2. stratified squamous, stratified cuboidal, stratified columnar, transitional; 3. no (the cells of stratified epithelium vary; basal cells, columnar or cuboidal in shape, change as they push toward the surface; the name of the stratified tissue is determined by the shape of the cells at the free surface); 4. basement membrane; 5. no, diffusion (from and to blood vessels found in underlying connective tissue); 6. simple; 7. stratified; 8. simple columnar; 9. pseudostratified (basal cell nuclei are small and darkly stained, apical cells are larger and stain lighter); 10a. pseudostratified; 10b. transitional; 10c. stratified squamous; 10d. transitional; 11. exocrine; 12. unicellular glands; 13. stratified cuboidal (other ducts may have other types of epithelial linings); 14. lysis; 15. exocytosis; (electron microscopy does not support the existence of apocrine secretion, sometimes mentioned as a third type of secretion); 16. endocrine; 17. blood vessels.

Epithelial Tissue

C. Connective Tissue

(*Color the same in all drawings)

Label figure 4.3a:
a. mesenchyme (embryonic tissue)

Color (trace) and label:
1. ○ mesenchymal cells
2. ○ fibers
3. ○ matrix*

Label figure 4.3b–f:
b. loose (also called areolar)
c. dense irregular
d. dense regular
e. elastic
f. reticular tissue

Color (or trace) and label:
1. ○ fibroblasts (fibroblast nuclei)*
2. ○ mast cells
3. ○ macrophage cells
4. ○ reticular cells
5. ○ collagen fibers*
6. ○ elastic fibers*
7. ○ reticular fibers
8. ○ matrix*

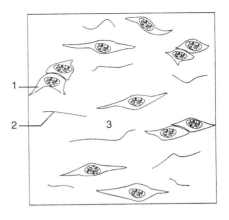

a _____

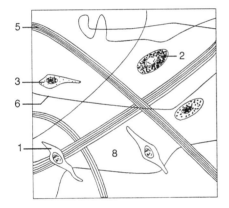

b _____

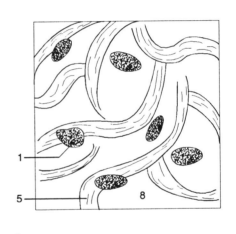

c _____

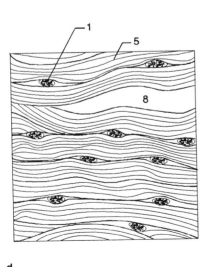

d _____

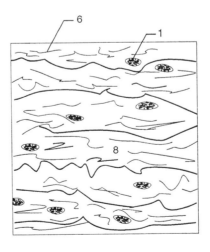

e _____

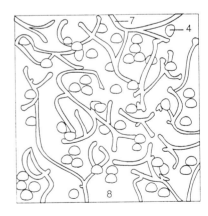

f _____

Figure 4.3 Types of connective tissue.

Label figure 4.3g:

g. adipose tissue

Color and label:

1. ○ nucleus
2. ○ cytoplasm
3. ○ fat deposits

Label figure 4.3h–j:

h. hyaline cartilage
i. elastic cartilage
j. fibrocartilage

Color (or trace) and label:

1. ○ chondrocytes
2. ○ lacunae
3. ○ collagen fibers*
4. ○ elastic fibers*
5. ○ matrix*

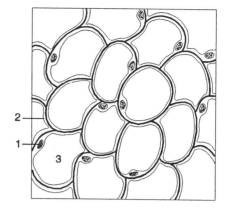

g _____

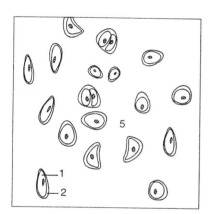

h _____

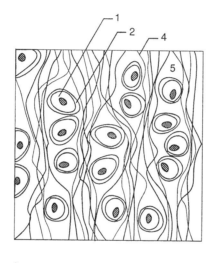

i _____

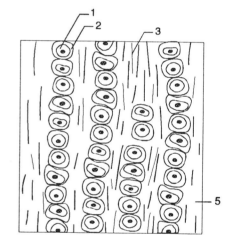

j _____

Label figure 4.3k:

k. bone

Color (or trace) and label:

1. ○ osteocytes
2. ○ Haversian canal
3. ○ lamellae
4. ○ canaliculi

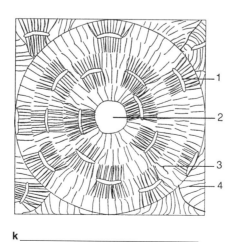

k _____

Connective Tissue 45

C. Connective Tissue continued

Label figure 4.3l:
1. blood

Color and label:
1. ○ erythrocytes
2. ○ leucocytes
3. ○ thrombocytes
4. ○ matrix* (plasma)

Exercise 4.3:

1. Which are more closely packed—epithelial or connective tissue cells? Which tissue has more intercellular material?
2. The matrix (intercellular material) contains _____ and _____ .
3. What general functions are served by connective tissue?
4. What functions are served by the following cells?
    a. fibroblasts
    b. mast cells
    c. macrophage cells
5. What functions are served by each type of fiber?
    a. collagen
    b. reticular
    c. elastic
6. Does hyaline cartilage contain fibers?
7. Are all types of connective tissue vascular?
8. Where are the following found in the body?
    a. hyaline cartilage
    b. elastic cartilage
    c. fibrocartilage
9. How do cartilage matrix and bone matrix differ?
10. Which connective tissue type matches each description?
    a. parallel collagenous fibers with nuclei between fiber bundles
    b. variety of cells and fibers, fibers run in all directions, fluid matrix
    c. contains chondrocytes in rows
    d. collagenous fibers run in various directions

Answers to Exercise 4.3: 1. epithelial cells, connective tissue; 2. ground substance (intercellular fluid and various large organic molecules), fibers; 3. bind organs together, support body, protect, nourish epithelium, insulate store energy, transport materials throughout body; 4a. make fibers; 4b. make heparin and histamine; 4c. engulf foreign and dead material; 5a. provide strength; 5b. create meshwork for holding cells and tissues in place; 5c. can stretch and return to original shape; 6. yes (only visible when the appropriate stain is used); 7. no (cartilage is avascular, and dense connective tissue has a poor blood supply); 8a. embryo, articular surface of bones, costal cartilages of ribs, nose, larynx, trachea, bronchi; 8b. external ear, epiglottis; 8c. intervertebral discs, pubic symphysis; 9. the presence of inorganic salts makes bone harder than cartilage; 10a. dense regular; 10b. loose; 10c. fibrocartilage; 10d. dense irregular.

46   Chapter 4   Tissues

D. Muscle Tissue

Label:
a. skeletal muscle
b. cardiac muscle
c. smooth muscle

Color (or trace) and label:
1. ○ nuclei
2. ○ striations
3. ○ sarcolemma
4. ○ intercalated discs

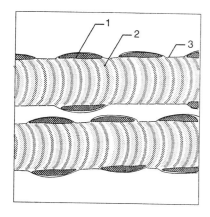 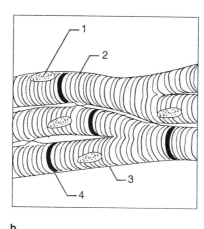 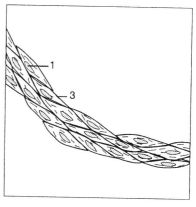

a _____ b _____ c _____

Figure 4.4. Types of muscle tissue.

Exercise 4.4:

_____ 1. Which muscle tissue shows the following characteristics?

_____   a. striations

_____   b. multinucleation

_____   c. fusiform (spindle shaped) cells

_____ 2. How many skeletal muscle cells do you see in figure 4.4a?

_____ 3. What cell surface modifications make up intercalated discs?

_____ 4. What functions are served by intercalated discs?

_____ 5. Single unit smooth muscle contains sheets of muscle that contract as a unit, while multiunit smooth muscle contains muscle cells that act independently. Which would you expect to have greater numbers of gap junctions?

_____ 6. Which muscle tissues are under involuntary control?

Answers to Exercise 4.4: 1a. skeletal, cardiac; 1b. skeletal; 1c. smooth muscle; 2. parts of 2 cells; 3. desmosomes, gap junctions; 4. serve to connect adjacent cells for strength and for movement of impulse; 5. single unit smooth muscle 6. cardiac, smooth.

E. Nervous Tissue

Color (or trace) and label:
1. ○ cell body
2. ○ nucleus
3. ○ dendrites
4. ○ axon
5. ○ myelin sheath
6. ○ terminal branches
    a. terminal boutons

Label:
7. node of Ranvier

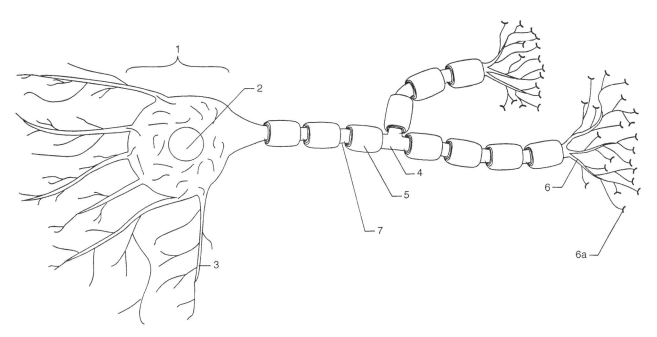

Figure 4.5. Motor neuron.

Chapter 5 Integument

A. General Anatomy

Color (trace) and label (save red and blue for blood vessels):
1. ○ epidermis
2. ○ dermis
3. ○ subcutaneous layer (hypodermis, superficial fascia)
4. ○ hair
5. ○ sebaceous gland
6. ○ arrector pili
7. ○ nerves
8. ○ sensory receptor
9. ○ arteries (red)
10. ○ veins (blue)
11. ○ eccrine sweat gland
12. ○ apocrine (odoriferous) sweat gland

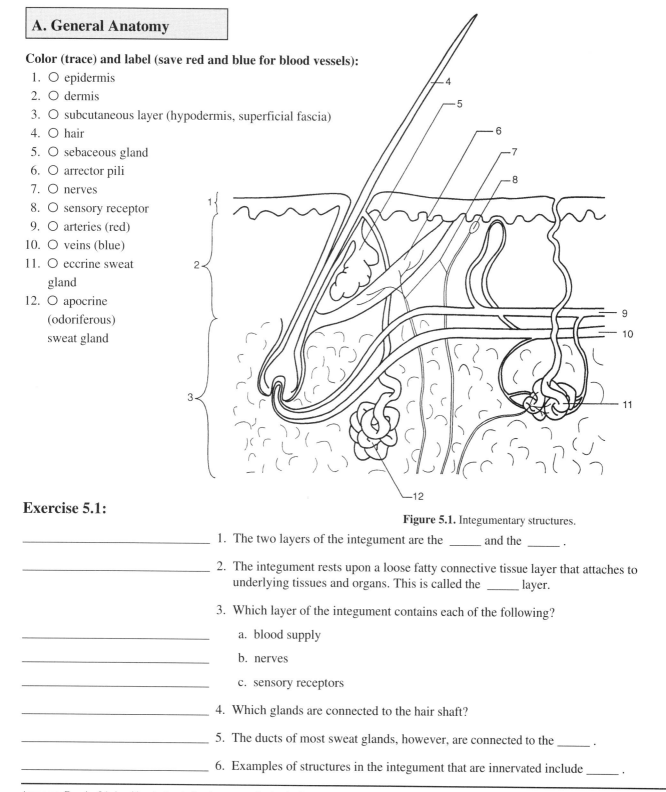

Figure 5.1. Integumentary structures.

Exercise 5.1:

1. The two layers of the integument are the _____ and the _____ .

2. The integument rests upon a loose fatty connective tissue layer that attaches to underlying tissues and organs. This is called the _____ layer.

3. Which layer of the integument contains each of the following?
   a. blood supply
   b. nerves
   c. sensory receptors

4. Which glands are connected to the hair shaft?

5. The ducts of most sweat glands, however, are connected to the _____ .

6. Examples of structures in the integument that are innervated include _____ .

Answers to Exercise 5.1: 1. epidermis, dermis; 2. subcutaneous; 3a. dermis; 3b. dermis; 3c. dermis, some receptors cross into epidermis; 4. sebaceous gland, apocrine sweat gland; 5. epidermis; 6. hair follicle, dermis, glands, arrector pili.

B. Histology of Integument

Label:
1. epidermis
2. dermis
3. subcutaneous layer
4. dermal papilla
5. epidermal pegs

Color and label:
6. ○ stratum corneum
7. ○ stratum lucidum
8. ○ stratum granulosum
9. ○ stratum germinativum
   a. ○ stratum spinosum (Malpighi)
   b. ○ stratum basale
10. ○ papillary layer of dermis
11. ○ reticular layer of dermis

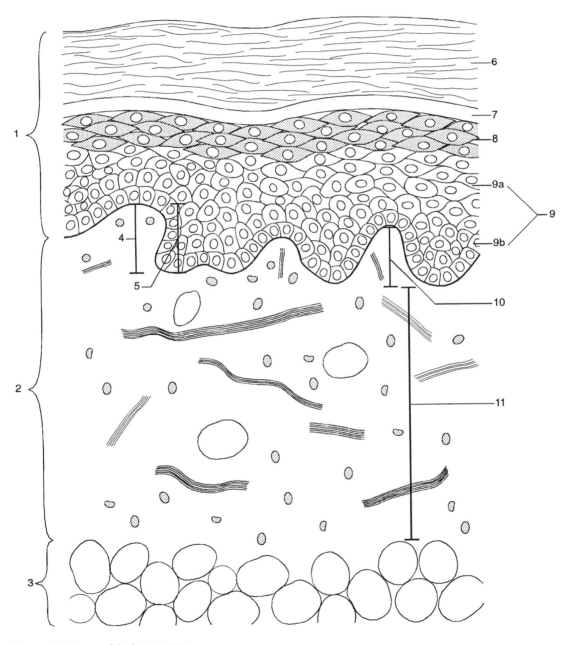

Figure 5.2. Layers of the integument.

50 Chapter 5 Integument

Exercise 5.2:

1. Match the layer with the description below.

   _____ a. clear, flat, dead cells containing eleidin

   _____ b. contains cells that continuously divide

   _____ c. loose connective tissue containing sensory receptors

   _____ d. cells make keratohyalin, nuclei begin to break down

   _____ e. polyhedral cells that have spiny processes

   _____ f. flat, dead cells containing keratin, constantly sloughing off

   _____ g. dense, irregular connective tissue containing hair follicles and sebaceous glands.

2. The following components contribute to integumentary functioning: melanin, nerve endings, stratified squamous, sweat glands, blood flow, Langerhans cells, keratin, dermal papillae, epidermal pegs. Match them with the list of functions below. (You may use the same component more than once.)

   _____ a. UV protection

   _____ b. sensation

   _____ c. excretion

   _____ d. temperature regulation

   _____ e. vitamin D production

   _____ f. immune response

   _____ g. prevention of water loss

   _____ h. protection from abrasion

   _____ i. prevention of epidermis and dermis from slipping laterally

   _____ 3. Do all samples of skin have the same number of layers?

   _____ 4. What is the boundary between the epidermis and dermis?

   _____ 5. Is there a sharp boundary between the dermis and hypodermis?

Answers to Exercise 5.2: 1a. stratum lucidum; 1b. stratum basale; 1c. dermis; 1d. stratum granulosum; 1e. stratum spinosum; 1f. stratum corneum; 1g. dermis; 2a. melanin; 2b. nerve endings; 2c. sweat glands; 2d. sweat glands, blood flow; 2e. stratified squamous; 2f. Langerhans cells; 2g. keratin; 2h. stratified squamous; 2i. dermal papillae and epidermal pegs interdigitate; 3. no (thin skin does not have a stratum lucidum, and the stratum corneum and stratum granulosum have fewer cell layers); 4. basement membrane; 5. no.

C. Hair

Color and label:
1. hair
   a. ○ hair shaft
   b. ○ hair root
2. hair follicle
   ○ a. epithelial root sheath
   ○ b. connective tissue root sheath
3. ○ sebaceous gland
4. ○ arrector pili
5. ○ blood vessel (red)

Label:
6. hair bulb
   a. matrix
   b. papilla

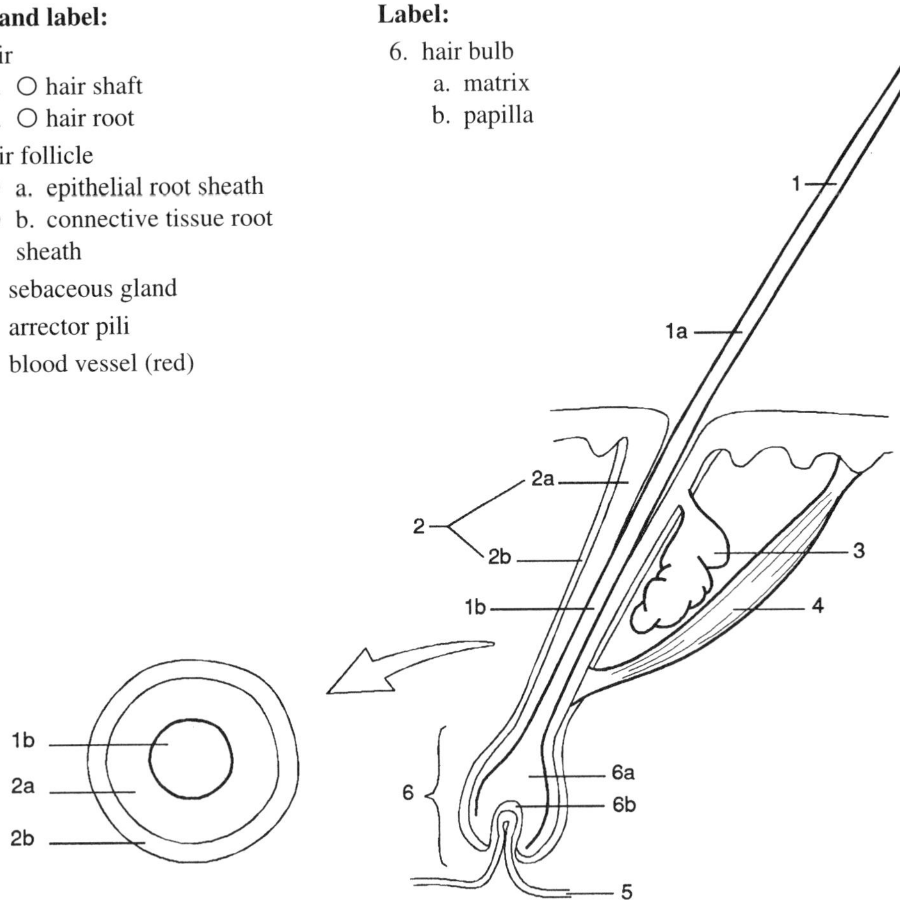

Figure 5.3. Section of hair and follicle.

Exercise 5.3:

_____ 1. What kinds of tissues make up the root sheath?

_____ 2. Where are the dividing cells that cause hair growth?

_____ 3. How is the hair root nourished?

_____ 4. In what layer of skin is the papilla?

_____ 5. What kind of tissue is the arrector pili?

_____ 6. What does the arrector pili do?

_____ 7. What does the sebaceous gland produce?

Answers to Exercise 5.3: 1. epithelial, connective; 2. matrix; 3. blood in papilla; 4. dermis; 5. smooth muscle; 6. causes "goose bumps"; 7. oil.

D. Structure of the Nail

Color and label:
1. ○ nail plate
2. ○ epidermis
3. ○ dermis
4. ○ hypodermis
5. ○ bone

Label:
6. nail bed
7. nail root
8. nail matrix
9. eponychium (cuticle)
10. hyponychium
11. nail wall
12. lunula

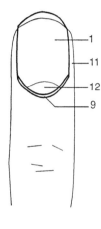
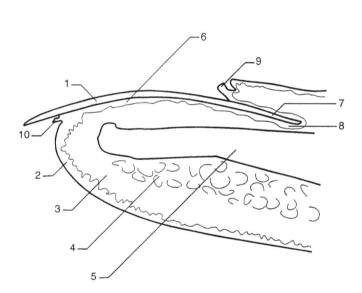

Figure 5.4. Nail anatomy.

Exercise 5.4:

_____ 1. Is the nail bed epidermis or dermis?

_____ 2. What cells divide in order for the nail to grow?

_____ 3. What protein is responsible for nail hardness? What layer of the integument contains this protein?

_____ 4. Is the tissue that connects with the upper surface of bone the epidermis, dermis, or hypodermis?

Answers to Exercise 5.4: 1. epidermis; 2. nail matrix; 3. keratin, stratum corneum; 4. dermis.

E. Burns

Color (trace) and label:
1. ○ epidermis
2. ○ dermis
3. ○ hypodermis
4. ○ hair
5. ○ nerve
6. ○ blood vessels (red)
7. ○ gland
8. ○ blister

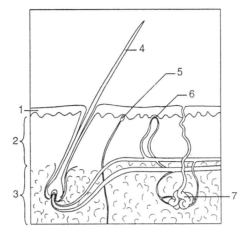

Figure 5.5a. Normal integument.

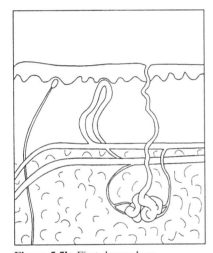

Figure 5.5b. First-degree burn.

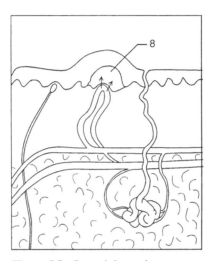

Figure 5.5c. Second-degree burn.

Figure 5.5d. Third-degree burn.

Exercise 5.5:

1. For first-degree burns, the blood vessels dilate, causing the skin to appear _____ and fluids to seep, resulting in _____ .

2. For second-degree burns, fluids seeping from blood vessels form _____ .

    a. What effect does a blister have on nutrient and waste product exchanges between epidermal cells and blood vessels?

    b. Therefore, the epidermis above a blister _____ .

    c. New epidermis grows from _____ .

3. Given the location of nerve endings, would first- and second-degree burns be painful?

4. For third-degree burns, the _____ and _____ (layers) are destroyed.

5. If hair follicles remain intact, can epidermis regenerate from them?

Answers to Exercise 5.5: 1. red, swelling (inflammation); 2. blisters; 2a. interferes with exchanges; 2b. dies; 2c. epidermis next to blister (and hair follicles); 3. yes; 4. epidermis, dermis; 5. yes.

Chapter 6 Skeletal System

A. Gross Anatomy of Bone

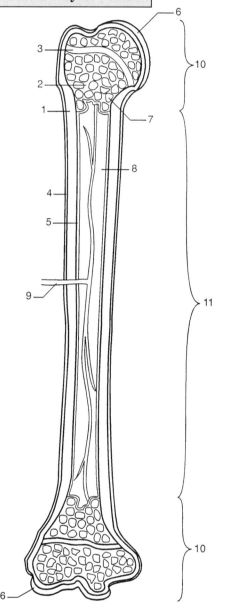

Figure 6.1a. Section of adult humerus.

Color and label:
1. ○ compact bone
2. ○ spongy (cancellous) bone
3. ○ epiphyseal line
4. ○ periosteum
5. ○ endosteum
6. ○ articular cartilage
7. ○ red marrow (red)
8. ○ marrow or medullary cavity (yellow marrow)
9. ○ nutrient artery (use red)

Label:
10. epiphysis
11. diaphysis
12. diploe

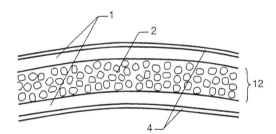

Figure 6.1b. Section of flat bone.

Exercise 6.1:

_____ 1. There are two types of bone, the denser _____ and the lighter _____ .

_____ 2. The outer surface of bone is _____ (compact, spongy).

_____ 3. The epiphyses of long bones are primarily _____ (compact, spongy bone).

_____ 4. The compact bone of the diaphysis is _____ (thicker, thinner) than that of the epiphyses.

55

_____ 5. What type of connective tissue is periosteum?

6. Which structures are described below?

_____ a. anchors blood vessels and nerves, contains bone forming cells, anchors tendons and ligaments

_____ b. lines cavities within bone, contains bone forming cells

_____ c. marks site of growth in length during childhood

_____ d. supplies blood to marrow

_____ e. site of blood cell production in fetus and young children

_____ f. site of blood cell production in adults

_____ g. contains fatty tissue in adult

Answers to Exercise 6.1: 1. compact bone, spongy bone; 2. compact; 3. spongy; 4. thicker; 5. dense, irregular, connective tissue; 6a. periosteum; 6b. endosteum; 6c. epiphyseal line; 6d. nutrient artery; 6e. spongy bone, medullary cavity; 6f. spongy bone; 6g. medullary cavity.

B. Microscopic Anatomy of Bone

Color and label:
1. ○ osteocytes
2. ○ calcified matrix
3. ○ periosteum
4. ○ blood vessels (red)

Label:
5. compact bone
6. spongy bone

7. lacuna
8. canaliculi
9. Volkmann's canal
10. Haversian canal
11. Haversian system (osteon)
12. lamella
13. trabecula
14. marrow spaces

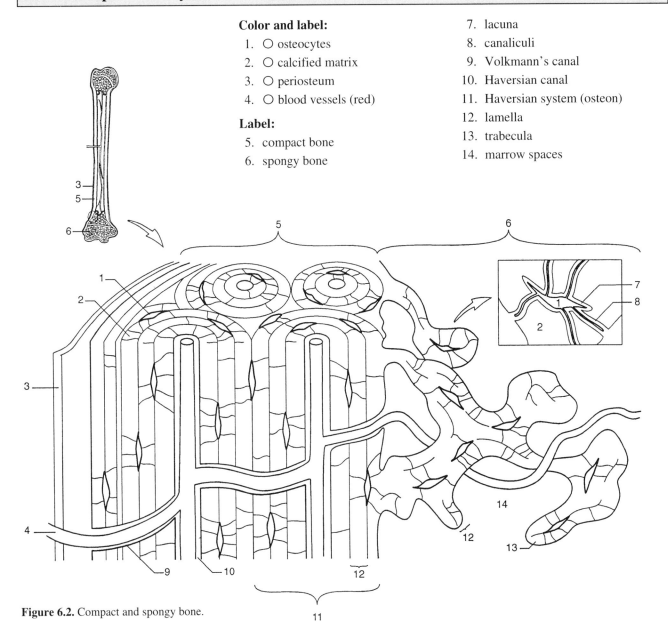

Figure 6.2. Compact and spongy bone.

Exercise 6.2:

_____ 1. Bone contains cells called _____ embedded in a matrix.

_____ 2. The organic portion of the matrix containing _____ and _____ is called the osteoid.

_____ 3. The inorganic portion of the matrix contains _____ salts.

_____ 4. What structures are the same in compact and spongy bone?

_____ 5. How are compact and spongy bone different?

_____ 6. The surface of every bone is _____ (compact, spongy bone) and is covered by a membranous _____ .

7. What structures match each of the following descriptions?

_____ a. spaces in bone matrix occupied by osteocytes

_____ b. space in bone matrix that provides for diffusion of nutrients and waste products

_____ c. location of blood vessels in compact bone

_____ d. location of blood vessels in spongy bone

_____ e. releases calcium from bone to maintain blood calcium levels, maintains organic matrix

_____ f. portion of matrix that provides rigidity

_____ g. portion of matrix that provides flexibility

_____ h. layered appearance of Haversian systems (due to collagen fiber orientation)

_____ 8. Which way do Haversian systems run within long bones?

Answers to Exercise 6.2: 1. osteocytes; 2. collagenous fibers, organic molecules; 3. calcium; 4. osteocytes and matrix; 5. organization (Haversian systems in compact bone and trabeculae in spongy bone); 6. compact, periosteum; 7a. lacuna and canaliculi; 7b. canaliculi; 7c. Haversian and Volkmann's canals; 7d. marrow spaces between trabeculae; 7e. osteocytes and osteoblasts; 7f. calcium salts; 7g. collagenous fibers; 7h. lamellae; 8. parallel to length of bone.

C. Types of Bone Development

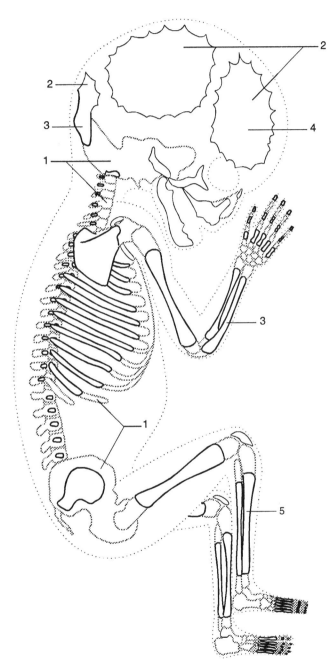

Figure 6.3a. Embryo at 10 weeks.

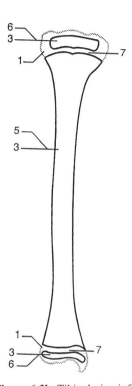

Figure 6.3b. Tibia during infancy.

58 Chapter 6 Skeletal System

Color and label:
1. ○ cartilage (thick, dashed line)
2. ○ intramembranous bone (thin, solid line)
3. ○ endochondral bone (thick, solid line)

Label:
4. ossification center
5. primary ossification center
6. secondary ossification center
7. epiphyseal cartilage

Exercise 6.3:

_____ 1. Bone always forms by replacement of preexisting connective tissue. Bone formation by replacement of mesenchyme is called _____ .

_____ 2. Intramembranous bone development in the embryo forms what bones?

_____ 3. Is there more than one center of ossification in the skull?

_____ 4. Do the bones of the skull fuse before birth, during childhood, or at puberty?

_____ 5. If the skull bones were to fuse before birth, what problems might result?

_____ 6. Bone formation by replacement of cartilage is called _____ .

_____ 7. During endochondral bone dvelopment, the primary ossification center provides strength for the cartilaginous skeleton by forming at the _____ (diaphysis, epiphysis).

_____ 8. Where are the secondary ossification centers located?

_____ 9. When do the secondary ossification centers form?

_____ 10. The area between the primary and secondary ossification centers is called the _____ .

_____ 11. The epiphyseal cartilage grows and bone replaces it. Is this called intramembranous or endochondral bone formation?

_____ 12. Where does growth in length occur in long bones?

_____ 13. After puberty, the epiphyseal cartilage is completely replaced by bone. This bone is the _____ .

_____ 14. Can growth in length occur at the epiphyseal line?

Answers to Exercise 6.3: 1. intramembranous; 2. most of the skull, clavicle, patella; 3. yes; 4. puberty; 5. lack of flexibility for birth and growth of brain; 6. endochondral; 7. diaphysis; 8. epiphyses; 9. at birth, during childhood (depends upon bone); 10. epiphyseal cartilage; 11. endochondral; 12. at the epiphyseal cartilage; 13. epiphyseal line; 14. no.

Types of Bone Development

D. Intramembranous Bone Formation

Color (trace) and label:
1. ○ mesenchymal cells
2. ○ osteogenic (osteoprogenitor) cells
3. ○ osteoblasts*
4. ○ osteocytes*
5. ○ collagenous fibers
6. ○ blood vessels (red)
7. ○ organic matrix (osteoid)
8. ○ inorganic matrix

Label:
9. lacuna
10. canaliculus

* cell processes only visible under high magnification

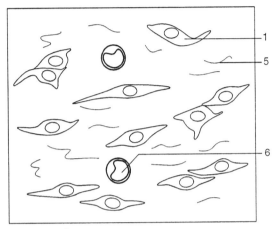
a. mesenchyme

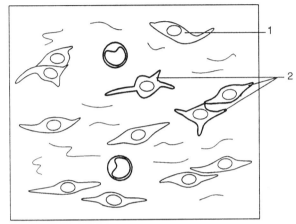
b. osteogenic cell formation

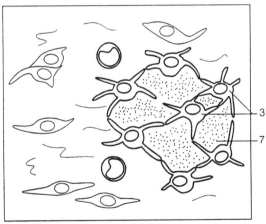
c. osteoblasts and osteoid formation

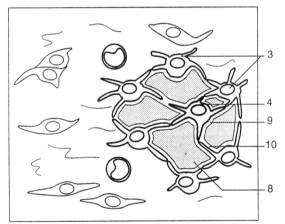
d. calcification of matrix

Figure 6.4. Steps in intramembranous bone formation.

Exercise 6.4:

_____ 1. Does mesenchyme contain blood vessels?

_____ 2. Mesenchymal cells become _____ cells.

_____ 3. Osteogenic cells divide to form _____ .

_____ 4. Osteoblasts first lay down the _____ matrix, then the _____ matrix.

_____ 5. When the osteoblasts are entrapped by calcium salts, they become _____ .

_____ 6. The spaces that remain around the cell bodies of osteocytes during calcification are called _____ . The interconnecting spaces around the processes of osteocytes are called _____ .

_____ 7. These ossification centers thicken when the surface cells, the _____ (cell type), continue to deposit new bone.

_____ 8. Spines of embryonic bone that form are called _____ .

_____ 9. Are blood vessels within or between these trabeculae?

_____ 10. In order for bone to increase in width (appositional growth), osteoblasts in the periosteum lay down new bone. Is this intramembranous or endochondral bone formation?

Answers to Exercise 6.4: 1. yes; 2. osteoprogenitor; 3. osteoblasts; 4. organic, inorganic; 5. osteocytes; 6. lacunae, canaliculi; 7. osteoblasts; 8. trabeculae; 9. between; 10. intramembranous.

Intramembranous Bone Formation

E. Endochondral Bone Formation

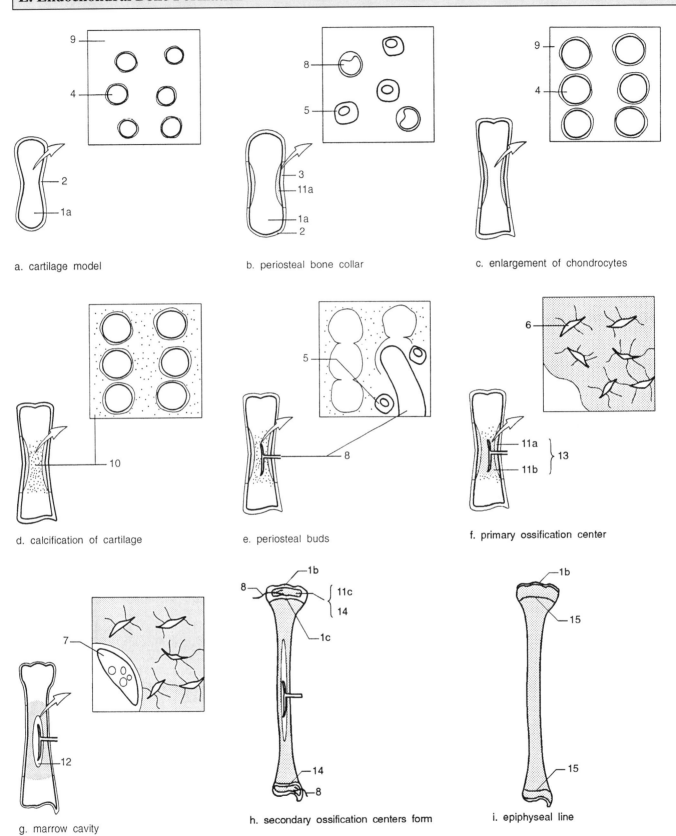

Figure 6.5. Steps in bone formation in a long bone.

Color and label:
1. ○ a. hyaline cartilage
   b. articular cartilage
   c. epiphyseal cartilage (plate)
2. ○ perichondrium (figure 6.5a–g)
3. ○ periosteum (figure 6.5 b–g)
4. ○ chondrocytes
5. ○ osteoblasts
6. ○ osteocytes
7. ○ osteoclast
8. ○ blood vessels (red)
9. ○ cartilaginous matrix
10. ○ calcified cartilaginous matrix
11. ○ bone
    a. periosteal bone collar
    b. center of diaphysis
    c. epiphysis
12. ○ marrow (medullary) cavity

Label:
13. primary ossification center
14. secondary ossification center
15. epiphyseal line

Exercise 6.5:

_____ 1. Does hyaline cartilage contain blood vessels? (see tissues, p. 45)

_____ 2. How do the chondrocytes of hyaline cartilage obtain nutrients and get rid of waste products?

_____ 3. During endochondral bone formation, what happens to the size of the chondrocytes? (figure 6.5c)

_____ 4. What happens to the cartilaginous matrix? (figure 6.5d)

_____ 5. How does calcification of the matrix alter the ability of materials to diffuse through the matrix?

_____ 6. What effect does this have on the chondrocytes? (figure 6.5e)

_____ 7. What happens to the calcified cartilaginous matrix?

_____ 8. When the matrix is weakened, what helps to support the structure?

_____ 9. Periosteal buds (blood vessels and osteoblasts) grow into the deteriorating cartilaginous matrix. The osteoblasts deposit _____ . (see figure 6.4e–f)

_____ 10. The primary ossification center forms in long bones at the _____ (epiphysis, diaphysis) and includes the _____ and the _____ . (figure 6.5f)

_____ 11. Is the formation of the periosteal collar *from periosteum* due to intramembranous or endochondral bone development? (figure 6.5b)

_____ 12. Is ossification in the center of the diaphysis, intramembranous, or endochondral development?

_____ 13. Does the diaphysis remain as solid bone?

_____ 14. What cells cause resorption of bone, creating the marrow cavity?

_____ 15. Is the entire embryonic hyaline cartilage model replaced?

_____ a. The secondary center of ossification forms in the _____ .

_____ b. What function is served by the epiphyseal cartilage?

_____ c. What function is served by the articular cartilage?

Answers to Exercise 6.5: 1. no; 2. diffusion through organic matrix; 3. they enlarge; 4. it calcifies; 5. prevents diffusion; 6. they die; 7. matrix deteriorates; 8. periosteal collar; 9. bone (see figure 6.4b–e for steps); 10. diaphysis, periosteal collar, center of the diaphysis; 11. intramembranous; 12. endochondral; 13. no (bone is resorbed to create medullary cavity); 14. osteoclasts; 15. no; 15a. epiphysis; 15b. growth in length; 15c. joint surface for articulation.

F. Growth and Remodeling

Color and label:
1. ○ osteoblasts
2. ○ osteocytes
3. ○ osteoclasts
4. ○ bone matrix

5. ○ a. blood vessels (red)
    b. marrow (red)

Label:
6. lamellae
7. trabecula

8. Haversian system
9. Haversian canal

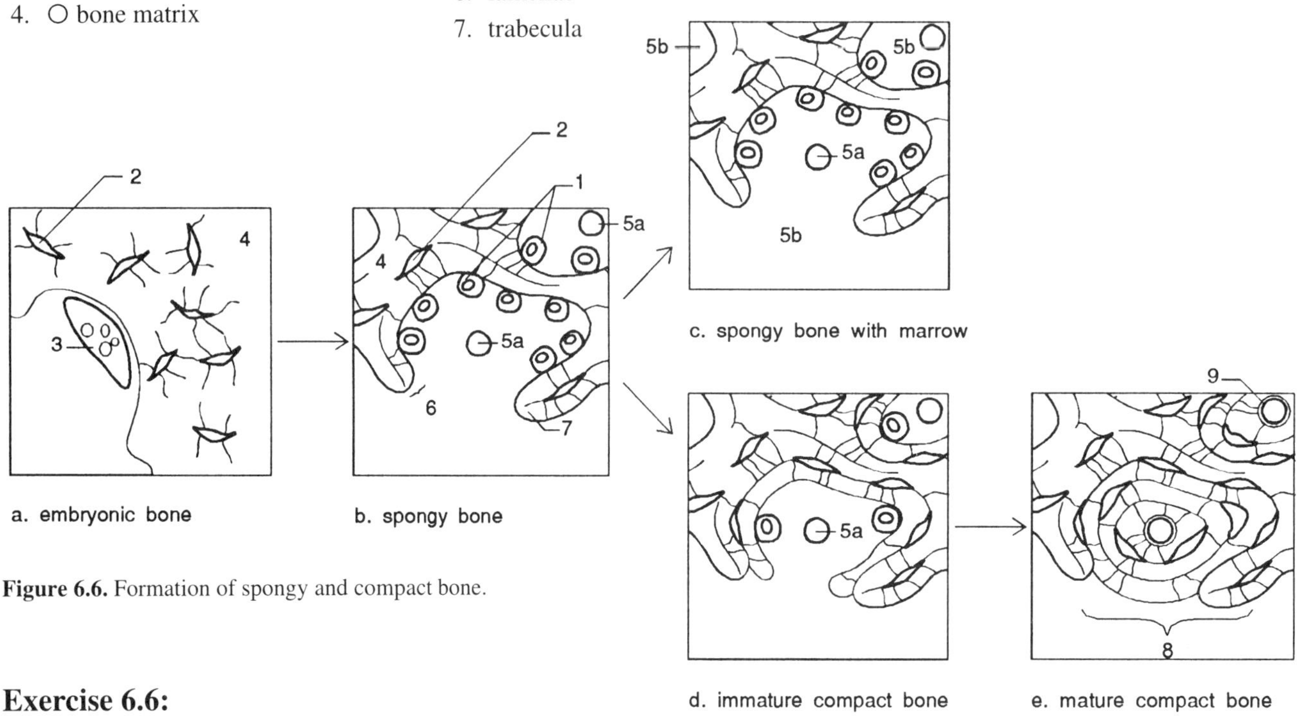

a. embryonic bone
b. spongy bone
c. spongy bone with marrow
d. immature compact bone
e. mature compact bone

Figure 6.6. Formation of spongy and compact bone.

Exercise 6.6:

_____ 1. Does embryonic (woven) bone form by intramembranous bone formation, endochondral bone formation, or both?

_____ 2. Does embryonic bone have lamellae?

_____ 3. In order for embryonic bone to develop into spongy and compact bone, the matrix (both inorganic and organic components) must first break down. What cells cause this resorption to occur? (figure 6.5g)

_____ 4. Remodeling of bone occurs at the inner and outer bony surfaces as well as at the epiphyseal cartilage. Where would you expect to find osteoclasts?

_____ 5. The new layered trabeculae that form are _____ (spongy, compact) bone.

_____ 6. The spaces between trabeculae can fill with _____ .

_____ 7. Trabeculae are covered with _____ (cell type).

_____ 8. What do you see in the space within the ring of osteoblasts? (figure 6.6b)

_____ 9. Bone formation around these blood vessels produces _____ (spongy, compact) bone.

_____ 10. Which layers of the Haversian system form first?

_____ 11. What happens to the blood vessel in the center?

_____ 12. As we have seen, osteoclasts break down embryonic bone. Can they also break down mature bone?

_____ 13. What functions might be served by resorption of mature bone?

_____ 14. To summarize, is the mature bone that forms by intramembranous and endochondral development the same or different?

_____ 15. Does growth and remodeling depend upon osteoblast activity, osteoclast activity, or both?

Answers to Exercise 6.6: 1. both; 2. no; 3. osteoclasts; 4. periosteum, endosteum, epiphyseal cartilage; 5. spongy; 6. marrow; 7. osteoblasts; 8. blood vessels; 9. compact; 10. outermost layers; 11. end up within Haversian canal; 12. yes; 13. remodel bone (during growth, in response to stress), repair; 14. the same; 15. both.

G. Repair

Color (trace) and label:
1. ○ periosteum
2. ○ endosteum
3. ○ a. compact bone
      b. spongy bone
4. ○ marrow cavity
5. ○ hematoma (blood clot) (red)
6. ○ blood vessels
7. ○ cartilage

Label:
8. internal callus
9. external callus

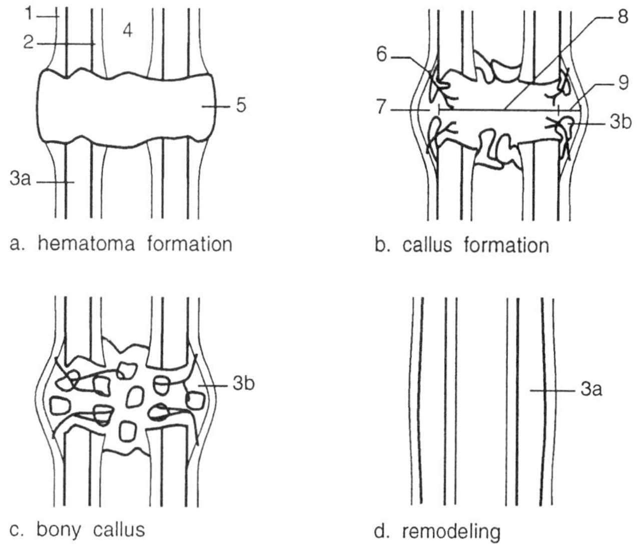

Exercise 6.7:

a. hematoma formation   b. callus formation

c. bony callus   d. remodeling

Figure 6.7. Repair of a fracture.

_____ 1. A fracture damages blood vessels in the _____ , _____ , and _____ , which causes a blood clot to form.

_____ 2. Callus formation occurs when chondroblasts and osteoblasts develop from cells in the periosteum and lay down _____ and _____ respectively.

_____ 3. As more blood vessels grow into the callus, the amount of _____ (tissue type) increases.

_____ 4. The bone that initially forms is _____ (compact, spongy).

_____ 5. Does the callus grow into "old" bone on either side of the injury?

_____ 6. Is this initial repair as strong as the original bone?

_____ 7. In order for bone remodeling to occur, spongy bone and the older neighboring bone are resorbed by _____ (cell type).

_____ 8. The new bone that bridges the injury site is _____ (compact, spongy).

Answers to Exercise 6.7: 1. periosteum, bone, endosteum; 2. cartilage, bone; 3. bone; 4. spongy; 5. no; 6. no; 7. osteoclasts; 8. compact.

H. Whole Skeleton

Color and label the bones of the axial skeleton:

1. ○ skull
2. ○ hyoid
3. ○ vertebral column
4. ○ sternum
   a. manubrium
   b. body
   c. xiphoid process
5. ○ ribs
6. ○ costal cartilage

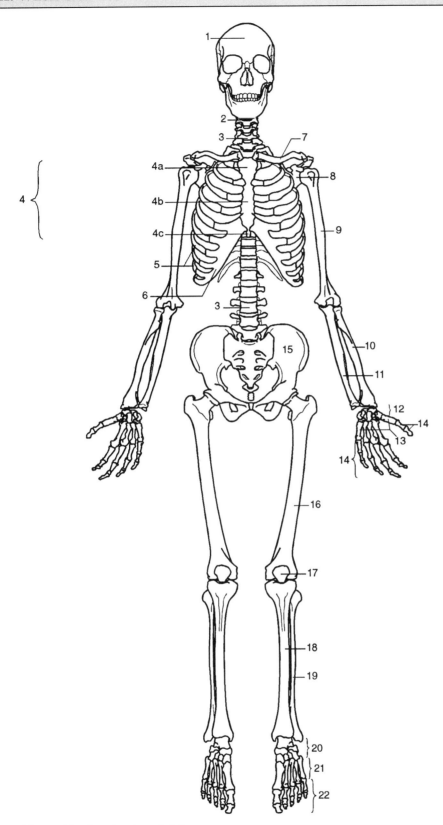

Figure 6.8a. Anterior view of skeleton.

Chapter 6   Skeletal System

**Color and label the bones of the appendicular skeleton:
(You will have to repeat some colors.)**

7. ○ clavicle
8. ○ scapula
9. ○ humerus
10. ○ radius
11. ○ ulna
12. ○ carpals
13. ○ metacarpals
14. ○ phalanges (of fingers)
15. ○ hip (coxal) bone
16. ○ femur
17. ○ patella
18. ○ tibia
19. ○ fibula
20. ○ tarsals
21. ○ metatarsals
22. ○ phalanges (of toes)

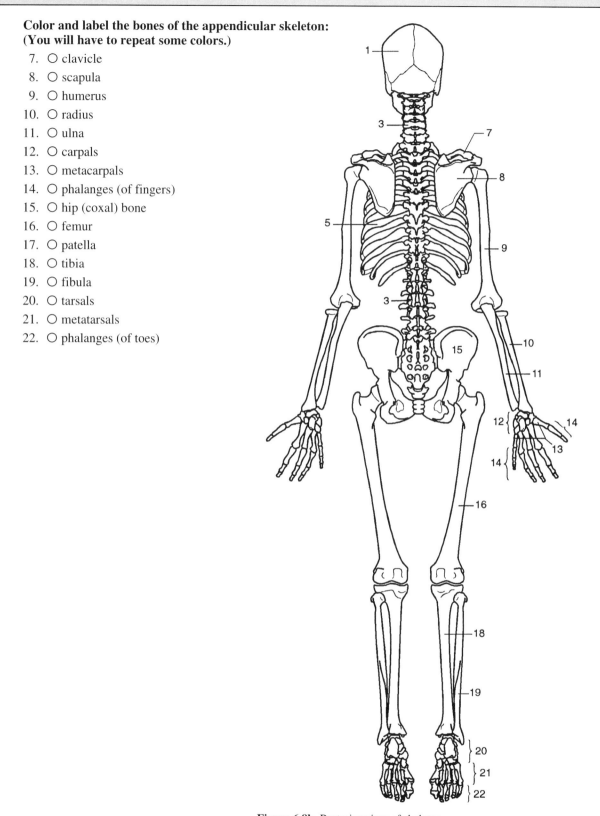

Figure 6.8b. Posterior view of skeleton.

Whole Skeleton 67

I. Skull

Color and label:
1. ○ frontal bone
   a. frontal sinus
2. ○ nasal bone
3. ○ vomer bone
4. ○ maxilla
5. ○ palatine bone
6. ○ mandible
   a. condyloid process
   b. coronoid process
7. ○ lacrimal bone
8. ○ ethmoid bone
   a. cribriform plate
9. ○ sphenoid bone
   a. sella turcica
10. ○ zygomatic bone
11. ○ temporal bone
    a. external auditory meatus
    b. mastoid process
    c. zygomatic process
    d. styloid process
12. ○ parietal bone
13. ○ occipital bone
    a. occipital condyles
    b. foramen magnum

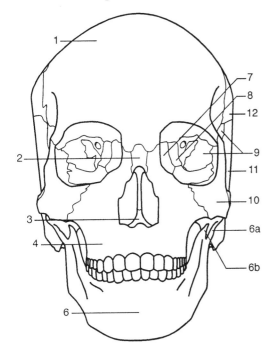

Figure 6.9a. Frontal view of skull.

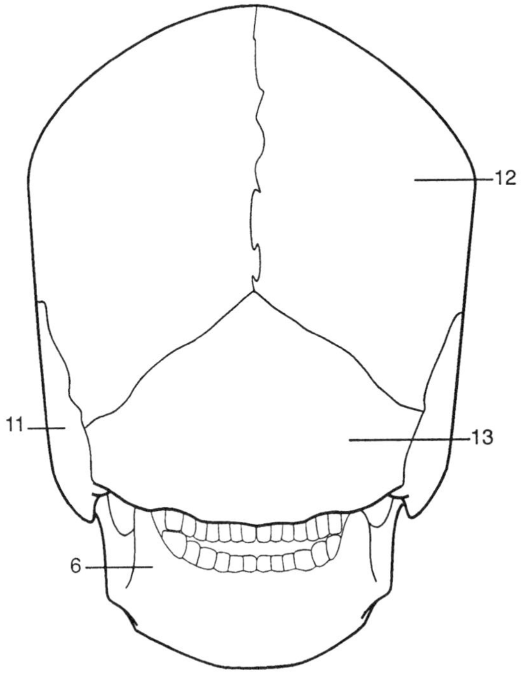

Figure 6.9b. Posterior view of skull.

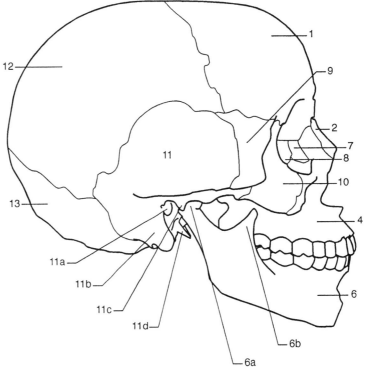

Figure 6.9c. Lateral view of skull.

68   Chapter 6   Skeletal System

Exercise 6.9:

1. Which bones are described below?

   _____ a. orbit of the eye

   _____ b. roof of the mouth (hard palate)

   _____ c. jaw

   _____ d. cheek bone

   _____ e. teeth embedded

   _____ 2. What bone contains the opening of the ear?

   _____ 3. Which bone articulates with vertebral column?

   4. What functions are served by the following?

   _____ a. occipital condyles

   _____ b. sella turcica

   _____ 5. The nerves of the spinal cord enter the brain through the opening called the _____ .

   _____ 6. The zygomatic arch includes parts of the _____ bone and the _____ bone.

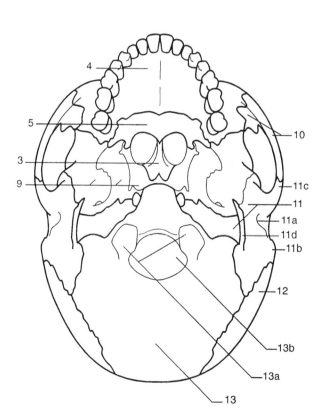

Figure 6.9d. Inferior view of skull.

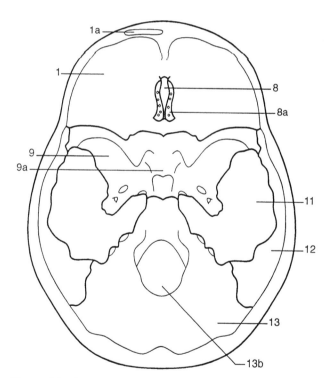

Figure 6.9e. Superior view of inside of skull.

Answers to Exercise 6.9: 1a. frontal, lacrimal, ethmoid, maxilla, zygomatic, sphenoid; 1b. maxilla, palatine; 1c. mandible; 1d. zygomatic, temporal; 1e. maxilla, mandible; 2. temporal; 3. occipital; 4a. articulates with first vertebra; 4b. location of pituitary gland; 5. foramen magnum; 6. zygomatic, temporal.

J. Vertebrae and Ribs

Color and label:
1. ○ cervical vertebra(e)
   a. atlas
   b. axis
2. ○ thoracic vertebra(e)
3. ○ lumbar vertebra(e)
4. ○ sacrum
5. ○ coccyx
6. ○ intervertebral discs

Label:
7. anterior arch (of atlas)
8. posterior arch (of atlas)
9. body
   a. odontoid process (dens)
10. vertebral arch
    a. lamina
    b. pedicle
11. spinous process
12. transverse process
13. superior articular surface
14. inferior articular surface
15. articular facets for ribs
16. vertebral foramen
17. intervertebral foramen
18. transverse foramen
19. sacral foramina

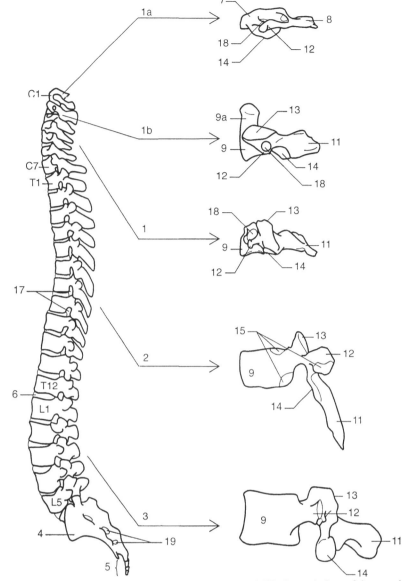

Exercise 6.10:

Figure 6.10a. Lateral view of vertebral column. Figure 6.10b. Lateral view of the vertebrae.

1. How many of each of the following are there?
   a. cervical vertebrae
   b. thoracic vertebrae
   c. lumbar vertebrae

2. Which vertebra is involved in the "yes" movement of the head? What vertebral articulating surface makes this possible?

3. Which vertebra contains the odontoid process (dens)? What function does it serve?

4. Which vertebrae articulate with ribs?

5. The ribs articulate with the body of the vertebrae, the transverse processes of the vertebrae, and the costal cartilage (sternum). Which of the following rib structures articulate with each of these?
   a. neck
   b. tubercle
   c. anterior end

Color and label (figure 6.10d):

20. ◯ rib
    a. tubercle
    b. neck
    c. head
    d. shaft
    e. anterior end
21. ◯ costal cartilage
22. ◯ sternum

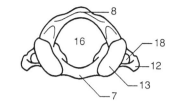
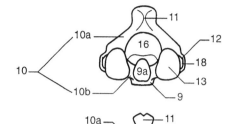
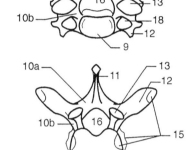
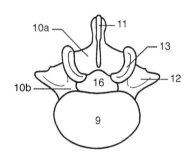
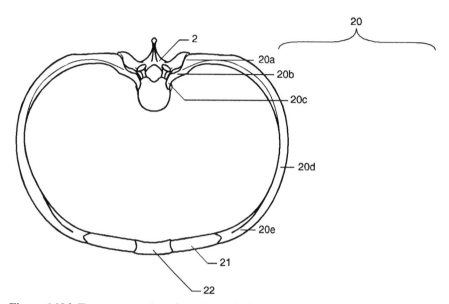

Figure 6.10d. Transverse section of vertebra and ribs.

Figure 6.10c. Superior view of vertebrae.

_____ 6. Muscles attach to the _____ and _____ of the vertebrae.

_____ 7. The greatest surfaces for muscle attachment are found on the _____ vertebrae.

_____ 8. Which vertebrae are fused?

_____ 9. What type of cartilage is the intervertebral discs?

10. Which openings are described below?

_____ a. location of spinal cord

_____ b. peripheral nerves enter and leave spinal cord

_____ c. vertebral arteries travel to brain

_____ d. nerves pass through sacrum

Answers to Exercise 6.10: 1a. 7; 1b. 12; 1c. 5; 2. atlas, superior articular surface; 3. axis, allows for "no" movement of head; 4. thoracic; 5a. body of vertebra; 5b. transverse process; 5c. costal cartilage (sternum); 6. spinous processes, transverse processes; 7. lumbar; 8. sacral; 9. fibrocartilage; 10a. vertebral foramen; 10b. intervertebral foramen; 10c. transverse foramen; 10d. sacral foramina.

Vertebrae and Ribs

K. Pectoral Girdle and Upper Arm

Color and label:

1. ○ clavicle
2. ○ scapula
   a. glenoid cavity
   b. acromion process
   c. coracoid process
   d. axillary border
   e. vertebral border
   f. spine
3. ○ humerus
   a. head of humerus
   b. greater tubercle
   c. lesser tubercle
   d. intertubercular groove
   e. deltoid tuberosity
   f. coronoid fossa
   g. olecranon fossa
   h. lateral epicondyle
   i. capitulum
   j. medial epicondyle
   k. trochlea
4. ○ radius
   a. radial tuberosity
   b. styloid process (of radius)
5. ○ ulna
   a. coronoid process
   b. olecranon process
   c. styloid process (of ulna)
6. ○ carpals
7. ○ metacarpals
8. ○ phalanges

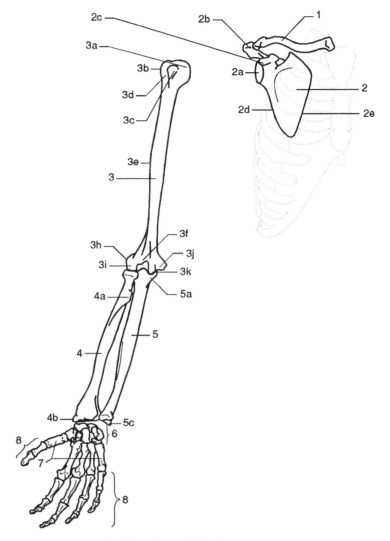

Figure 6.11a. Anterior view of pectoral girdle and upper arm.

Exercise 6.11:

1. Which bones correspond to the following surface structures?

   a. wrist

   b. palm

   c. elbow

   d. upper arm

   e. forearm

   f. shoulder blade

2. What bones make up the pectoral girdle?

3. The clavicle articulates with the _____ and _____ .

72   Chapter 6   Skeletal System

_____ 4. The humerus articulates with the _____ of the _____ .

_____ 5. Does the pectoral girdle articulate with the ribs?

_____ 6. Is the ulna lateral or medial to the radius?

_____ 7. Which articulates with the humerus—the radius or ulna?

_____ 8. Which articulates with the carpals—the radius or ulna?

_____ 9. When the arm is straight, where is the olecranon process?

_____ 10. How many phalanges make up the thumb? the fingers?

_____ 11. Projections that allow for muscle attachments include _____ .

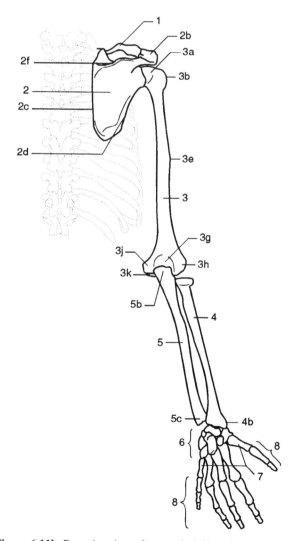

Figure 6.11b. Posterior view of pectoral girdle and upper arm.

Answers to Exercise 6.11: 1a. carpals; 1b. metacarpals; 1c. ulna; 1d. humerus; 1e. radius, ulna; 1f. scapula (spine); 2. scapula, clavicle; 3. scapula, manubrium (of sternum); 4. glenoid cavity, scapula; 5. no; 6. medial; 7. ulna; 8. radius; 9. in the olecranon fossa; 10. 2, 3; 11. epicondyles, tubercles, tuberosities, processes.

Pectoral Girdle and Upper Arm

L. Pelvic Girdle and Lower Leg

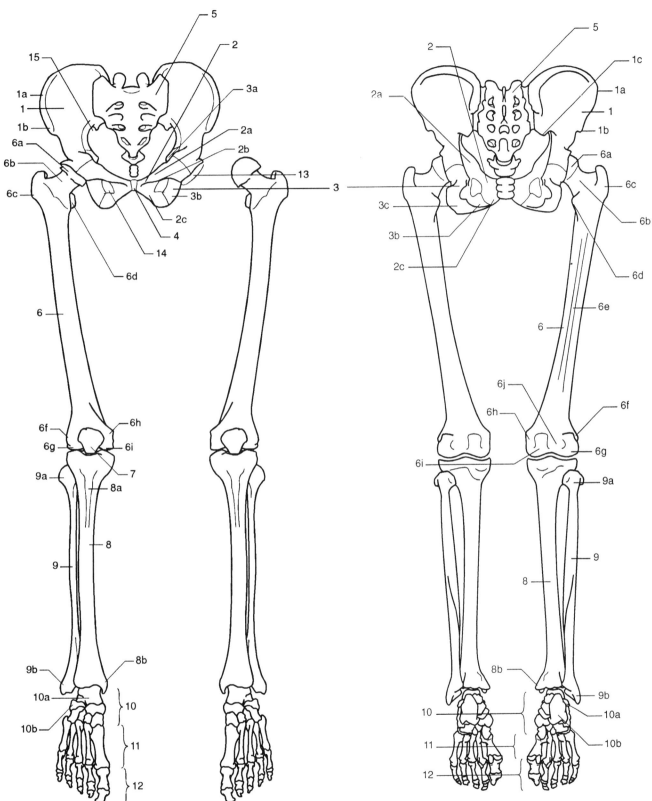

Figure 6.12a. Anterior view of pelvic girdle and lower limb.

Figure 6.12b. Posterior view of pelvic girdle and lower limb.

Color and label the parts of the hip (coxal) bone:
1. ○ ilium
   a. iliac crest
   b. anterior superior iliac spine
   c. posterior superior iliac spine
2. ○ pubis
   a. superior ramus (of pubis)
   b. pubic tubercle
   c. inferior ramus (of pubis)
3. ○ ischium
   a. spine (of ischium)
   b. ramus (of ischium)
   c. ischial tuberosity
4. ○ symphysis pubis

Color and label:
5. ○ sacrum
6. ○ femur
   a. head of femur
   b. neck of femur
   c. greater trochanter
   d. lesser trochanter
   e. linea aspera
   f. lateral epicondyle
   g. lateral condyle
   h. medial epicondyle
   i. medial condyle
   j. intercondylar fossa
7. ○ patella
8. ○ tibia
   a. tibial tuberosity
   b. medial malleolus
9. ○ fibula
   a. head of fibula
   b. lateral malleolus
10. ○ tarsals
    a. talus
    b. calcaneous
11. ○ metatarsals
12. ○ phalanges

Label:
13. acetabulum
14. obturator foramen
15. greater sciatic notch
16. true pelvis

Exercise 6.12:

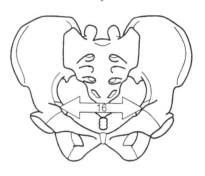

Figure 6.12c. Male vs. female pelvis.

1. Which bones correspond to the following surface structures?
   a. thigh _____
   b. hip _____
   c. lower leg _____
   d. heel _____
2. What bones make up the pelvic girdle?
3. What type of cartilage is the symphysis pubis? What advantage does this type of cartilage offer? (see tissues, p. 44)
4. You sit on your _____ .
5. When you put your hands on your hips you are touching the _____ .
6. Embryologically, the ischium, ilium, and pubis are separate bones. Are they separate in the adult?
7. Is the tibia lateral or medial to the fibula?
8. The fibula articulates with the _____ .
9. The bump on the lateral surface of the ankle is the _____ of the _____ (bone).
10. The bump on the medial surface of the ankle is the _____ of the _____ (bone).
11. The tibia articulates with which tarsal?
12. How many phalanges make up the big toe? the other toes?
13. Which is wider—the male or female pelvis?
14. Is the male or female pelvis deeper?

Answers to Exercise 6.12: 1a. femur; 1b. coxal bone; 1c. tibia, fibula; 1d. calcaneous; 2. coxal bones, sacrum, coccyx; 3. fibrocartilage; strength, resilience; 4. ischium; 5. ilium; 6. no; 7. medial; 8. tibia; 9. lateral malleolus, fibula; 10. medial malleolus, tibia; 11. talus; 12. 2, 3; 13. female; 14. male.

M. Articulations

Color and label:
1. ○ fibrous tissue
2. cartilaginous tissue
   a. ○ hyaline
   b. ○ fibrocartilage
3. ○ synovial fluid
4. ○ bone
5. ○ articular cartilage
6. ○ capsular ligaments
7. ○ synovial membrane
8. ○ articular discs
9. ○ muscle
10. ○ tendon
11. ○ bursa

Label:
12. interosseous membrane
13. epiphyseal cartilage
14. intervertebral discs
15. pubic symphysis
16. articular capsule
17. suture
18. gomphosis
19. syndesmosis
20. synchondrosis
21. symphysis

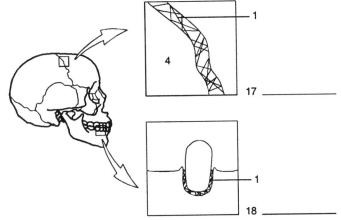

Figure 6.13a. Fibrous joints.

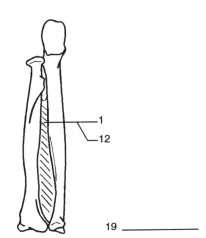

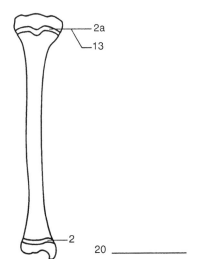

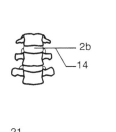

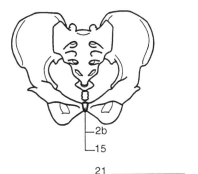

Figure 6.13b. Cartilaginous joints.

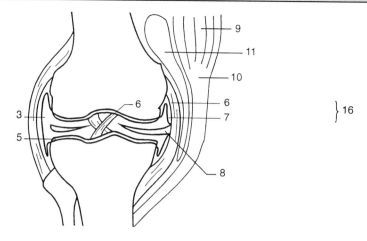

Figure 6.13c. Synovial joints.

Exercise 6.13:

_____
_____
_____
_____
_____
_____
_____
_____
_____
_____
_____
_____
_____
_____
_____
_____
_____
_____
_____
_____
_____
_____

1. Which joints are
   a. fibrous?
   b. cartilaginous?
   c. synovial?
2. Which fibrous joint has the most fibrous connective tissue? How does this affect movement?
3. Which fibrous joint is replaced by bone in adults?
4. What type of cartilage is in a synchondrosis?
5. An example of a synchondrosis is the _____ .
6. What type of cartilage is in a symphysis?
7. Two examples of symphyses are _____ and _____ .
8. The shape of the articular surface in a synovial joint is determined by _____ , _____ , and _____ .
9. Do all synovial joints have articular discs?
10. What function is served by the articular discs?
11. The articular capsule includes the outer _____ and the inner _____ .
12. A dense, regular, connective tissue, bone-to-bone attachment is called a _____ .
13. Where are ligaments found—inside or outside the synovial cavity?
14. Synovial fluid is made by the _____ .
15. The avascular articular cartilage receives nutrients from the _____ .
16. The range of motion of a joint is determined by the joint structures _____ and _____ .
17. Outside the joint capsule, muscles and their connective tissue attachments to bone (called _____) cross the joint.
18. Do muscles and tendons lend support to the joint?

Articulations

Articulations continued

_____ 19. Ligaments and tendons tend to shorten with inactivity. What effect will this have on the motion of a joint?

_____ 20. A fluid-filled sac that prevents friction damage between moving structures is called a _____ .

Answers to Exercise 6.13. 1a. suture, gomphosis, syndesmosis; 1b. synchondrosis, symphysis; 1c. synovial joint; 2. syndesmosis; movement possible; 3. suture; 4. hyaline; 5. epiphyseal cartilage; 6. fibrocartilage; 7. intervertebral discs, pubic symphysis; 8. bone shape, articular cartilage, articular discs; 9. most, but not all; 10. helps to prevent slipping; 11. capsular ligaments, synovial membrane; 12. ligament; 13. both; 14. synovial membrane; 15. synovial fluid; 16. bone, ligaments; 17. tendons; 18. yes; 19. decrease range of motion; 20. bursa.

Chapter 7: Muscular System

A. Gross Structure of Skeletal Muscle

Color and label:
1. ○ tendon (origin)
   a. ○ muscle belly
2. ○ synergists
3. ○ bone
4. ○ joint
5. ○ tendon (insertion)
6. muscle (antagonist of muscles 2 and 3)

Exercise 7.1:

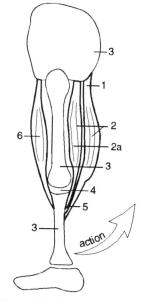

Figure 7.1. Skeletal muscle.

_____ 1. A skeletal muscle is attached at both of its ends to _____ .

_____ 2. The connective tissue endings are called _____ .

_____ 3. When a muscle contracts, its length becomes _____ .

_____ 4. The end of the muscle that usually remains fixed (does not move) is called the _____ .

_____ 5. The end of the muscle that usually is pulled toward the fixed end is called the _____ .

_____ 6. Muscles that act together to produce an action are called _____ .

_____ 7. Muscles that perform opposite actions are called _____ .

_____ 8. In order to move a bone, each muscle must cross at least one _____ .

Answers to Exercise 7.1: 1. bones; 2. tendons; 3. shorter; 4. origin; 5. insertion; 6. synergists; 7. antagonists; 8. joint.

B. Skeletal Muscle Cell

Color and label:
1. ○ sarcoplasmic reticulum
2. ○ T (transverse) tubules
3. ○ mitochondrion
4. ○ sarcolemma
5. ○ glycogen granules
6. ○ nucleus
7. ○ myofibril

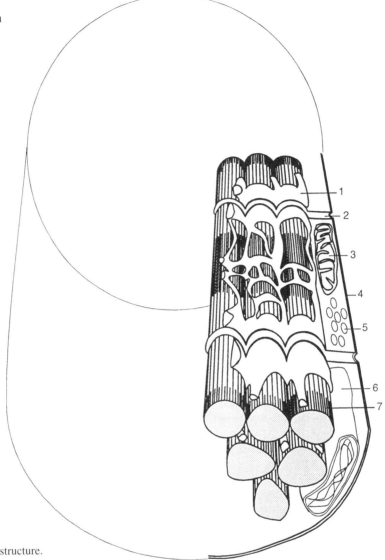

Figure 7.2. Skeletal muscle cell structure.

Exercise 7.2:

_____ 1. The thick and thin filaments are organized into parallel bundles called _____ .

_____ 2. The myofibrils are wrapped in endoplasmic reticulum which, in skeletal muscle, is called _____ .

_____ 3. Projecting into the muscle cell from the surface are the _____ .

_____ 4. The T tubules also wrap around the _____ and are adjacent to the _____ .

_____ 5. Two structures shown in this figure that provide energy for contraction are _____ and _____ .

_____ 6. Information for the synthesis of muscle proteins and other molecules is found in the _____ .

Answers to Exercise 7.2: 1. myofibrils; 2. sarcoplasmic reticulum; 3. T tubules; 4. myofibrils, sarcoplasmic reticulum; 5. mitochondrion, glycogen granules; 6. nucleus.

C. Structure of the Sarcomere

Color and label:
1. ○ thick filament
2. ○ thin filament
3. ○ z line

Label:
4. actin
5. tropomyosin
6. troponin
7. myosin head

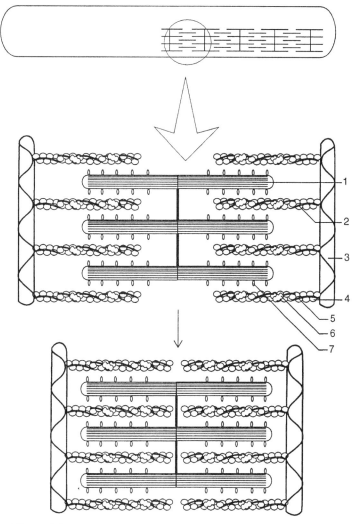

Figure 7.3. The relaxed and contracted sarcomere.

Exercise 7.3:

_____ 1. Myofibrils are made of alternating bundles of _____ and _____ filaments.

_____ 2. The filaments between two z lines are collectively known as the _____ .

_____ 3. The thin filaments are connected at one end to the _____ .

_____ 4. The thick filaments are joined to each other at their _____ .

_____ 5. The thin filament is made up of two chains of the globular protein _____ and two chains of alternating proteins called _____ and _____ .

_____ 6. The thick filament is made up of a bundle of the protein _____ .

_____ 7. When the muscle contracts, the sarcomere becomes (longer, shorter) _____ .

Answers to Exercise 7.3: 1. thick, thin; 2. sarcomere; 3. z line; 4. centers; 5. actin, tropomyosin, troponin; 6. myosin; 7. shorter.

Structure of the Sarcomere 81

D. Contraction of Skeletal Muscle

Color and label:
1. ○ sarcoplasmic reticulum
2. ○ calcium ion
3. ○ z line
4. ○ actin
5. ○ tropomyosin
6. ○ troponin
7. ○ myosin head
8. ○ thin filament
9. ○ thick filament

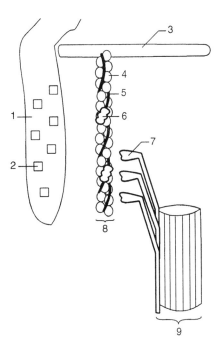

Figure 7.4a. Unattached cross-bridges.

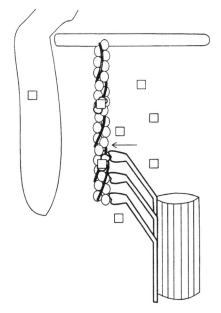

Figure 7.4b. Cross-bridge formation.

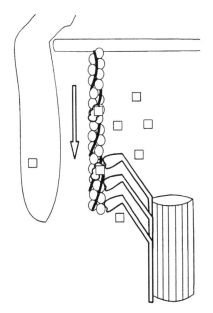

Figure 7.4c. Cross-bridge swiveling.

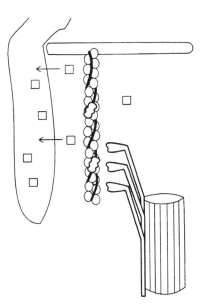

Figure 7.4d. Cross-bridge release.

Exercise 7.4:

_____ 1. When the sarcoplasmic reticulum is stimulated it releases _____ .

_____ 2. Calcium ions diffuse to the thin filament where they bind with _____ .

_____ 3. Troponin then uncovers the binding sites on _____ .

_____ 4. Actin can then attach to the heads of _____ .

_____ 5. The heads of myosin protrude from the thick filaments and form the _____ .

_____ 6. The myosin heads cause ATP to hydrolyze, thus storing _____ .

_____ 7. This energy allows the heads of myosin on the thick filament to pull the thin filament toward the center of the thick filament, causing the sarcomere to _____ .

_____ 8. When the nerve impulses stop, the sarcoplasmic reticulum is again able to bind _____ .

_____ 9. This calcium is removed from the _____ by active transport.

_____ 10. Troponin then causes the myosin binding sites to be covered on the _____ .

_____ 11. Actin and myosin no longer form cross-bridges and the muscle can _____ .

_____ 12. The muscle will relax only if the cross-bridges are broken by _____ .

_____ 13. If ATP is not present, then the muscle will remain stiff in the state of _____ .

Answers to Exercise 7.4: 1. calcium; 2. troponin; 3. actin; 4. myosin; 5. cross-bridges; 6. energy; 7. shorten; 8. calcium; 9. troponin; 10. actin; 11. relax; 12. ATP; 13. rigor mortis.

E. Energetics

Color:
- ○ ATP (and corresponding arrows)
- ○ ADP
- ○ CP (creatine phosphate)
- ○ C (creatine)

- ○ glycolysis
- ○ lactic acid
1. ○ myosin (thick filaments)
2. ○ thin filaments

Color and label:
3. ○ glycogen granules
4. ○ mitochondrion

Label:
5. cross-bridge

Exercise 7.5:

_____ 1. Energy is needed during muscle contraction for _____, _____, and _____.

_____ 2. The source of this energy is _____.

_____ 3. As ATP is used, which way does the reaction C + ATP ↔ CP + ADP proceed? (see reversible reactions, p. 13)

_____ 4. Another source of energy for ATP production is _____.

_____ 5. Glycogen is converted to _____, which is broken down during _____.

_____ 6. The end product of glycolysis, _____, can be further metabolized or converted to lactic acid.

_____ 7. During less strenuous activity, the slower metabolism of pyruvic acid by oxygen, called aerobic metabolism, occurs in the _____ (organelle).

_____ 8. During strenuous activity, glycolysis proceeds faster and extra pyruvic acid is converted to _____. This is anaerobic metabolism.

_____ 9. Which ATP source (creatine phosphate, aerobic metabolism, or anaerobic metabolism) supplies the most ATP?

_____ 10. As lactic acid accumulates during anaerobic metabolism, the pH of muscle cells becomes _____ (higher, lower). (see pH, p.15)

_____ 11. With this low pH, what happens to enzyme activity? (see enzymes, p. 24)

_____ 12. Decreased ability to contract is called _____.

_____ 13. The oxygen needed to metabolize lactic acid and return the muscle to normal function is called the _____.

_____ 14. When at rest, which way does the reaction C + ATP ↔ CP + ADP proceed?

_____ 15. Therefore, the levels of creatine phosphate will _____ (increase, decrease).

Answers to Exercise 7.5: 1. cross-bridge swiveling, bridge breakage, calcium ion reuptake (by sarcoplasmic reticulum); 2. ATP; 3. to the left; 4. glycogen; 5. glucose, glycolysis; 6. pyruvic acid; 7. mitochondria; 8. lactic acid; 9. aerobic metabolism; 10. lower; 11. decreases; 12. fatigue; 13. oxygen debt; 14. to the right; 15. increase.

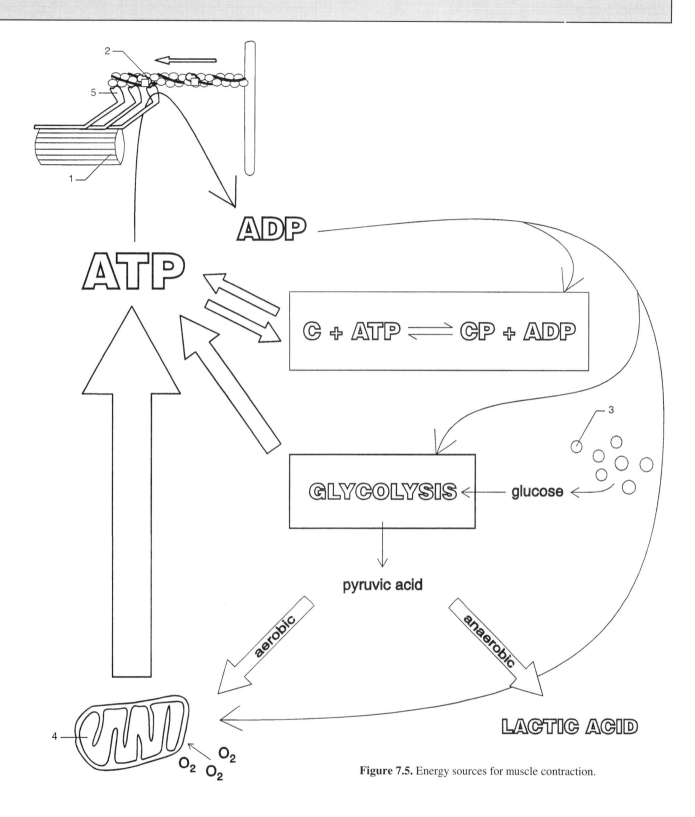

Figure 7.5. Energy sources for muscle contraction.

F. The Neuromuscular Junction

Color and label:
1. ○ axon
   a. axon terminal
   b. axon membrane
2. ○ mitochondrion
3. ○ synaptic vesicle
4. ○ muscle cell
   a. ○ muscle cell membrane
5. ○ acetylcholine esterase
6. ○ acetylcholine receptor
7. ○ acetate (acetyl group)
8. ○ choline

Label:
9. synaptic cleft
10. acetylcholine

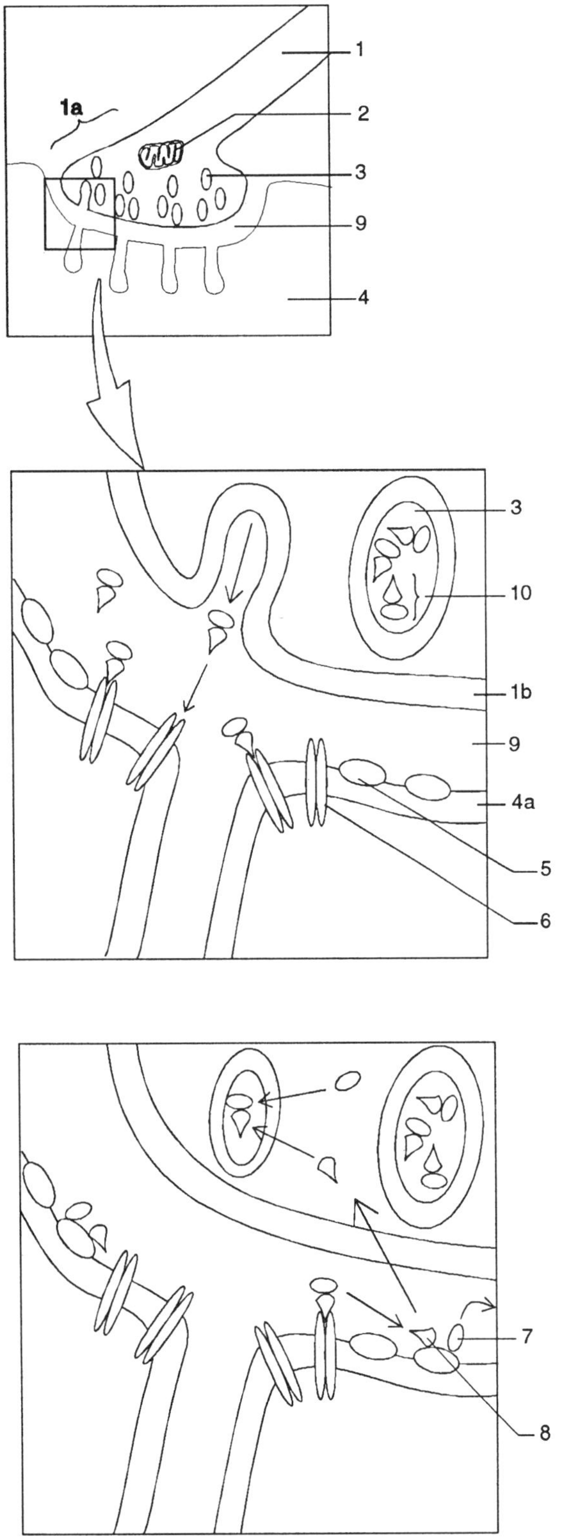

Figure 7.6. Neurotransmitter release and action.

Exercise 7.6:

_____ 1. Each branch (collateral) of the axon ends in a structure called the _____ .

_____ 2. This axon terminal rests in a(n) _____ on the surface of the muscle fiber (cell).

_____ 3. The axon terminal contains many membranous sacs called _____ .

_____ 4. Each of these synaptic vesicles contains _____ .

_____ 5. When the nerve impulse (action potential) travels along the axon and reaches the axon terminal, it causes the synaptic vesicles to release _____ into the _____ .

_____ 6. When acetylcholine reacts with _____ on the muscle membrane, the muscle membrane is then stimulated to transmit the action potential.

_____ 7. In order to stop the stimulation, the acetylcholine must be broken down by _____ , which is located on the _____ .

_____ 8. Acetylcholine is broken down to _____ and _____ .

_____ 9. Most of the choline is transported back into the _____ .

_____ 10. It is then resynthesized into _____ and returned to the _____ .

_____ 11. On the muscle cell, the action potential sweeps across the sarcolemma and into the cell on the membrane of the _____ .

Answers to Exercise 7.6: 1. axon terminal; 2. synaptic trough; 3. synaptic vesicles; 4. acetylcholine; 5. acetylcholine, synaptic cleft; 6. receptors; 7. acetylcholine esterase, sarcolemma; 8. acetate, choline; 9. axon terminal; 10. acetylcholine, synaptic vesicle; 11. T tubules.

The Neuromuscular Junction

G. The Motor Unit

Color and label:
1. ○ central nervous system
2. ○ motor neuron
   a. ○ axon
   b. ○ terminal branches
3. ○ muscle cells

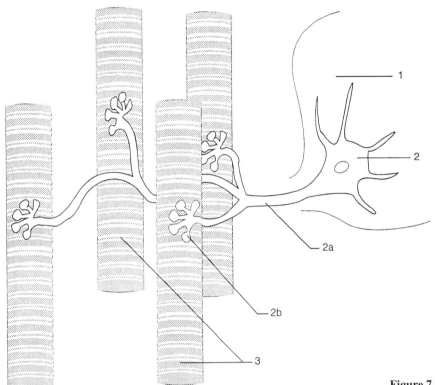

Figure 7.7. Neuron and muscle fibers of the motor unit.

Exercise 7.7:

_____ 1. Each muscle fiber (cell) maintains contact with the nervous system by one _____ .

_____ 2. This neuromuscular junction is at one of the axon endings of the _____ .

_____ 3. Each motor neuron ends in many _____ .

_____ 4. The motor neuron and all of the muscle cells that it innervates is called the _____ .

_____ 5. When the motor neuron transmits an impulse, what happens to the muscle cells in that motor unit?

_____ 6. If all of the muscle cells in a whole muscle were part of one motor unit, then the muscle would be either fully relaxed or _____ .

_____ 7. To be able to exert many different amounts of force, the whole muscle must be made up of many _____ .

_____ 8. Which muscle would be able to exert finer control—one with few large motor units or one with many small motor units?

Answers to Exercise 7.7: 1. neuromuscular junction; 2. motor neuron; 3. terminal branches; 4. motor unit; 5. they contract; 6. fully contracted; 7. motor units; 8. one with many small motor units.

H. Muscle Activity

Color the arrows:

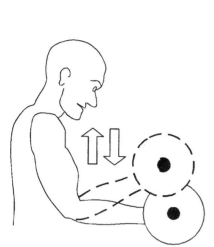

Figure 7.8a. Muscle twitch.

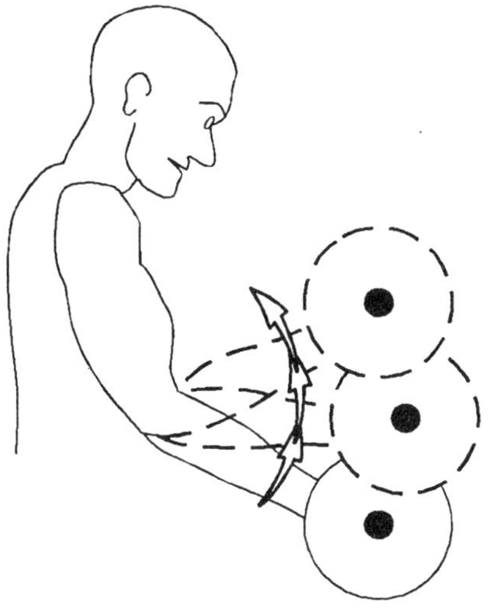

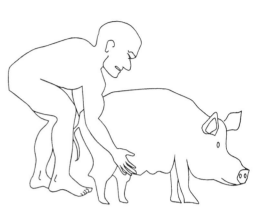

Figure 7.8b. Sustained contraction.

Figure 7.8c. Unsuccessful attempt to lift pig.

Exercise 7.8:

_____ 1. The response of a muscle to a single, brief stimulus is called a _____ . (see figure 7.8a)

_____ 2. Before any movement is observed, excitation-contraction coupling occurs during the _____ .

3. During the latent period, after the muscle is stimulated but before movement is observed,

_____ a. the stimulus crosses the _____ onto the sarcolemma.

_____ b. Then it travels into the muscle cell via the _____

_____ c. and stimulates the _____ to release the ion _____ ,

_____ d. which attaches to _____ and allows _____ .

_____ 4. When the thick and thin filaments move, the muscle _____ .

_____ 5. After the stimulus ceases, the muscle _____ .

_____ 6. If a second stimulus enters the muscle before it has completely relaxed from the previous one, the muscle continues to shorten. This process is called _____ .

_____ 7. If the next contraction occurs at the peak of the previous one, the process is called _____ .

_____ 8. If the muscle shortens and moves the load, the contraction is called _____ . (see figure 7.8b)

_____ 9. If the load is greater than the force that the muscle can generate, then no movement occurs and the contraction is _____ . (see figure 7.8c)

Answers to Exercise 7.8: 1. twitch; 2. latent period; 3a. myoneural junction; 3b. T tubules; 3c. sarcoplasmic reticulum, calcium; 3d. troponin, cross-bridge formation; 4. contracts; 5. relaxes; 6. incomplete tetanus; 7. complete tetanus; 8. isotonic; 9. isometric.

I. Skull

Color and label:
1. ○ orbicularis oculi
2. ○ frontalis
3. ○ temporalis
4. ○ orbicularis oris
5. ○ levator labii superioris
6. ○ levator anguli oris
7. ○ masseter
8. ○ risorius

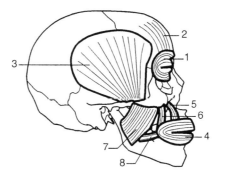

Figure 7.9. Lateral view of the skull.

J. Trunk

Color and label:
1. ○ sternocleidomastoid
2. ○ external oblique
3. ○ internal oblique
4. ○ transverse abdominis
5. ○ rectus abdominis

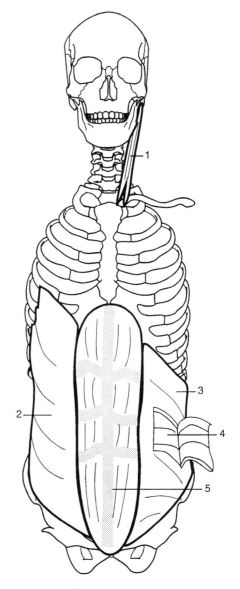

Figure 7.10. Anterior view of the trunk.

K. Anterior Pectoral Girdle and Upper Arm

Color and label:
1. ○ pectoralis major
2. ○ coracobrachialis
3. ○ biceps brachii
4. ○ bracialis

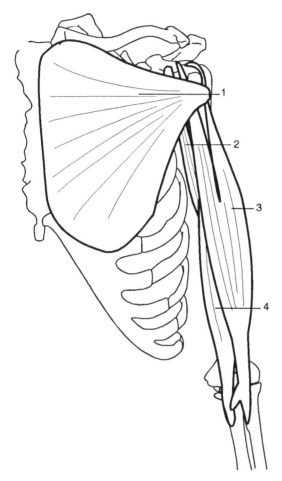

Figure 7.11. Anterior view of the shoulder.

L. Posterior Pectoral Girdle and Upper Arm

Color and label the superficial muscles:
1. ○ trapezeus
2. ○ deltoid
3. ○ latissimus dorsi
4. ○ triceps brachii

Color and label the deep muscles:
5. ○ levator scapuli
6. ○ rhomboid minor
7. ○ rhomboid major
8. ○ teres major
9. ○ supraspinatus
10. ○ infraspinatus
11. ○ teres minor

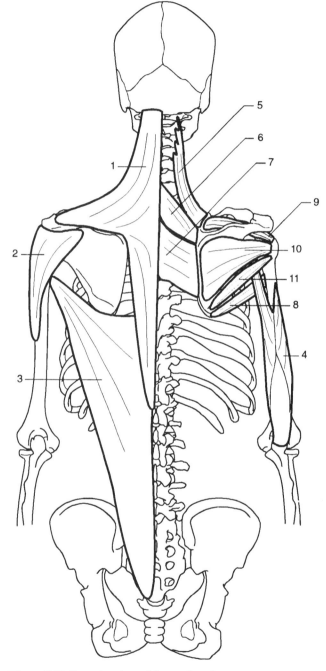

Figure 7.12. Posterior view of the trunk and upper arm.

M. Anterior Forearm

Color and label:
1. ○ flexor carpi radialis
2. ○ palmaris longus
3. ○ flexor carpi ulnaris
4. ○ flexor digitorum superficialis and profundus
5. ○ supinator
6. ○ pronator teres
7. ○ pronator quadratus

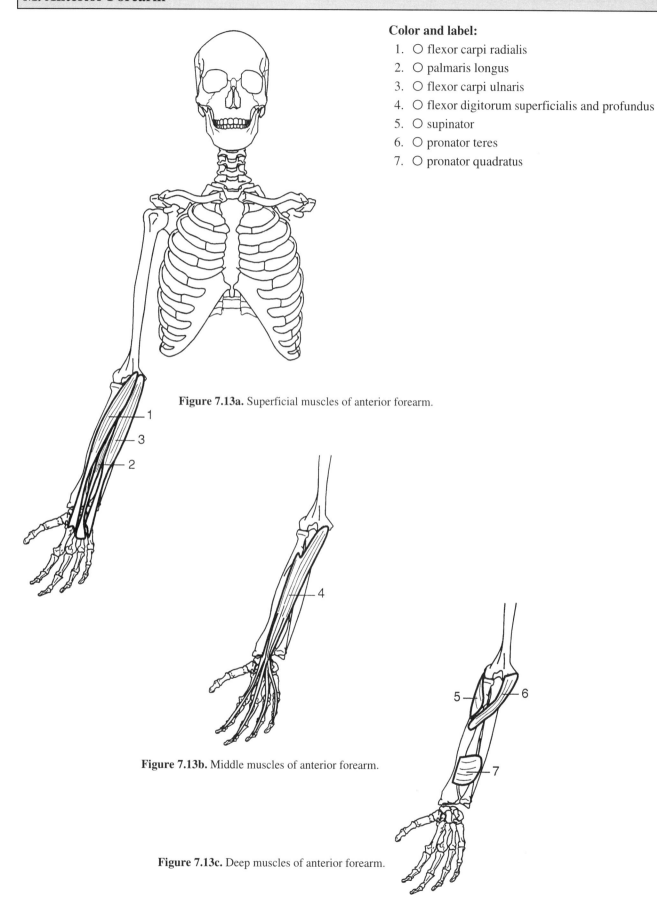

Figure 7.13a. Superficial muscles of anterior forearm.

Figure 7.13b. Middle muscles of anterior forearm.

Figure 7.13c. Deep muscles of anterior forearm.

Anterior Forearm

N. Posterior Forearm

Color and label:
1. ◯ extensor carpi radialis longus
2. ◯ extensor carpi radialis brevis
3. ◯ extensor carpi ulnaris
4. ◯ extensor digitorum communis

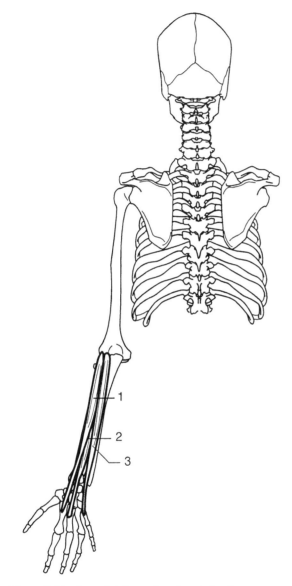

Figure 7.14a. Superficial muscles of posterior forearm.

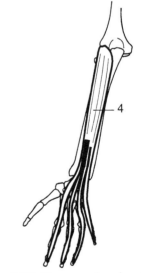

Figure 7.14b. Deep muscles of posterior forearm.

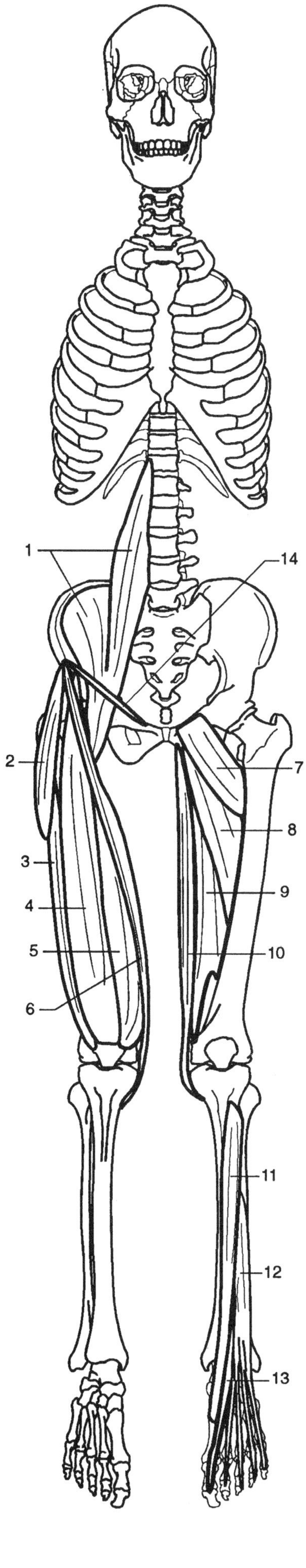

O. Anterior Pelvic Girdle and Leg

Color and label:
1. ○ iliopsoas
2. ○ tensor fasciae latae
3. ○ vastus lateralis
4. ○ rectus femoris
5. ○ vastus medialis
6. ○ sartorius
7. ○ pectineus
8. ○ adductor longus
9. ○ adductor magnus
10. ○ gracilis
11. ○ tibialis anterior
12. ○ extensor digitorum longus
13. ○ extensor hallucis longus
14. ○ inguinal ligament

Figure 7.15. Anterior view of the hip, thigh, and leg.

P. Posterior Pelvic Girdle and Leg

Color and label:
1. ○ gluteus maximus
2. ○ semimembranosus
3. ○ semitendinosus
4. ○ biceps femoris (long head)
5. ○ biceps femoris (short head)
6. ○ gastrocnemius
7. ○ soleus
8. ○ iliotibial tract

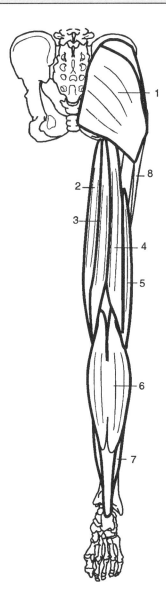

Figure 7.16. Posterior view of the hip, thigh, and leg.

96   Chapter 7   Muscular System

Q. Anatomical Position

Outline and label:
1. ○ midsagittal plane
2. ○ frontal (coronal) plane
3. ○ transverse plane

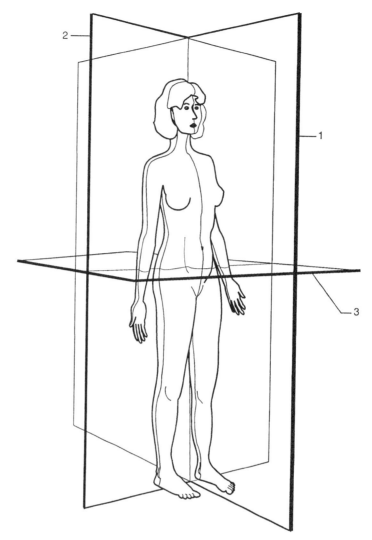

Figure 7.17. Anterolateral view of the body.

Exercise 7.17:

_____ 1. Someone standing in the anatomical position is standing erect with arms _____ .

_____ 2. The palms of the hands are facing _____ .

_____ 3. The head and toes are facing _____ .

_____ 4. The body is divided into right and left sides by the _____ plane.

_____ 5. The _____ plane separates the body into superior and inferior sections.

_____ 6. The _____ plane separates the anterior from the posterior section.

Answers to Exercise 7.17: 1. straight down; 2. forward; 3. forward; 4. midsagittal; 5. transverse; 6. frontal.

R. Movement

Color the arrow and label:
1. ○ flexion of the arm
2. ○ extension of the arm
3. ○ flexion of the elbow
4. ○ extension of the elbow
5. ○ hyperextension of the arm
6. ○ flexion of the thigh
7. ○ extension of the thigh
8. ○ plantar flexion of the foot
9. ○ dorsiflexion of the foot
10. ○ abduction of the thigh
11. ○ adduction of the thigh
12. ○ abduction of the arm
13. ○ adduction of the arm
14. ○ rotation of the arm
15. ○ pronation of the forearm
16. ○ supination of the forearm
17. ○ inversion of the foot
18. ○ eversion of the foot
19. ○ protraction of the shoulder
20. ○ retraction of the shoulder
21. ○ elevation of the shoulder
22. ○ depression of the shoulder

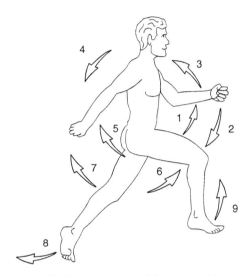
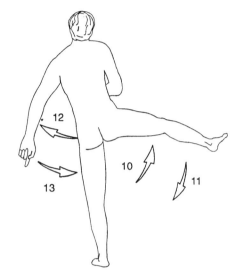

Figure 7.18a. Movements of the arms and legs.

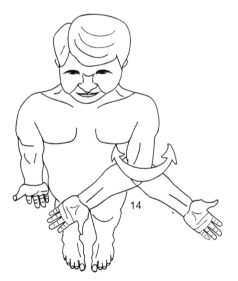
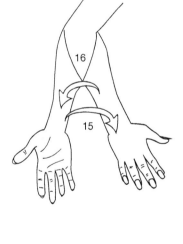
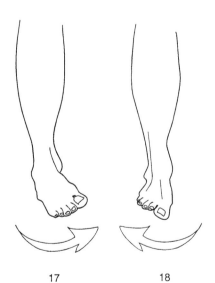

Figure 7.18b. Movements of the hands and feet.

Exercise 7.18:

_____ 1. Flexion and extension occur in the _____ plane.

_____ 2. Adduction and abduction occur in the _____ plane.

_____ 3. Dorsiflexion and plantar flexion refer to movements of the _____ .

_____ 4. The palm rotates medially out of anatomical position when the forearm _____ .

_____ 5. The forearm rotates back into anatomical position in the movement of _____ .

_____ 6. Shrugging the shoulders results in the motion of _____ .

_____ 7. At the end of a push-up exercise, the shoulder _____ .

_____ 8. If a person's ankle turns so that the sole of the foot faces inward, it is _____ .

_____ 9. Spreading the fingers is the motion of _____ .

_____ 10. When a person's trunk bends forward, the body is _____ .

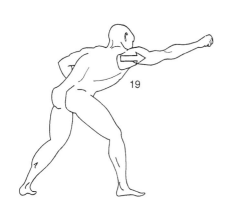
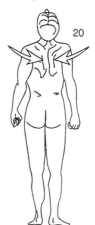
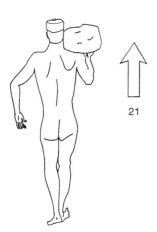
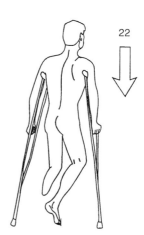

Figure 17.18c. Movements of the shoulder.

Answers to Exercise 7.18: 1. sagittal; 2. frontal; 3. foot; 4. pronates; 5. supination; 6. elevation; 7. protracts; 8. inverted; 9. abduction; 10. flexed.

S. Levers

Color:
○ E (effort)
○ R (resistance)
○ A (axis)

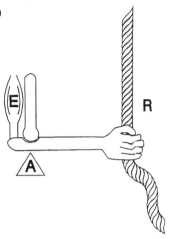

Figure 7.19a. First class lever.

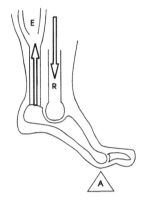

Figure 7.19b. Second class lever.

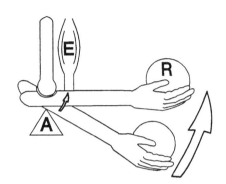

Figure 7.19c. Third class lever.

Exercise 7.19:

_____ 1. A lever is a rigid structure. In the body, a lever is usually a(n) _____ .

_____ 2. The lever pivots around an axis (fulcrum). In the body, this axis is usually a(n) _____ .

_____ 3. The axis is not always a joint. It may be the point of contact when the body is pushing away from something. Which drawing shows this?

_____ 4. Effort must be expended to move the lever against resistance. In the body, this effort (force) is usually provided by _____ .

_____ 5. In the body, the effort is considered to be applied at the insertion of the _____ .

_____ 6. In the above drawings, the resistance is provided by a. _____ , b. _____ , c. _____ .

_____ 7. If the effort and resistance are the same distance from the axis (fulcrum), such as when two people of equal weight sit on a seesaw, then how much effort must be expended to move a particular resistance?

_____ 8. Which drawing shows the effort arm (the distance the effort is away from the axis) as longer than the resistance arm?

9. When the resistance arm is longer than the effort arm,

_____ a. will the effort have to be greater or less than the resistance in order to cause movement?

_____ b. will the effort have to move through a longer or shorter distance than the resistance?

_____ 10. In a limb, what will be the effect on strength of moving the insertion of a muscle farther away from the joint?

_____ 11. In the above case, what will be the effect on range of motion?

Answers to Exercise 7.19: 1. bone; 2. joint; 3. b; 4. muscle; 5. muscle; 6. the rope, the weight of the body, the ball; 7. the same amount of effort; 8. b; 9a. greater, 9b. shorter; 10. increased; 11. decreased.

Chapter 8 Nervous System

A. Anatomical Organization

Color and label:
1. central nervous system
    a. ○ brain
    b. ○ spinal cord

Label:
2. peripheral nervous system
    a. cranial nerves
    b. spinal nerves

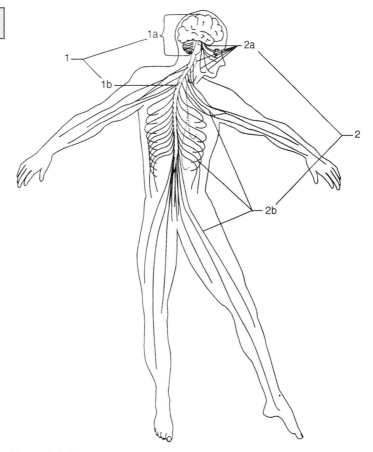

Figure 8.1. Central vs. peripheral nervous systems.

Exercise 8.1:

_____ 1. The central nervous system is made up of the _____ and _____ .

_____ 2. The nerves that connect to the central nervous system are called the _____ nervous system.

_____ 3. The nerves that are connected to the brain are called _____ nerves.

_____ 4. The nerves that are connected to the spinal cord are called _____ nerves.

Answers to Exercise 8.1: 1. brain, spinal cord; 2. peripheral; 3. cranial; 4. spinal.

101

B. Functional Organization

Color and label:
1. ○ central nervous system
2. ○ sensory pathway
3. motor pathway
   a. ○ somatic motor
   b. ○ autonomic

Label:
4. salivary gland
5. heart
6. skeletal muscle
7. intestine

Figure 8.2. Sensory and motor pathways.

Exercise 8.2:

_____ 1. Information entering the central nervous system from the internal or external environment is called _____ .

_____ 2. Information traveling to muscles or glands from the central nervous system is called _____ .

_____ 3. Motor output to skeletal muscles is called _____ .

_____ 4. Motor output to smooth muscle, cardiac muscle, or glands is called _____ .

Answers to Exercise 8.2: 1. sensory; 2. motor; 3. somatic; 4. autonomic.

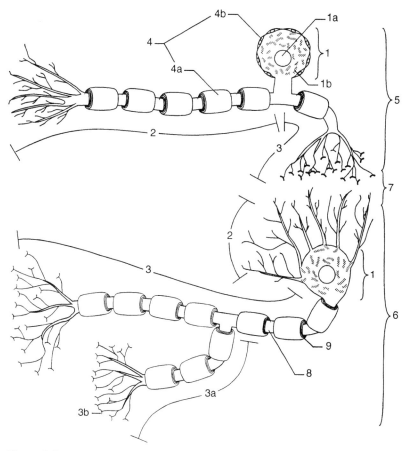

Figure 8.3. Sensory and motor neurons, peripheral neuroglia.

C. Neurons and Neuroglia

Color (trace) and label:

1. cell body
   a. ○ nucleus
   b. ○ Nissl bodies
      (endoplasmic reticulum
      and ribosomes)

2. ○ dendrites
3. ○ axon
   a. collateral
   b. terminal button
4. neuroglia (supporting cells)
   a. ○ Schwann cell
   b. ○ satellite cell

Label:
5. sensory neuron
6. motor neuron
7. synapse
8. node of Ranvier
9. myelin layers

Exercise 8.3:

1. The nervous system contains two classes of cells:
   a. cells that transmit action potentials called _____ .
   b. supporting cells called _____ .

2. The part of the neuron that contains the nucleus is called the _____ .

3. The cell body contains large numbers of _____ , which are made of _____ .

4. Therefore, you would expect a major function of neuron cell bodies to be _____ . (see cell structure, p. 32)

5. Many neurons (multipolar) have large numbers of _____ projecting from their cell bodies.

Neurons and Neuroglia 103

6. All functioning neurons have one _____, which may branch into one or more _____.

7. These collaterals end in _____.

8. The terminal buttons of an axon form part of the communicating structure between neurons (or neuron and muscle) called the _____.

9. The two types of support cells found in the peripheral nervous system are called _____ and _____.

10. Fill in the following table:

support cells	location	function
a.	wrapped around peripheral neuron fibers	form myelin sheath in PNS, phagocytic, nerve fiber regeneration
b.	around peripheral neuron cell bodies	provide nutrients for cell bodies

Answers to Exercise 8.3: 1a. neurons; 1b. neuroglia; 2. cell body; 3. Nissl bodies, endoplasmic reticulum and ribosomes; 4. protein synthesis; 5. dendrites; 6. axon, collaterals; 7. terminal buttons; 8. synapse; 9. Schwann cells, satellite cells; 10a. Schwann cells; 10b. satellite cells.

D. Neuroglia in Central Nervous System

Color (trace) and label:
1. ○ cell body
2. ○ dendrites
3. ○ axon

4. neuroglia
    a. ○ microglia
    b. ○ astrocytes
    c. ○ oligodendrocytes
    d. ○ ependymal cells
5. ○ blood vessel (red)

Label:
6. myelin
7. ventricles
8. choroid plexus

Exercise 8.4:

1. The support cells found in the central nervous system are _____, _____, _____, and _____.

2. Schwann cells (see figure 8.3) and oligodendrocytes have multiple layers of membrane that wrap around axons and are called _____.

3. The spaces between adjacent Schwann cells (and oligodendrocytes) are the _____.

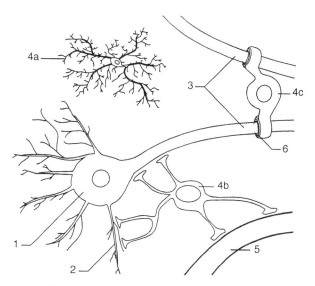

Figure 8.4a. Oligodendrocytes, astrocytes, microglia.

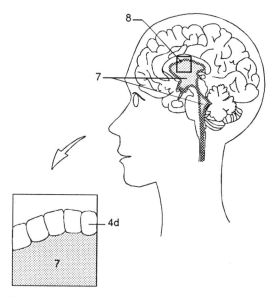

Figure 8.4b. Ependymal cells of choroid plexus.

4. Fill in the following table :

support cell	location	function
a.	throughout CNS, attached to neurons and capillaries	control chemical environment around neurons (control ions, recapture neurotransmitters)
b.	throughout CNS	macrophages, engulf invading microorganisms and dead tissue
c.	line ventricles of brain and central cavity of spinal cord	form part of blood-brain barrier between cerebrospinal fluid and interstitial fluid, secrete cerebrospinal fluid (see meninges, p. 119)
d.	wrapped around CNS neuron fibers	form myelin sheaths in CNS

Answers to Exercise 8.4: 1. microglia, astrocyte, oligodendrocytes, ependymal cells; 2. myelin; 3. nodes of Ranvier; 4a. astrocytes; 4b. microglia; 4c. ependymal; 4d. oligodendrocyte.

E. Membrane Potentials

Color:
- sodium ions (Na$^+$)
- potassium ions (K$^+$)
- calcium ions (Ca^{++})
- phosphate ions (HPO$_4^=$)
- chloride ions (Cl$^-$)
- proteins

Color and label:
1. sodium channels (color the same as sodium ions)
2. potassium channels (color the same as potassium ions)
3. sodium-potassium pump
4. activation gate
5. inactivation gate
6. potassium gate

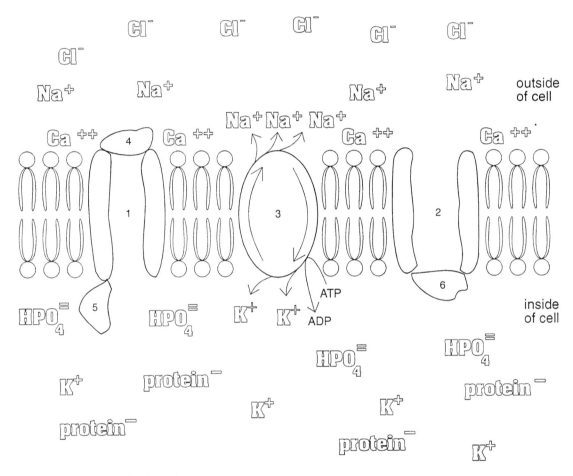

Figure 8.5. Membrane with channels.

Exercise 8.5:

1. The nerve cell membrane is a double layer of _____ with many _____ imbedded in it. (see membrane structure, p. 27)

2. Membrane permeability to ions is controlled by proteins organized as _____ .

3. The major ions that interact with the membrane are _____ , _____ , _____ , and _____ .

   a. The ion associated with the gates on sodium channels is _____ .

   b. Sodium is most concentrated _____ (inside, outside) the membrane.

   c. Potassium is most concentrated _____ (inside, outside) the membrane.

   d. Chloride is most concentrated _____ (inside, outside) the membrane.

4. Due to the unequal distribution of ions, the outside of the membrane is _____ (positive, negative) in relation to its inside.

5. This electrical charge (voltage) is called the _____ .

6. Would the ions remain distributed unequally if the membrane were freely permeable to all ions?

7. Which gate prevents the free diffusion of Na^+ into the cell?

8. Since there is some leakage and movement of ions, how does the membrane maintain the gradients (unequal distributions) of sodium and potassium?

9. In order to move these ions against a gradient, _____ is required.

10. When ATP hydrolyzes, _____ is released.

11. The nerve cell must expend _____ to maintain its membrane potential.

Answers to Exercise 8.5: 1. phospholipids, proteins; 2. channels; 3. calcium, sodium, potassium, chloride; 3a. calcium; 3b. outside; 3c. inside; 3d. outside; 4. positive; 5. resting potential; 6. no; 7. sodium activation gate; 8. the sodium-potassium pump; 9. ATP; 10. energy; 11. energy.

F. Action Potential

Color:
- ○ sodium ions (Na+)
- ○ potassium ions (K+)
- ○ stimulus (S)

1. ○ sodium channels (color the same as sodium ions)
2. ○ potassium channels (color the same as potassium ions)
3. ○ sodium-potassium pump
4. ○ activation gate
5. ○ inactivation gate
6. ○ potassium gate

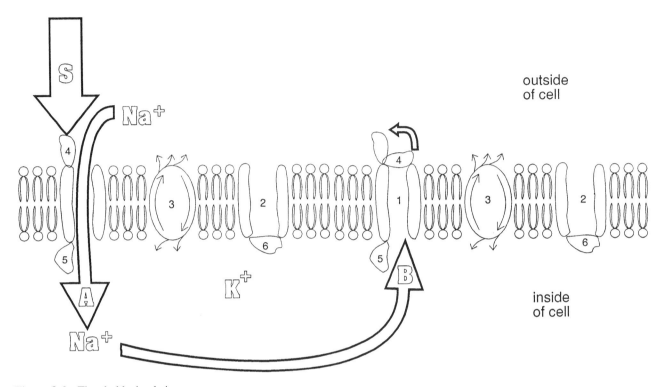

Figure 8.6a. Threshold stimulation.

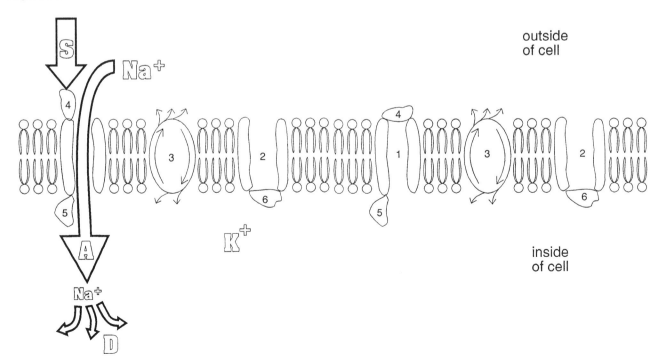

Figure 8.6b. Subthreshold stimulation.

Exercise 8.6:

1. When the membrane is at equilibrium, it is polarized. This means that its inside and outside surfaces carry opposite _____ .

2. The charge on the outside of the membrane is _____ . (see 8.5, question 4)

3. Since the sodium ion carries a _____ (positive, negative) charge, it would be attracted to the _____ (inside, outside) of the membrane.

4. Since the sodium ion is more concentrated _____ (inside, outside) the membrane, its tendency is to flow to the _____ (inside, outside) of the membrane.

5. Even though these two gradients (electrical and concentration) predict that sodium will flow into the cell, it does not flow in the resting membrane because the _____ are closed on the sodium channels.

6. If something (a stimulus) causes the activation gates on sodium channels to open, then sodium will _____ , as shown by arrow _____ .

7. When sodium enters the cell, the polarity of the membrane will _____ (increase, decrease).

8. When the membrane depolarizes, the gates in the immediate area _____ (open, close). This is shown by arrow _____ .

9. As more sodium gates open, the polarity (voltage) of the membrane continues to _____ (increase, decrease).

10. If the membrane depolarizes enough to open large numbers of adjacent gates (threshold), then these parts of the membrane will also _____ .

11. This then becomes a self-sustaining wave of depolarization called the _____ .

12. Small amounts of sodium (subthreshold) can enter the cell without causing a(n) _____ . This is represented by arrow _____ .

13. The action potential conforms to the all-or-none principle. This means that when compared to an action potential produced by a threshold stimulus

    a. the action potential produced by a stimulus of greater magnitude would be _____ (the same, larger).

    b. a subthreshold stimulus would produce _____ (no, a smaller) action potential.

Answers to Exercise 8.6: 1. charges; 2. positive; 3. positive, inside; 4. outside, inside; 5. gates; 6. flow into the cell, A; 7. decrease; 8. open, B; 9. decrease; 10. depolarize; 11. action potential; 12. action potential, D; 13a. the same; 13b. no.

Action Potential

G. Repolarization and Hyperpolarization

Color:
○ sodium ions (Na⁺)
○ potassium ions (K⁺)

1. ○ sodium channels (color the same as sodium ions)
2. ○ potassium channels (color the same as potassium ions)
3. ○ sodium-potassium pump
4. ○ activation gate
5. ○ inactivation gate
6. ○ potassium gate

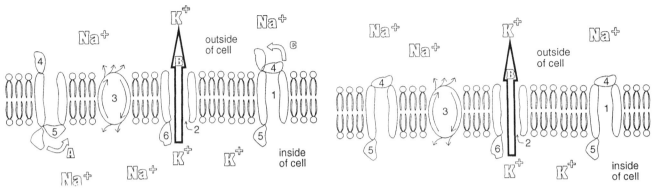

Figure 8.7a. Repolarization.

Figure 8.7b. Hyperpolarization.

Exercise 8.7:

1. Potassium gates are also opened by voltage changes but at a slower rate. When potassium gates open, this ion will flow (into, out of) _____ the cell
   a. because the concentration gradient for potassium is _____ (in, out).
   b. because the electrical gradient for potassium is _____ (in, out).

2. Potassium carries a _____ charge.

3. As potassium flows out of the cell, it will cause the membrane to _____ (depolarize, repolarize). This is shown by arrow _____ .

4. As the potassium gates open, sodium channel _____ (activation, inactivation) gates close on previously depolarized membrane, causing the membrane to enter its refractory period. This is shown by arrow _____ .

5. Even if activation gates on sodium channels are reopened during the refractory period, sodium cannot flow through the channels until the _____ gates reopen.

6. Therefore, a new action potential _____ (can, cannot) be generated during the refractory period.

7. In figure 8.7a, the activation gate on the sodium channel to the right is opening. (arrow C)
   a. This means that this part of the membrane is _____ (depolarizing, repolarizing).
   b. Is this part of the membrane in its refractory period?

8. If potassium gates open on a resting membrane, the outside of the membrane becomes more (positive, negative) _____ . (see figure 8.7b)

9. If the outside of the membrane becomes more positive, the membrane is (hyperpolarized, depolarized).

10. Hyperpolarization makes it _____ (easier, more difficult) for the membrane to generate a new action potential.

Answers to Exercise 8.7: 1. out of; 1a. out; 1b. out; 2. positive; 3. repolarize, B; 4. inactivation, A; 5. inactivation; 6. cannot; 7a. depolarizing; 7b. no; 8. positive; 9. hyperpolarized; 10. more difficult.

H. Propagation

Label:
a. unmyelinated axon
b. myelinated axon
○ Trace the arrows representing the movement of an action potential on a and b.

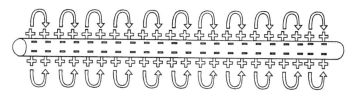

a _____

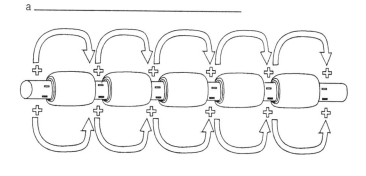

b _____

Figure 8.8. Movement of an action potential.

Exercise 8.8:

_____ 1. When a section of the membrane reaches threshold voltage, the activation gates on the adjacent channels _____ .

_____ 2. This causes a rapid inflow of _____ , which brings that section of membrane to its _____ voltage.

_____ 3. This continues from section to section on the membrane and is called the _____ potential.

_____ 4. If sections of nerve membrane are covered by myelin, those sections do not contain _____ .

_____ 5. On myelinated fibers, the action potential propagates (jumps) from _____ to _____ .

_____ 6. On the above illustration, which axon has less exposed membrane to depolarize? _____

_____ 7. Since much membrane is skipped on myelinated fibers, the action potential moves _____ (faster, slower) than on unmyelinated fibers.

_____ 8. Since a section of myelinated axon contains fewer channels and pumps than an unmyelinated axon, does it consume as much energy (ATP)?

_____ 9. The conduction of an action potential on a myelinated axon is called _____ conduction.

Answers to Exercise 8.8: 1. open; 2. sodium, threshold; 3. action; 4. channels; 5. node of Ranvier, node of Ranvier; 6. B (myelinated axon); 7. faster; 8. no; 9. saltatory.

I. Synapse

Color:
- ○ neurotransmitter (NT)
- ○ calcium ions (Ca++)
- ○ sodium ions (Na+)
- ○ potassium ions (K+)

Label:
1. terminal button of presynaptic neuron
2. synaptic cleft
3. postsynaptic neuron
4. ion channels
5. synaptic vesicles
   a. excitatory postsynaptic potential (EPSP)
   b. inhibitory postsynaptic potential (IPSP)

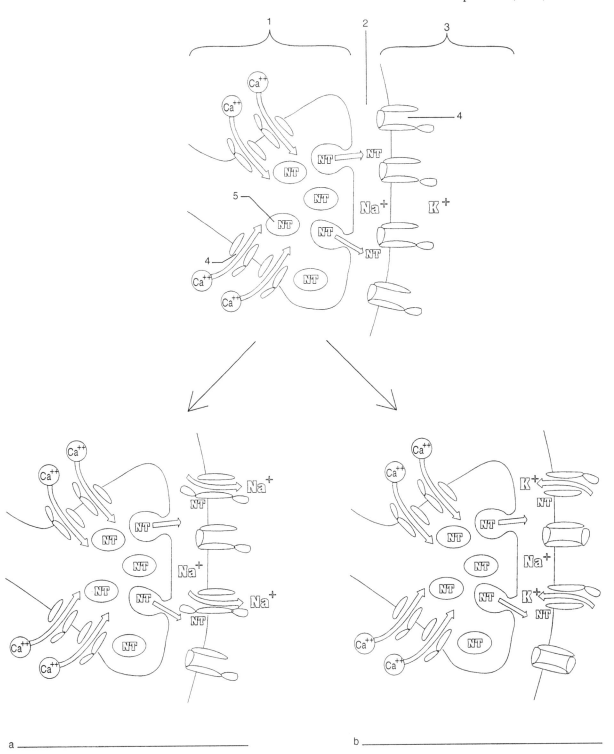

Figure 8.9. Neurotransmitter release.

Exercise 8.9:

_____ 1. When the action potential reaches the terminal button it stimulates the opening of _____ channels.

_____ 2. The inflow of calcium ions causes the _____ to fuse with the plasma membrane.

_____ 3. This results in the release of _____ .

_____ 4. The transmitter substance diffuses across the _____ .

_____ 5. The postsynaptic membrane contains channels whose gates open in response to _____ .

_____ 6. If these channels conduct sodium ions, then sodium will flow _____ (into, out of) the postsynaptic cell.

_____ 7. This inflow will cause an excitatory postsynaptic potential (EPSP) that tends to _____ (depolarize, hyperpolarize) the postsynaptic cell.

_____ 8. If the channels on the postsynaptic membrane conduct other ions such as potassium or chloride, then the result of opening their gates will be hyperpolarization, which is the opposite of _____ .

_____ 9. If the postsynaptic cell is hyperpolarized, then it will be _____ (easier, more difficult) to reach threshold.

_____ 10. An inhibitory postsynaptic potential (IPSP) would be the result of the opening of _____ channels.

_____ 11. Transmitter substance can be removed from the synaptic cleft either by being broken down (see figure 7.6) or by _____ .

Answers to Exercise 8.9: 1. calcium; 2. synaptic vesicles; 3. transmitter substance; 4. synaptic cleft; 5. transmitter substance; 6. into; 7. depolarize; 8. depolarization; 9. more difficult; 10. potassium (or chloride); 11. diffusion.

J. Summation

Color:
- ○ presynaptic neurons A, B, C
- ○ presynaptic neuron D
- ○ sodium ions
- ○ potassium ions
- ○ arrow S (summation)
- ○ arrow AP (action potential)

Label:
1. postsynaptic neuron
2. axon hillock
3. axon

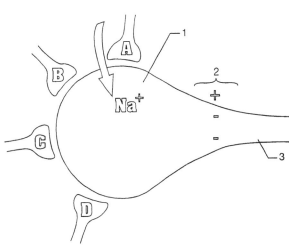

Figure 8.10a. Subthreshold stimulus at synapse.

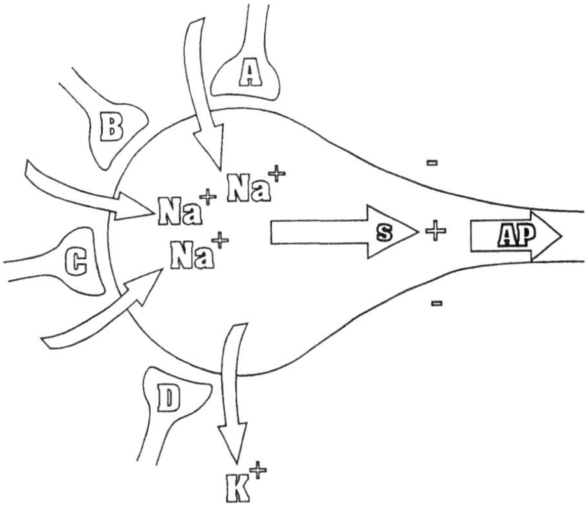

Figure 8.10b. Summation at synapse.

Exercise 8.10:

1. A single excitatory postsynaptic potential (EPSP) is a subthreshold stimulus. This means that it does not cause enough sodium inflow to produce a(n) _____ .

2. If many EPSP's are induced rapidly, then enough _____ may enter the cell to produce a(n) _____ .

3. This addition of the effects of subthreshold stimuli (EPSPs) is called _____ .

4. Summation can occur from many presynaptic neurons acting together, which is called _____ summation, or by a few presynaptic neurons producing EPSPs at high frequencies, which is called _____ summation.

5. If this summation results in the postsynaptic cell reaching the threshold voltage, then is an action potential propagated?

6. In figure 8.10b, summation results in action potentials starting at the _____ .

7. In figure 8.10b, the inhibitory postsynaptic potentials (IPSPs) result from the outflow of _____ .

8. An IPSP should counteract the effect of a(n) _____ .

9. Summation is the adding of the depolarizations of all the _____ and subtracting the hyperpolarizations of all of the _____ .

Answers to Exercise 8.10: 1. action potential; 2. sodium, action potential; 3. summation; 4. spatial, temporal; 5. yes; 6. axon hillock; 7. potassium; 8. EPSP; 9. EPSPs, IPSPs.

K. Generator Potential

Color and label:
1. ○ sensory receptor
2. ○ sensory neuron
- ○ S (stimulus arrow)
- ○ GP (generator potential arrow)
- ○ AP (action potential arrow)

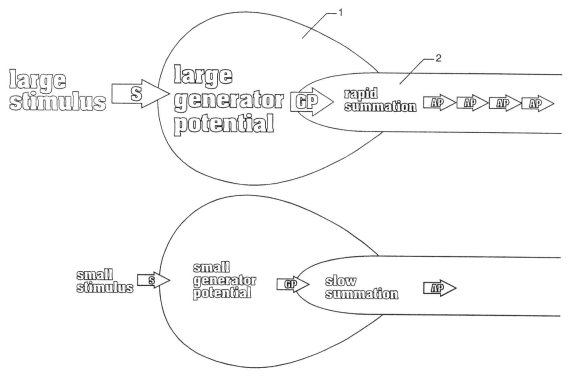

Figure 8.11. Effect of stimulus strength on generator potential.

Exercise 8.11:

_____ 1. The generator potential (receptor potential) is produced by a _____.

2. The generator potential is proportional to the stimulus.

_____ a. A large stimulus causes the sensory receptor to produce a _____ (large, small) generator potential.

_____ b. A small stimulus causes the sensory receptor to produce a _____ (large, small) generator potential.

_____ c. Therefore, does the generator potential conform to the all-or-none principle?

_____ 3. Generator potentials in the sensory receptors cause _____ on the sensory neuron.

_____ a. A large generator potential will result in _____ (many, few) action potentials on the sensory neuron.

_____ b. A small generator potential will result in _____ (many, few) action potentials on the sensory neuron.

_____ c. If the generator potential is very small (subthreshold), will it cause the sensory neuron to produce action potentials?

Answers to Exercise 8.11: 1. sensory receptor; 2a. large; 2b. small; 2c. no; 3. action potentials; 3a. many; 3b. few; 3c. no.

L. Sensory Receptors—Exteroceptors

Color (trace) and label:
1. ○ free nerve endings
2. ○ root hair plexuses
3. ○ organs of Ruffini
4. ○ Meissner's corpuscle
5. ○ Merkel's corpuscle
6. ○ Pacinian corpuscles
7. ○ bulb of Krause
8. ○ sensory (afferent) neuron

Label:
9. epidermis
10. dermis
11. hypodermis

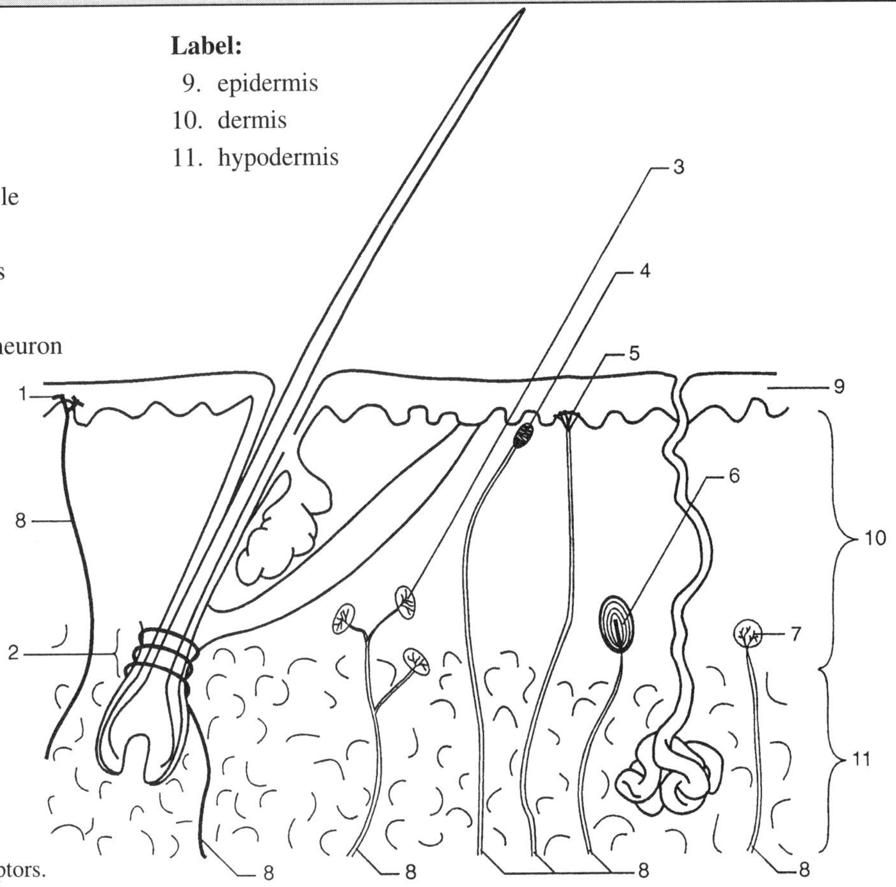

Figure 8.12. Cutaneous receptors.

Exercise 8.12:

_____ 1. Receptors can be classified by the location of the stimulus: external environment, body position, and internal environment. What do exteroceptors detect?

_____ 2. Cutaneous receptors are a particular type of exteroceptor. Where are they located?

3. Receptors convert energy forms (mechanical, thermal, light, and chemical) into action potentials.

   a. What type of receptor is the root hair plexus?

   b. Organs of Ruffini, Meissner's corpuscle, Merkel's corpuscle, Pacinian corpuscles, and bulbs of Krause are _____ (type of receptor).

_____ 4. Some areas of the integument contain *only* free nerve endings and can respond to touch, warmth, cold, and pain. What sensory receptor(s) would, therefore, detect these environmental stimuli?

_____ 5. Meissner's corpuscles and Pacinian corpuscles are rapidly adapting touch receptors. Under continued stimulation, they stop triggering action potentials. Which stimuli would demonstrate this adaptation?

   a. sitting on a chair

   b. an insect walking on your skin

   c. the feel of your clothing on your body

_____ 6. Would you expect pain receptors to shown adaptation?

_____ 7. Which cutaneous receptors extend into the epidermis?

_____ 8. What other types of exteroceptors are there?

Answers to Exercise 8.12: 1. external environment; 2. epidermis, dermis, hypodermis; 3a. mechanical; 3b. mechanical; 4. free nerve endings; 5. a, c; 6. no; 7. free nerve endings (pain receptors), Merkel's corpuscles; 8. taste, smell, vision, hearing (see special senses, chp. 9).

M. Sensory Receptors—Proprioceptors and Interoceptors

Color (trace) and label:
1. ○ muscle spindle
2. ○ Golgi tendon organ
3. ○ joint kinesthetic receptor
4. ○ free nerve ending
5. ○ afferent neuron
6. ○ muscle
7. ○ tendon
8. ○ joint capsule
9. ○ small intestine

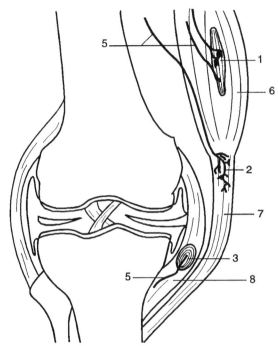

Figure 8.13a. Muscle, tendon, and joint.

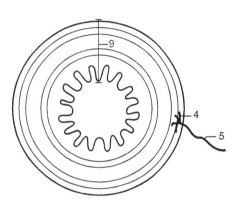

Figure 8.13b. Stretch receptors in the intestine.

Exercise 8.13:

_____ 1. Proprioceptors, which detect body position, are shown in figure _____ .

2. Which proprioceptors detect each of the following?

_____ a. change in position of joint

_____ b. muscle stretch (lengthening)

_____ c. tendon stretch (primarily due to strong muscle contraction)

_____ 3. Interoceptors, which monitor the internal environment, are shown in figure _____ .

_____ 4. In addition to stretch receptors, other examples of interoceptors include _____ .

Answers to Exercise 8.13: 1. 8.13a; 2a. joint kinesthetic; 2b. muscle spindle; 2c. Golgi tendon organ; 3. 8.13b; 4. pain receptors, chemoreceptors (CO_2, H^+, O_2, glucose).

N. Structure of the Brain

Color and label:
1. ◯ cerebrum
   a. cerebral cortex
   b. central sulcus
   c. corpus callosum
   d. gyres (convolutions)
2. ◯ thalamus
3. ◯ hypothalamus
4. ◯ optic chiasma
5. ◯ stalk of pituitary
6. ◯ pineal gland
7. ◯ midbrain
8. ◯ cerebellum
9. ◯ pons
10. ◯ medulla
11. ◯ spinal cord

Label:
12. third ventricle
13. fourth ventricle

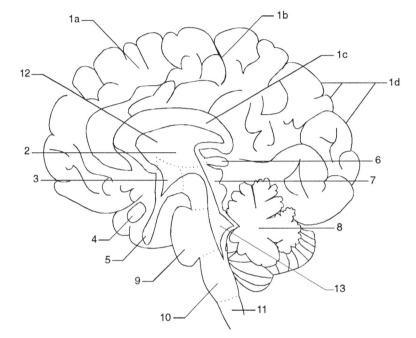

Figure 8.14. Midsagittal view of brain.

Exercise 8.14:

_____ 1. The largest part of the brain is the _____ .

_____ 2. The cerebral cortex is folded into _____ .

_____ 3. The large crease in the coronal plane is called the central _____ .

_____ 4. A major connection between the left and right sides of the cerebral cortex is the _____ .

_____ 5. In this diagram, cerebrospinal fluid is found in the _____ and _____ .

_____ 6. The pituitary gland is joined to the _____ .

_____ 7. The part of the brain that is joined to the spinal cord is the _____ .

_____ 8. The two parts of the brain that make up the lateral walls of the third ventricle are the _____ and _____ .

Answers to Exercise 8.14: 1. cerebrum; 2. gyres (convolutions); 3. sulcus; 4. corpus callosum; 5. third, fourth ventricles; 6. hypothalamus; 7. medulla; 8. thalamus, hypothalamus.

O. Meninges and Cerebrospinal Fluid

Trace in color and label the meninges:
1. ○ pia mater
2. ○ arachnoid
3. ○ dura mater

Color the areas containing cerebrospinal fluid the same and label accordingly:
4. choroid plexus
5. third ventricle
6. fourth ventricle
7. central canal
8. subarachnoid space
9. arachnoid villus

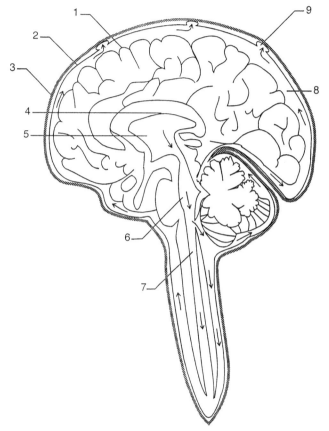

Figure 8.15. Midsagittal section of central nervous system.

Exercise 8.15:

_____ 1. Which of the meninges is closest to the nervous tissue of the central nervous system?

_____ 2. Which of the meninges is closest to the bones of the skull?

_____ 3. Cerebrospinal fluid is produced by the _____ (see neuroglia, p. 105).

_____ 4. The cerebrospinal fluid flows through the central nervous system in the _____ of the brain and the _____ of the spinal cord.

_____ 5. The cerebrospinal fluid flows around the central nervous system in the _____ space.

_____ 6. Cerebrospinal fluid flows from the subarachnoid space into veins under the dura mater through the _____ .

Answers to Exercise 8.15: 1. pia mater; 2. dura mater; 3. choroid plexus; 4. ventricles, central canal; 5. subarachnoid; 6. arachnoid villus.

Meninges and Cerebrospinal Fluid

P. Structure of the Spinal Cord and Spinal Nerves

Color and label:

1. ○ spinal nerve
   a. ○ dorsal root
   b. ○ dorsal root ganglion
   c. ○ ventral root
2. ○ gray matter
   a. dorsal horn
   b. ventral horn
3. ○ central canal
4. ○ white matter

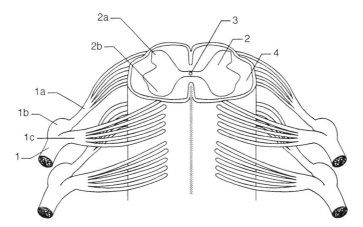

Figure 8.16. Ventral view of spinal cord.

Exercise 8.16:

_____ 1. Each spinal nerve connects to the spinal cord by a _____ and a _____ .

_____ 2. The dorsal root contains portions of _____ (sensory, motor) neurons. (see functional organization, p. 102)

_____ 3. The dorsal root ganglion is a group of cell bodies that belong to _____ neurons.

_____ 4. The ventral root contains portions of _____ (sensory, motor) neurons.

_____ 5. Since gray matter contains dendrites and cell bodies, what type of nerve activity occurs there?

_____ 6. In the spinal cord, the gray matter is _____ (deep, superficial) to the white matter.

_____ 7. The central canal of the spinal cord contains _____ .

_____ 8. The white matter contains bundles of axons (tracts) that carry information between the spinal cord and the _____ .

Answers to Exercise 8.16: 1. dorsal root, ventral root; 2. sensory; 3. sensory; 4. motor; 5. synaptic; 6. deep; 7. cerebrospinal fluid; 8. brain.

Q. Spinal Reflex

Color (trace) and label:
1. ○ stimulus
2. ○ sensory receptor
3. ○ sensory neuron
4. ○ association neuron
5. ○ motor neuron
6. ○ motor effector
7. spinal cord
   a. ○ gray matter
8. synapse
9. dorsal root ganglion

Label:

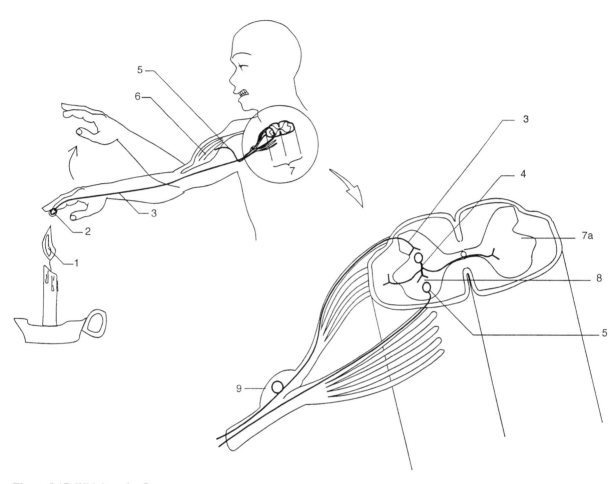

Figure 8.17. Withdrawal reflex.

Exercise 8.17:

_____ 1. The reflex is initiated when a stimulus activates a(n) _____ .

_____ 2. Action potentials from the sensory receptor travel directly to the _____ .

_____ 3. At the spinal cord, the sensory neuron synapses with a(n) _____ .

_____ 4. The association neuron activates the _____ .

_____ 5. Is the brain necessary for the completion of a spinal reflex?

_____ 6. Does the sensory input of a reflex have a predictable motor response?

Answers to Exercise 8.17: 1. sensory receptor; 2. sensory neuron; 3. association neuron; 4. motor neuron; 5. no; 6. yes.

R. Functions of the Brain

Color and Label:
1. ⭕ reticular activating system (dotted line)
2. ⭕ limbic system (dotted line)

Label:
3. brain stem
   a. medulla oblongata
   b. pons
   c. midbrain
4. diencephalon
   a. thalamus
   b. hypothalamus
5. cerebral cortex
6. cerebellum

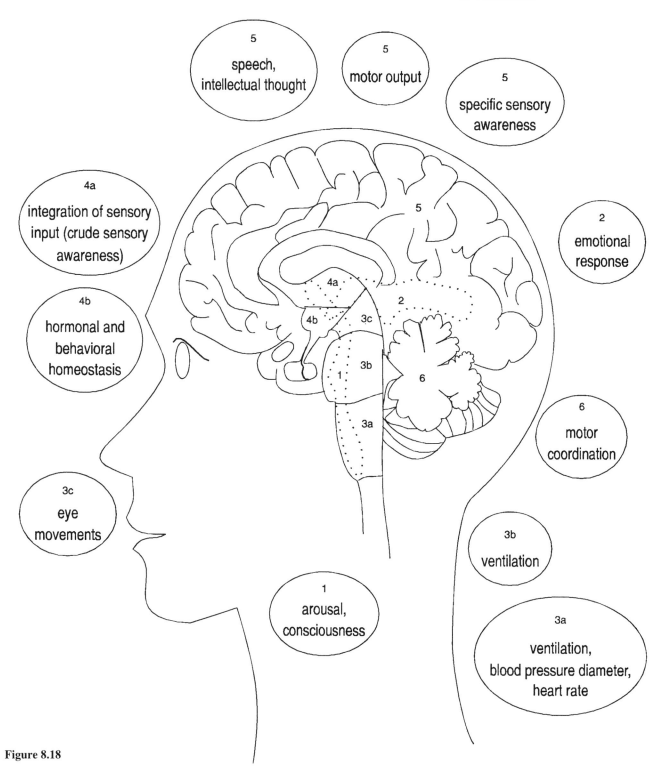

Figure 8.18

Exercise 8.18:

_____ 1. The brain stem includes the _____, _____, and _____.

_____ 2. The diencephalon includes the _____ and _____.

_____ 3. The reticular activating system is spread throughout the _____ and _____.

_____   a. Its function is to maintain _____.

_____ 4. Centers in the pons and medulla oblongata control basic vegetative functions such as _____, _____, and _____.

_____   a. Do these functions generally require consciousness?

_____ 5. Motor control of eye movements is found in the _____.

_____ 6. The limbic system is contained in the _____, _____, and _____.

_____   a. Behaviors such as reproduction and self-defense require emotional motivation, which may be centered in the _____ system.

_____ 7. Most somatic sensory input is first perceived by the _____.

_____ 8. Much homeostatic activity requiring hormonal and/or behavioral response is mediated by the _____.

_____ 9. Specific localization of sensory input and fine motor output is controlled by the _____.

_____ 10. Memory, planning, and speech are centered in the _____.

_____ 11. The coordination of skeletal muscles and balance are mediated by the _____.

_____ 12. In summary, most activities that do not require conscious participation are controlled by the _____.

_____   a. Activities that require behaviors such as eating or drinking are controlled by the _____.

_____   b. Fear, rage, sexual performance, and other emotionally based activities may be controlled by the _____ system.

_____   c. Speech and the intellectual processes connected with it are controlled by the _____.

Answers to Exercise 8.18: 1. medulla, pons, midbrain; 2. hypothalamus, thalamus; 3. brain stem, diencephalon; 3a. consciousness; 4. ventilation (breathing), heart rate, blood vessel diameter; 4a. no; 5. midbrain; 6. thalamus, hypothalamus, cerebral cortex; 6a. limbic; 7. thalamus; 8. hypothalamus; 9. cerebral cortex; 10. cerebral cortex; 11. cerebellum; 12. brain stem; 12a. hypothalamus; 12b. limbic; 12c. cerebral cortex.

S. Cranial Nerves

Color and label the cranial nerves.
- I. olfactory nerve
- II. optic nerve
- III. oculomotor nerve
- IV. trochlear nerve
- V. trigeminal nerve
- VI. abducent nerve
- VII. facial nerve
- VIII. statoacoustic (vestibulocochlear) nerve
- IX. glossopharyngeal nerve
- X. vagus nerve
- XI. accessory nerve
- XII. hypoglossal nerve
- **Using the same color, color and label structures containing receptors.**

1. olfactory receptors
2. retina
3. dura mater covering brain
4. facial structures (from skin into mucous membrane)
5. taste receptors on tongue
6. semicircular canals
7. cochlea
8. pharynx
9. carotid sinus
10. auricle
- **Using one color, color and label muscles.**
11. upper eyelid
12. iris
13. ciliary body
14. muscles moving eyes
15. muscles of mastication
16. muscles of expression
17. muscles of pharynx
18. muscles of larynx
19. sternocleidomastoid
20. trapezius
21. muscles of tongue
- **Using one color, color and label glands.**
22. lacrimal
23. nasal mucosa
24. salivary gland
    a. parotid
- **Color and label structures containing receptors, muscles, and glands.**
25. thoracic and abdominal organs

Exercise 8.19:

Which cranial nerve(s):

_____ 1. control muscles moving the eyes? (3 nerves)

_____ 2. detects odors? (1)

_____ 3. control muscles involved in chewing of food? (3)

_____ 4. control swallowing? (3)

_____ 5. detects light? (1)

_____ 6. innervates structures outside the head or neck? (1)

_____ 7. innervate salivary glands? (2)

Answers to Exercise 8.19: 1. III, IV, VI; 2. I; 3. V, VII, XII; 4. IX, XI, XII; 5. II; 6. X; 7. VII, IX.

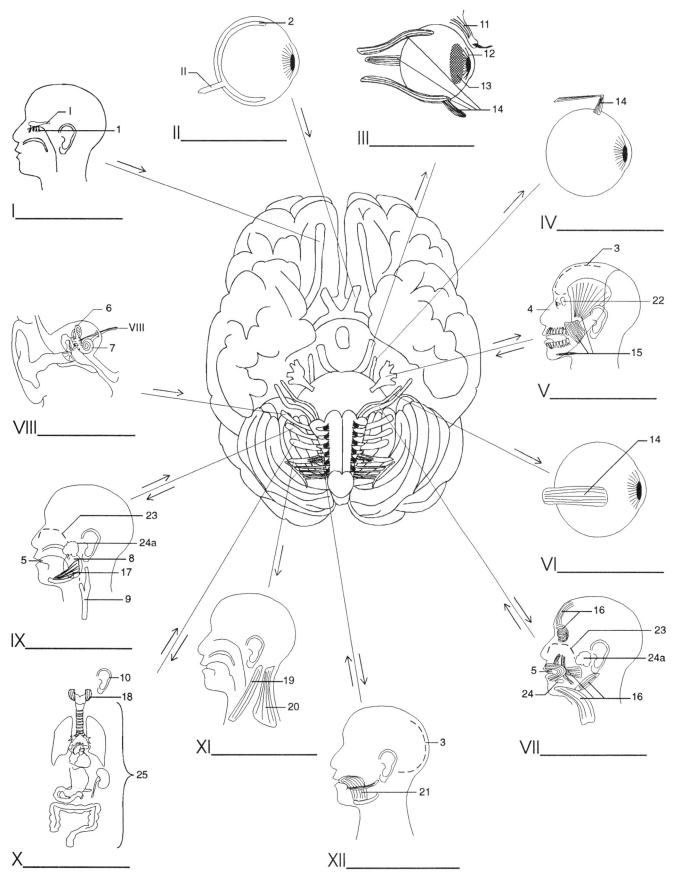

Figure 8.19. Inferior surface of brain and structures innervated by cranial nerves.

T. Autonomic Nervous System

Color (trace) and label:
1. ○ motor neuron
2. ○ preganglionic neurons (dashed line in figure 8.20b)
3. ○ postganglionic neurons (solid line in figure 8.20b)
4. ○ brain
5. ○ spinal cord
6. ○ skeletal muscle
7. ○ heart
8. ○ intestines

Label:
9. somatic nervous system
10. autonomic nervous system
    a. sympathetic division
    b. parasympathetic division
11. ganglion (synapse of pre- and postganglionic neurons)

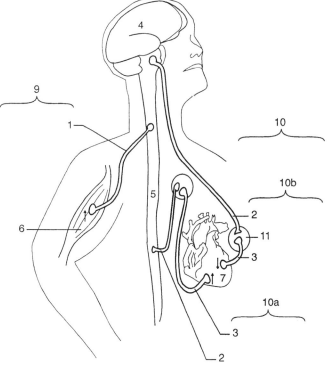

Exercise 8.20:

1. Compare the somatic and autonomic nervous systems by filling in this table

Figure 8.20a. Somatic vs. autonomic nervous system.

	Somatic	Autonomic
a. What is innervated?		
b. Do neurons synapse after leaving CNS?		
c. Does excitation of neurons always activate effector?		

_____ 2. The autonomic nervous system is divided into two parts, the _____ and _____ .

3. Fill in the table:

	Sympathetic	Parasympathetic
a. Where do preganglionic neurons leave the CNS?		
b. Are preganglionic neurons short or long?		
c. Are ganglia interconnected or discrete?		

_____ 4. In emergencies, a coordinated response of the whole body is needed. Which division of the ANS is designed for this type of response?

_____ 5. During restorative activities, individual organs can function as needed. Which division is designed for this?

_____ 6. Which division would be most active during physical exercise?

_____ 7. Look at the effects of the ANS on the heart and intestines as shown in figure 8.20b.

_____   a. Is the sympathetic division always excitatory and the parasympathetic division always inhibitory?

    b. As a general rule, do the two augment each other or do they have reciprocal actions?

Answers to Exercise 8.20: 1a. skeletal muscle; smooth muscle, cardiac muscle and glands; 1b. no, yes; 1c. yes, no; 2. sympathetic, parasympathetic; 3a. all thoracic and the first two lumbar; cranial nerves III, VII, IX, X, sacral nerves 2, 3, 4; 3b. short, long; 3c. interconnected, discrete; 4. sympathetic; 5. parasympathetic; 6. sympathetic; 7a. no; 7b. reciprocal actions.

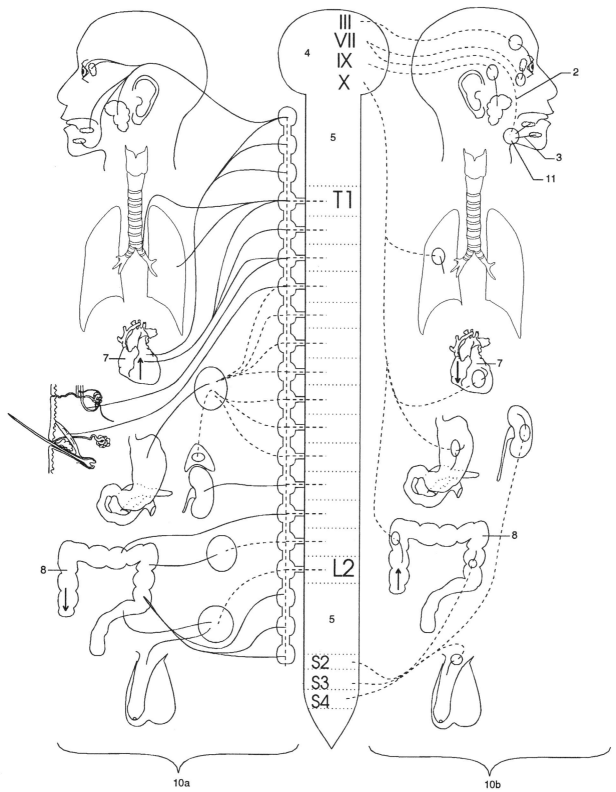

Figure 8.20b. Divisions of ANS.

Autonomic Nervous System

U. Functions of Hypothalamus

Color each center of the hypothalamus differently. Use the same color for input to each center and for output.

Exercise 8.21:

_____ 1. Hypothalamic temperature centers will react to decreased _____ by stimulating skeletal muscles to _____ .

_____ 2. Exposure to cold weather will cause the metabolism to _____ (increase, decrease).

_____ 3. Cutaneous vasoconstriction will result in _____ (more, less) blood flow to the skin.

_____ 4. Since the skin is usually cooler than core body temperature, increasing blood flow to the skin usually results in _____ (increasing, decreasing) body temperature.

_____ 5. A decreased blood glucose level leads to a(n) _____ (increase, decrease) in the motility of the digestive system.

_____ 6. The sight, smell, or thought of food involves what part of the brain? _____

_____ 7. Since the hypothalamus contains osmoreceptors, increased salt would produce an effect similar to _____ (increased, decreased) water.

_____ 8. In addition to controlling the adjustment of internal physiological conditions, the hypothalamus initiates _____ , which cause the organism to seek out necessities from the external environment.

_____ 9. These behaviors include _____ , _____ , _____ , and _____ .

_____ 10. The hypothalamus is a major controller of the _____ system.

Answers to Exercise 8.21: 1. blood temperature, shiver; 2. increase; 3. less; 4. decreasing; 5. increase; 6. cerebral cortex; 7. decreased; 8. behavior; 9. food seeking, drinking, sexual behavior, defensive behavior; 10. endocrine.

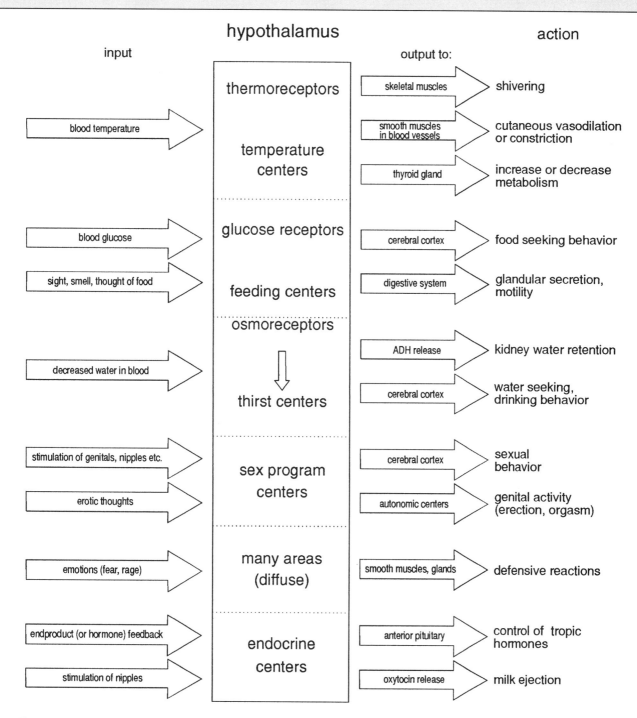

Figure 8.21. Centers of the hypothalamus.

V. Somatic Sensory System

Color (trace) and label:

1. ○ sensory receptor
2. sensory neuron
   a. ○ in spinal nerve
   b. ○ in cranial nerve
3. ○ spinal cord
4. ○ neuron in ascending tract
5. ○ thalamus
6. ○ postcentral gyrus of cerebral cortex
   a. somatotopic map

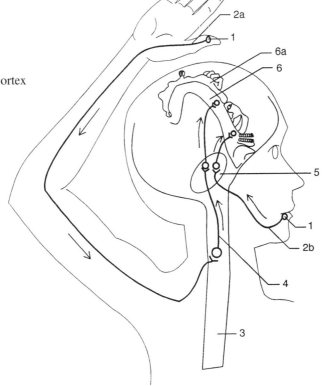

Figure 8.22. Sensory pathway.

Exercise 8.22:

_____ 1. Sensory information enters the nervous system through _____ .

_____ 2. This information is transmitted into the central nervous system on _____ .

_____ 3. The spinal cord pathway, which carries sensory information to the brain, is a(n) _____ .

_____ 4. This ascending tract is located in the _____ (white, gray) matter of the spinal cord.

_____ 5. Most somatic sensory information synapses in the _____ .

_____ 6. From the thalamus this information is projected to the _____ of the cerebral cortex.

_____ 7. The orderly arrangement of neurons of the postcentral gyrus that correspond to the parts of the body is called a(n) _____ map.

_____ 8. The size of the areas in the somatotopic map is not proportional to the anatomical size of the body areas, but to the number of _____ that each area contains.

_____ 9. According to this diagram, what part of the brain first processes sensory input?

Answers to Exercise 8.22: 1. sensory receptors; 2. sensory neurons; 3. ascending tract; 4. white; 5. thalamus; 6. postcentral gyrus; 7. somatotopic; 8. sensory receptors; 9. thalamus.

W. Somatic Motor System—Motor Cortex

Color (trace) and label:

1a. ○ primary motor cortex (somatotopic map)
 b. ○ premotor cortex
 c. ○ supplementary motor cortex
2. ○ descending tract
3. ○ peripheral motor pathway
4. ○ muscle

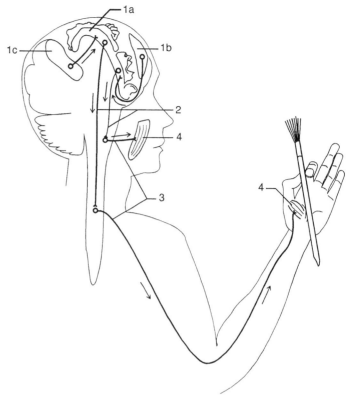

Figure 8.23. Skeletal muscle control.

Exercise 8.23:

_____ 1. Neurons controlling specific motor units in muscles are located in the _____ .

_____ 2. The neurons of the primary motor cortex are organized into patterns that represent the particular muscles of the body. This organization is called the _____ .

_____ 3. The primary motor cortex gets information from the _____ and the _____ .

_____ 4. Neurons leaving the primary motor cortex form _____ (ascending, descending) tracts.

_____ 5. These descending tracts form the _____ (gray, white) matter of the brain and spinal cord.

_____ 6. These neurons synapse with neurons in the _____ (gray, white) matter of the spinal cord.

Answers to Exercise 8.23: 1. primary motor cortex; 2. somatotopic map; 3. premotor cortex, supplementary motor cortex; 4. descending; 5. white; 6. gray.

X. Somatic Motor System—Basal Ganglia

Color and label:
1. ○ basal ganglia
2. ○ thalamus
3. ○ motor cerebral cortex

Color input and output arrows to basal ganglia using different colors.

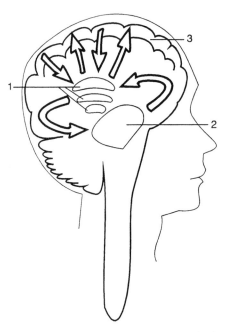

Figure 8.24. Role of basal ganglia in skeletal muscle control.

Exercise 8.24:

_____ 1. The masses of neurons at the base of the cerebrum are called _____ .

_____ 2. The basal ganglia receive information from the _____ and _____ .

_____ 3. The basal ganglia send information to many areas of the _____ and to the _____ .

_____ 4. Since the basal ganglia contain patterns of motor activity, should they fire before or after the primary motor area?

_____ 5. Since disease of the basal ganglia may cause muscle rigidity or uncontrolled movements, is it likely that the motor cortex needs them for normal muscle activity?

Answers to Exercise 8.24: 1. basal ganglia; 2. motor cerebral cortex, thalamus; 3. cerebral cortex, thalamus; 4. before; 5. yes.

Y. Function of the Cerebellum

Color and label:
1. ○ sensory receptors
   a. vestibular apparatus
   b. eye
   c. muscle spindle
   d. joint receptor
2. ○ muscle

Label:
3. cerebral cortex
4. thalamus
5. cerebellum
6. ascending tract
7. sensory neuron

8. descending tract
   a. nucleus in descending tract
9. motor neuron

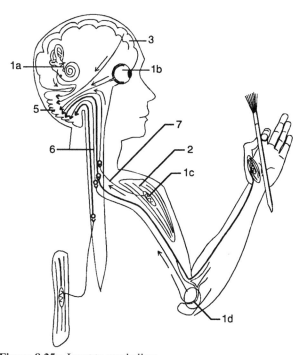

Figure 8.25a. Input to cerebellum.

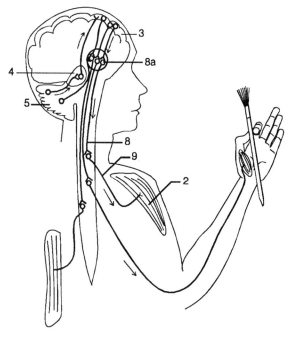

Figure 8.25b. Output from cerebellum.

Exercise 8.25:

_____ 1. The cerebellum gets information from the ____ , ____ , ____ , ____ , and ____ .

_____ 2. The cerebellum affects the action of muscles by sending information out to the ____ and ____ .

_____ 3. The vestibular apparatus senses ____ .

_____ 4. As the center of gravity shifts during movements, the cerebellum adjusts the tension of the postural ____ .

_____ 5. Since the cerebellum stores information about the sequence of somatic motor actions, which activity would be most directly affected by cerebellar damage? a. breathing, b. organized movement of the intestines, c. playing the piano.

Answers to Exercise 8.25: 1. cerebral cortex, vestibular apparatus, spinal cord, joint receptors, muscles; 2. cerebral cortex, spinal cord; 3. equilibrium; 4. muscles; 5. c.

Chapter 9: Special Senses

A. Gustatory and Olfactory Receptors

Color (trace) and label:
1. ○ bitter
2. ○ sour
3. ○ salty
4. ○ sweet
5. ○ gustatory receptors
6. ○ supporting cells
7. ○ olfactory neurons (receptors)
8. ○ a. afferent (sensory) neurons (for taste)
   b. olfactory bulb
   c. olfactory tract
9. ○ a. cribriform plate (of ethmoid bone)
   b. palate
10. ○ a. nasal cavity
    b. nasopharynx

Label:
11. taste pore
12. epithelium

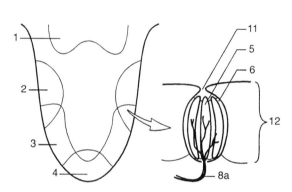

Figure 9.1a. Taste regions of tongue, structure of taste bud.

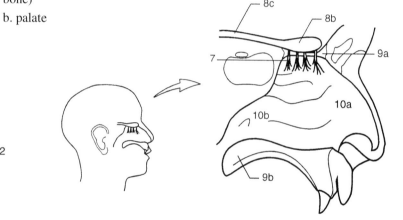

Figure 9.1b. Olfactory receptors.

Exercise 9.1:

1. Which receptors detect chemicals in the external environment?
2. In order for chemoreceptors to fire, the chemical must be dissolved in water. Where is the "water" for gustatory and olfactory receptor function?
3. When mucus covers the aqueous olfactory surfaces during a cold, what happens to the sense of smell?
4. How many primary tastes can the gustatory receptors detect? Which have more primary sensations—olfactory or gustatory receptors?
5. Sensory information from the taste buds travels to the brain via cranial nerves _____, _____, and _____ .
6. Gustatory information registers on the _____ of the _____ (part of the brain).
7. What cranial nerve transmits olfactory information?
8. Olfactory information registers on the _____ of the _____ (part of the brain).

Answers to Exercise 9.1: 1. gustatory, olfactory; 2. saliva, mucous membrane; 3. it decreases; 4. 4, olfactory; 5. VII (facial), IX (glossopharyngeal), X (vagus); 6. postcentral gyrus (sensory cortex), cerebrum (see brain, p. 130); 7. I (olfactory); 8. olfactory cortex, cerebrum.

B. General Anatomy of the Eye

Color and label:
1. ○ cornea
2. ○ sclera
3. ○ choroid
4. ciliary body
    ○ a. ciliary muscle
    ○ b. ciliary process
5. ○ iris
6. ○ lens
7. ○ suspensory ligament
8. ○ retina
9. ○ optic nerve

Label:
10. pupil
11. macula lutea
    a. fovea centralis
12. optic disk

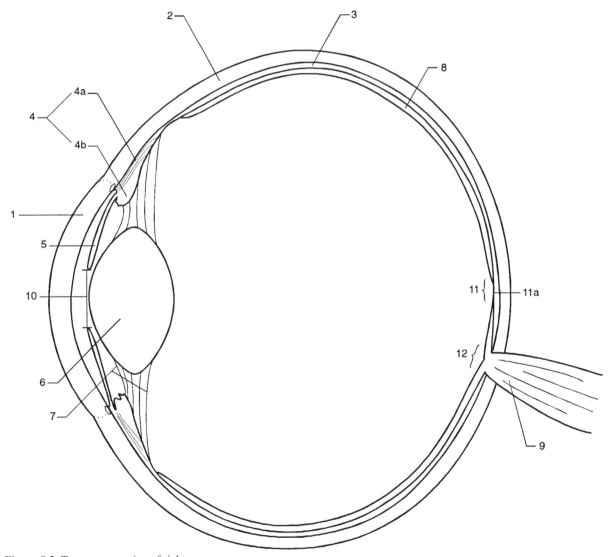

Figure 9.2. Transverse section of right eye.

Exercise 9.2:

_____ 1. The outermost layer of the eye includes the _____ and _____ .

_____ 2. The middle layer includes the _____ , _____ , and _____ .

_____ 3. The innermost layer is the _____ .

4. Name the structures of the eye described below:

_____ a. tough, white connective tissue outer layer, prevents light from entering eye other than anteriorly

_____ b. central space where light passes through iris

_____ c. part of retina with no photoreceptors (blind spot), optic nerve leaves retina

_____ d. concentration of cones

_____ e. densest concentration of cones within macula lutea, vision clearest here

_____ f. site of aqueous humor production

_____ g. contains many blood vessels, melanin, prevents scattering of light inside the eye

_____ h. contains light receptors

_____ i. transparent surface structure that allows light to enter eye, main site of light refraction (bending)

_____ j. regulates amount of light entering eye

_____ k. focuses light onto retina

_____ l. sends sensory information from retina to brain

_____ m. smooth muscle that changes shape of lens

_____ n. attaches lens to ciliary body

Answers to Exercise 9.2: 1. sclera, cornea; 2. choroid, ciliary body, iris; 3. retina; 4a. sclera; 4b. pupil; 4c. optic disk; 4d. macula lutea; 4e. fovea centralis; 4f. ciliary process; 4g. choroid; 4h. retina; 4i. cornea; 4j. iris; 4k. lens; 4l. optic nerve; 4m. ciliary muscle; 4n. suspensory ligaments.

C. Chambers of the Eye

Color and label:
1. ○ aqueous humor
2. ○ vitreous humor
3. ○ cornea
4. ○ iris
5. ○ lens
6. ○ ciliary process
7. ○ retina
8. ○ canal of Schlemm

Label:
9. anterior cavity (aqueous chamber)
   a. anterior chamber
   b. posterior chamber
10. posterior cavity (vitreous chamber)

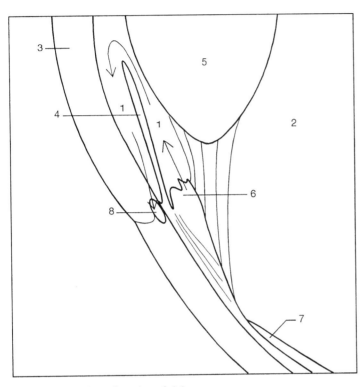
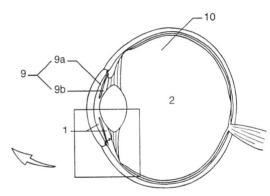

Figure 9.3. Horizontal section of right eye.

Exercise 9.3:

_____ 1. The avascular cornea and lens get nutrients and oxygen from the _____ .

_____ 2. Where is the aqueous humor made?

_____ 3. The aqueous humor first drains through the _____ chamber, then the _____ chamber.

_____ 4. How does the aqueous humor drain out of the eye?

_____ 5. The vitreous humor is found in the _____ .

_____ 6. Is vitreous humor continuously produced?

_____ 7. The function of vitreous humor is _____ .

Answers to Exercise 9.3: 1. aqueous humor (cornea gets oxygen directly from air); 2. ciliary process; 3. posterior, anterior; 4. canal of Schlemm; 5. vitreous chamber; 6. no (it's made prenatally); 7. to help maintain intraocular pressure.

D. Accommodation

Color (trace) and label:
1. ○ cornea
2. ○ ciliary body
3. suspensory ligaments
4. ○ lens
5. ○ retina
6. ○ light rays

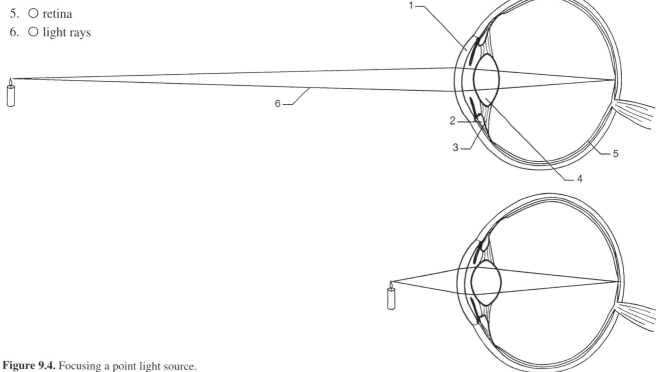

Figure 9.4. Focusing a point light source.

Exercise 9.4:

_____ 1. When light rays enter the eye, the greatest refraction (bending) occurs at the _____ . Light is also refracted by the _____ .

_____ 2. The light rays traveling from a distant point that reach the eye are _____ (nearly parallel, diverging).

_____ 3. When a person looks at distant objects, the ciliary muscles are relaxed, causing the ciliary body to pull on the suspensory ligaments. The tensed suspensory ligaments, in turn, pull on the lens, causing it to _____ (flatten, thicken).

_____ 4. The light then converges (focuses) onto the _____ .

_____ 5. The light rays traveling from a near point that reach the eye are _____ (nearly parallel, diverging).

_____ 6. When looking at near objects, the ciliary muscles contract, reducing tension on the suspensory ligaments. This reduced tension, in turn, allows the elastic lens to _____ (flatten, thicken).

_____ 7. The thickened lens refracts the light so that it focuses on the _____ .

_____ 8. Therefore, the _____ (lens, cornea) allows for accommodation, the focusing of near and far objects.

Answers to Exercise 9.4: 1. cornea, lens; 2. nearly parallel; 3. flatten; 4. retina; 5. diverging; 6. thicken; 7. retina; 8. lens.

E. Rods and Cones

Color (trace) and label:
- V ○ violet
- B ○ blue
- G ○ green
- Y ○ yellow
- O ○ orange
- R ○ red

1. ○ rods (use gray or black)
2. cones
   a. blue (use blue)
   b. green (use green)
   c. red (use red)
3. ○ bipolar neurons
4. ○ ganglion neurons

Label:
5. light
6. retina
7. choroid
8. sclera
9. optic nerve

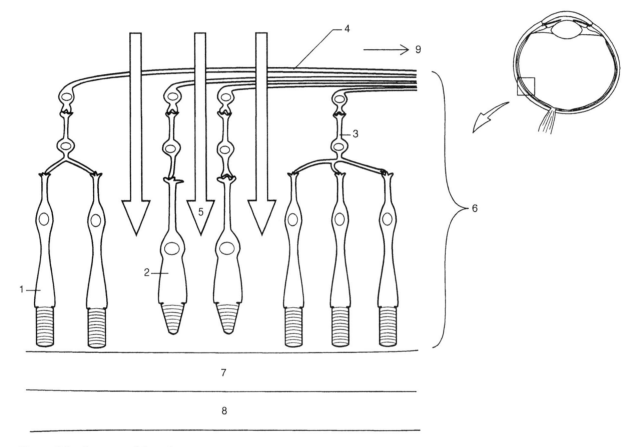

Figure 9.5a. Structure of the retina.

Exercise 9.5:

_____ 1. The two types of receptors in the retina are ____ and ____.

_____ 2. Cones are concentrated at the ____ and rods at the ____ of the retina.

_____ 3. Rods detect the range of light wavelengths called the ____.

_____ 4. The three types of cones are ____, ____, and ____.

140 Chapter 9 Special Senses

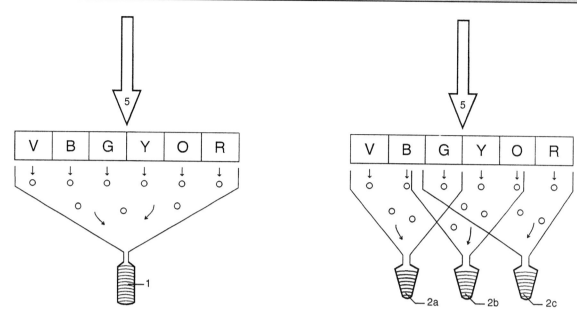

Figure 9.5b. Visible spectrum.

_____ 5. Each type cone detects _____ (a portion of, the whole) visible spectrum.

_____ 6. Rods are more sensitive to light than cones. Which photoreceptor responds to dim light?

_____ 7. The pigment in rods bleaches out in bright light. Which photoreceptor responds to bright light?

_____ 8. The brain interprets rod stimulation as black and white, and cone stimulation as color. What do you see in dim light? In bright light?

_____ 9. Stimulated photoreceptors form a(n) _____ (generator, action) potential.

_____ 10. Photoreceptors synapse with _____ neurons, which synapse with _____ neurons.

_____ 11. Ganglion neurons, which become the _____ nerve, convey sensory information to the brain as _____ potentials.

_____ 12. Convergence at the synapses in the retina is more likely to occur from _____ (rods, cones).

_____ 13. If three rods converge onto one bipolar neuron, can the brain tell which rod triggered the action potentials?

_____ 14. Does convergence affect visual acuity? Is vision in dim light sharp or blurry?

_____ 15. Is vision in bright light sharp or blurry? Why?

Answers to Exercise 9.5: 1. rods, cones; 2. fovea centralis, periphery; 3. visible spectrum; 4. blue, green, red; 5. a portion of; 6. rods; 7. cones; 8. black and white, color; 9. generator; 10. bipolar, ganglion; 11. optic, action; 12. rods; 13. no; 14. yes, blurry; 15. sharp, less convergence.

Rods and Cones

F. Visual Pathway to the Brain

Color (trace) and label:
1. ○ (solid line)
   a. visual field of left eye
   b. neural pathway from retina of left eye
2. ○ (dashed line)
   a. visual field of right eye
   b. neural pathway from retina of right eye

Label:
3. optic nerve
   a. optic chiasma
   b. optic tract
4. lateral geniculate body (of thalamus)
5. midbrain
6. primary visual cortex (occipital lobe of cerebrum)

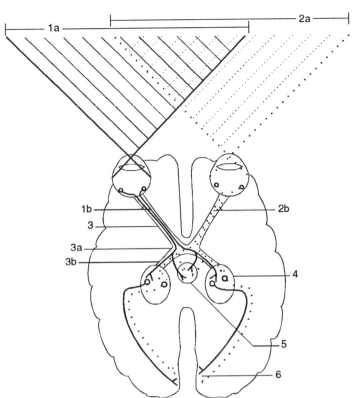

Figure 9.6. Transverse view of brain and visual pathways.

Exercise 9.6:

1. Each eye receives light from visual fields that are _____ (the same, overlapping, completely separate).

2. Light from the right side of each visual field focuses on the _____ (right, left) side of the retina, while light from the left side of each visual field focuses on the _____ (right, left) side of the retina.

3. Sensory information from the retina travels along the _____ .

4. The right and left optic nerves meet at the _____ .
   a. Fibers from the lateral retina _____ (remain on the same side, cross over).
   b. Fibers from the medial retina _____ (remain on the same side, cross over).

5. Therefore, each optic tract contains information from _____ (one eye only, both eyes).

6. Most of the fibers from the optic nerve synapse at the _____ , relaying sensory information to the _____ , where conscious perception of visual information occurs.

7. Collaterals from the optic nerve synapse at the _____ and control visual reflexes.

8. The lateral geniculate body and visual cortex integrate sensory information from the lateral retina on the _____ (same side, opposite side) and the medial retina on the _____ (same side, opposite side).

9. Interpretation of the differences between the sensory information on the two sides of the brain gives _____ (monocular, binocular) vision.

Answers to Exercise 9.6: 1. overlapping; 2. left, right; 3. optic nerve; 4. optic chiasma; 4a. remain on the same side; 4b. cross over; 5. both eyes; 6. lateral geniculate body, visual cortex; 7. midbrain; 8. same side, opposite side; 9. binocular.

G. General Anatomy of the Ear

Color (trace) and label:
1. ○ tympanic membrane
2. ossicles
   a. ○ malleus (hammer)
   b. ○ incus (anvil)
   c. ○ stapes (stirrup)
3. ○ semicircular canals
4. ○ vestibule
   a. utricle
   b. saccule
5. ○ cochlea
6. ○ vestibulocochlear nerve
   a. vestibular nerve
   b. cochlear nerve

Color:
7. ○ temporal bone

Label:
8. outer ear
   a. pinna
   b. external auditory meatus
9. middle ear
10. inner ear
11. eustachian tube

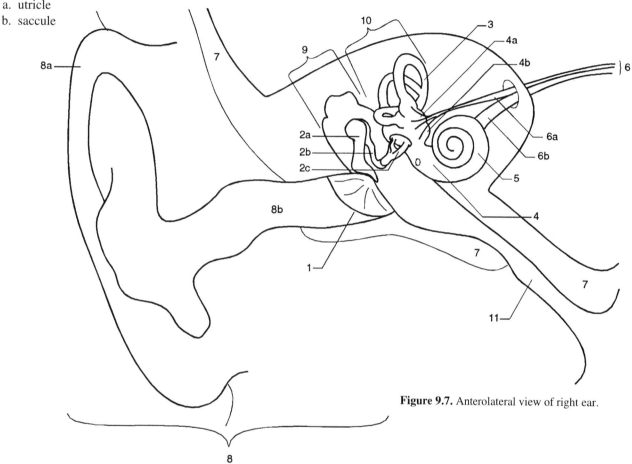

Figure 9.7. Anterolateral view of right ear.

Exercise 9.7:

_____ 1. The outer ear includes the _____ , _____ , and _____ .

_____ 2. The middle ear includes the _____ .

_____ 3. The inner ear includes the _____ , _____ , and _____ .

_____ 4. What material surrounds the ossicles of the middle ear?

_____ 5. When atmospheric pressure against the tympanic membrane changes, the air pressure in the middle ear equalizes. How does air enter or leave the middle ear?

_____ 6. Which inner ear structure is most anterior?

_____ 7. The vestibulocochlear nerve is a _____ (sensory, motor, mixed) nerve.

Answers to Exercise 9.7: 1. pinna, external auditory meatus, tympanic membrane; 2. ossicle (malleous, incus, stapes); 3. semicircular canals, vestibule, cochlea; 4. air; 5. through the eustachian tube; 6. cochlea; 7. sensory.

H. Hearing

Color (trace) and label:

1. ○ tympanic membrane
2. ○ ossicles
3. perilymph
   a. ○ scala vestibuli
   b. ○ scala tympani
4. ○ endolymph (cochlear duct, also called scala media)
5. ○ vestibular membrane
6. ○ basilar membrane
7. ○ tectorial membrane
8. ○ hair cells
9. ○ cochlear nerve

Label:

10. cochlea
    a. oval window
    b. round window
11. organ of Corti

Exercise 9.8:

_____ 1. Sound waves (air compression) represented by arrow _____ enter the ear and push against the _____ .

_____ 2. When the tympanic membrane moves, the _____ of the middle ear move. This is represented by arrow _____ .

_____ 3. The stapes vibrates against the _____ , which moves the fluid (perilymph) in the scala _____ . Perilymph movement is represented by arrow _____ .

_____ 4. The pressure of the perilymph moves in two directions. It is transmitted forward through the scala _____ , returns through the scala _____ , and is dissipated through the _____ .

_____ 5. In addition, sideways pressure, when great enough, is transmitted to the endolymph of the cochlear duct and moves the _____ membrane. Endolymph movement is represented by arrow _____ .

_____ 6. When the basilar membrane moves, it pulls on the auditory receptors, the _____ , which are anchored to the _____ .

_____ 7. Mechanical deformation of the hair cells causes a(n) _____ to form.

_____ 8. Generator potentials in the hair cells can initiate a(n) _____ potential on the _____ .

Answers to Exercise 9.8: 1. a, tympanic membrane; 2. ossicles, b; 3. oval window, vestibuli, c; 4. vestibuli, tympani, round window; 5. basilar, d; 6. hair cells, tectorial membrane; 7. generator potential; 8. action, cochlear nerve.

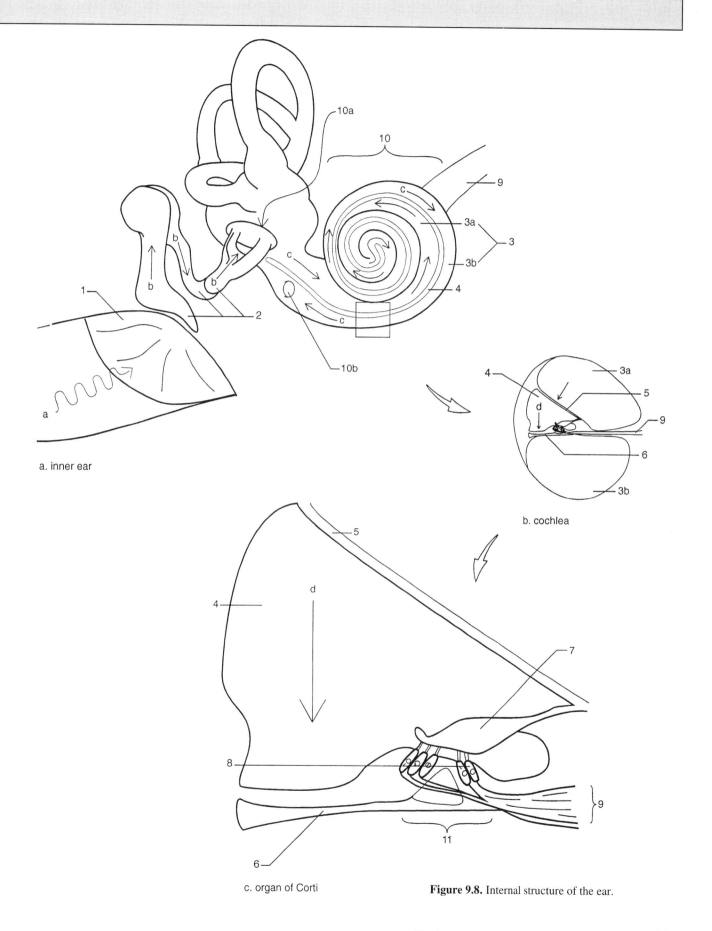

a. inner ear

b. cochlea

c. organ of Corti

Figure 9.8. Internal structure of the ear.

Hearing

I. Sound Perception

Color (trace) and label:
1. ○ stapes

Color the sound waves:
2. ○ high frequency
3. ○ low frequency
4. ○ high amplitude
5. ○ low amplitude

Label:
6. oval window
7. scala vestibuli
8. cochlear duct
9. basilar membrane
10. scala tympani
11. round window

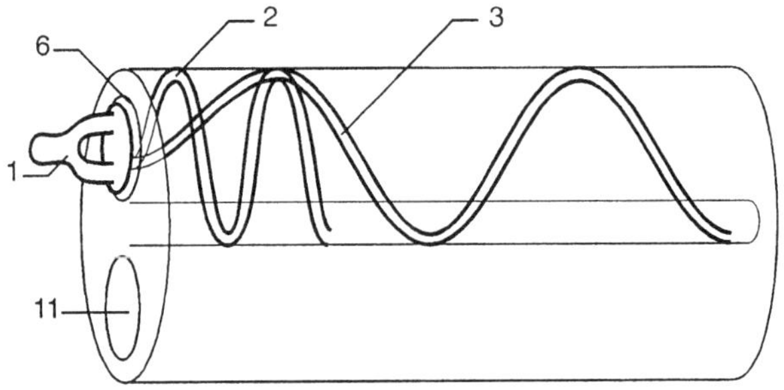

Figure 9.9a. Pitch.

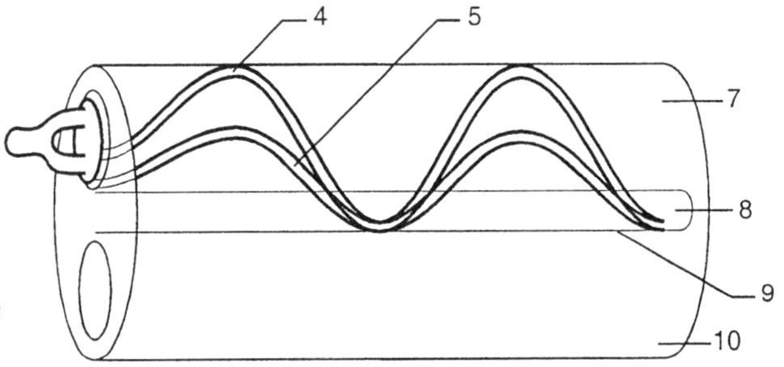

Figure 9.9b. Intensity.

Exercise 9.9:

1. Two characteristics of sound waves are _____ and _____ .

2. Pitch depends upon the _____ of the sound wave
   a. High pitch equals _____ frequency.
   b. Low pitch equals _____ frequency.

3. High frequency sound stimulates hair cells _____ (closer to, farther from) the oval window.

4. Low frequency sound stimulates hair cells _____ (closer to, farther from) the oval window.

5. Thus, the nervous system codes for pitch by _____ (stimulus location, amount of stimulation).

6. The intensity (loudness) depends upon the _____ of the sound waves.
   a. Loud equals _____ amplitude.
   b. Soft equals _____ amplitude.

7. High amplitude sounds cause _____ (great, less) basilar membrane deflection, which _____ (increases, decreases) the pull on the hair cells.

8. Low amplitude sounds cause _____ (great, less) basilar membrane deflection, which _____ (increases, decreases) the pull on the hair cells.

9. As basilar deflection increases, the amplitude of the generator potential _____ (increases, remains the same, decreases) and the number of action potentials _____ (increases, remains the same, decreases).

10. For very loud sounds, muscles contract to pull the ossicles away from the tympanic membrane and oval window.
    a. How does this affect the transmission of sound vibrations to the inner ear?
    b. What benefit does this acoustic reflex provide?

11. Timbre (tone quality) depends upon a pattern of _____ and _____ .

Answers to Exercise 9.9: 1. pitch, intensity; 2. frequency; 2a. high; 2b. low; 3. closer to; 4. farther from; 5. stimulus location; 6. amplitude; 6a. high; 6b. low; 7. great, increases; 8. less, decreases; 9. increases, increases; 10a. decreases it; 10b. helps prevent damage to hair cells; 11. pitch, intensity.

J. Dynamic Equilibrium

Color (trace) and label:
1. ○ semicircular canals
   a. superior
   b. posterior
   c. lateral
   d. ampulla
2. ○ hair cells
3. ○ cupula
4. ○ endolymph
5. ○ vestibular nerve

Label:
6. crista (ampullaris)

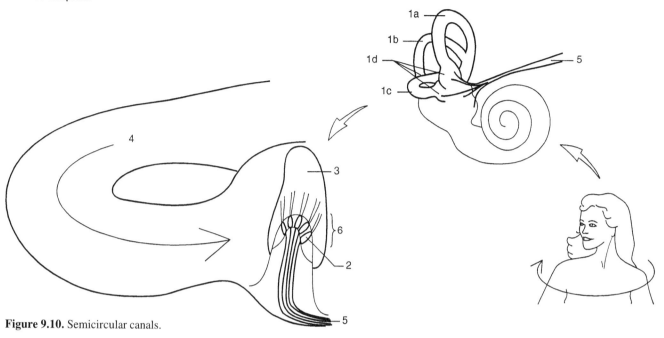

Figure 9.10. Semicircular canals.

Exercise 9.10:

1. The three semicircular canals point roughly in the following directions: ____, ____, and ____.

2. They contain the fluid ____.

3. Each canal has an enlarged area called the ____ containing receptor cells called ____.

4. Hair cells are embedded in the gelatinous ____.

5. Movement of the endolymph can deflect the cupula, which pulls on the ____ and alters their membrane potentials.

6. Generator potentials on the hair cells can trigger an action potential on the ____ nerve.

7. The person in figure 9.10 is turning to her ____ (right, left).

8. As she turns, the endolymph in the ____ (superior, lateral, posterior) canal lags behind and effectively spins in the ____ (same, opposite) direction.

9. The moving endolymph puts pressure on the ____, which causes the ____ cells to bend.

10. When the head is stationary (in any position), is the endolymph moving? Therefore, do the semicircular canals detect head position (static equilibrium)?

Answers to Exercise 9.10: 1. superior, posterior, lateral; 2. endolymph; 3. ampulla, hair cells; 4. cupula; 5. hair cells; 6. vestibular; 7. right; 8. lateral, opposite; 9. cupula, hair; 10. no, no.

K. Static Equilibrium

Color (trace) and label:
1. ○ macula
   a. hair cells
2. ○ otolith membrane
3. ○ otolith crystals
4. ○ endolymph
5. ○ vestibular nerve

Label:
6. utricle
7. saccule

Figure 9.11a. Head upright.

148 Chapter 9 Special Senses

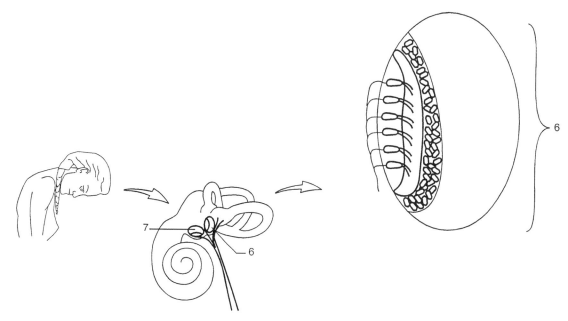

Figure 9.11b. Head bent (utricle enlarged).

Exercise 9.11:

_____ 1. The utricle and saccule each contain the fluid _____ and a sensory area called the _____ .

_____ 2. When the head is upright, the utricular macula is oriented _____ (vertically, horizontally) and the saccular macula is oriented _____ (vertically, horizontally).

_____ 3. When the head is bent, the utricular macula now orients _____ (vertically, horizontally) and the saccular macula orients _____ (vertically, horizontally).

_____ 4. The receptor cells that comprise the maculae are _____ cells.

_____ 5. The hair cells are embedded in the gelatinous _____ membrane, which contains _____ made of calcium carbonate.

_____ 6. Since otolith crystals are denser than endolymph, gravity has a greater pull on _____ .

7. Note the changes in the utricle when the head is bent. (see figure 9.11b)

_____ a. The otolith crystals pull on the _____ .

_____ b. The sagging otolith membrane can cause mechanical deformation of the _____ cells, thus altering their membrane potentials.

_____ c. A generator potential on the hair cells can trigger a(n) _____ potential on the _____ nerve.

_____ 8. To summarize, dynamic equilibrium depends upon movement of _____ , which deforms the hair cells of the semicircular canals. Static equilibrium depends upon the pull of gravity on _____ , which deform the hair cells of the macula.

Answers to Exercise 9.11: 1. endolymph, macula; 2. horizontally, vertically; 3. vertically, horizontally; 4. hair; 5. otolith, otolith crystals; 6. otolith crystals; 7a. otolith membrane; 7b. hair; 7c. action, vestibular; 8. endolymph, otolith crystals.

Chapter 10 | Endocrine System

A. Characteristics of Hormones

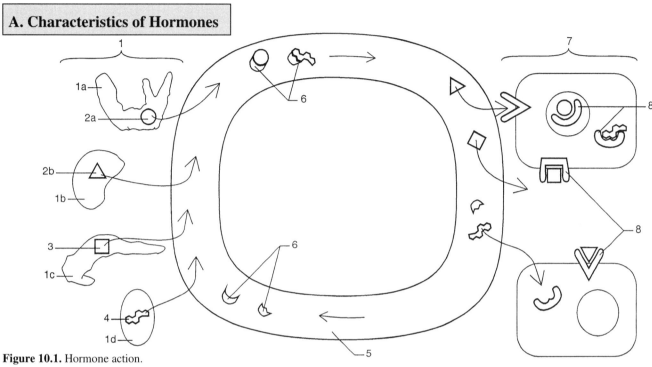

Figure 10.1. Hormone action.

Color and label:

1. ○ endocrine glands
   a. thyroid
   b. adrenal gland
   c. pancreas
   d. ovary

2. amine hormones (amino acid derivatives)
   a. ○ thyroid hormones
   b. catecholamines (epinephrine, norepinephrine)

3. ○ protein and polypeptide hormones

4. ○ steroid hormones
5. ○ blood (use red)
6. ○ carrier protein
7. ○ target cells
8. ○ receptors

Exercise 10.1

_____ 1. Molecules made by specialized cells and usually transported by the blood that act upon specific target cells are called _____ .

_____ 2. The ductless glands that produce hormones are called _____ glands.

_____ 3. The hormones that generally attach to carrier proteins for transport in the blood are _____ and _____ .

_____ 4. The cell's ability to respond to a hormone depends upon the presence of specific _____ .

_____ 5. Hormones that are nonpolar, lipid soluble _____ (can, cannot) cross the membrane. (see membrane structure, p. 27)

_____ a. Their receptors are found _____ (on the membrane, inside the cell).

_____ b. Examples of these hormones are _____ and _____ .

_____ 6. Hormones that are large and/or lipid insoluble _____ (can, cannot) cross the membrane.

_____ a. Their receptors are found _____ (on the membrane, inside the cell).

_____ b. Examples of these hormones are _____ .

Answers to Exercise 10.1: 1. hormones; 2. endocrine; 3. steroids, thyroid hormones; 4. receptors; 5. can; 5a. inside the cell; 5b. steroids, thyroid hormones; 6. cannot; 6a. on the membrane; 6b. proteins, polypeptides, catecholamines.

B. Intracellular Mechanism of Action

Color (trace) and label:
1. ○ hormone
   a. thyroid hormones
   b. steroid hormones
2. ○ receptor
3. ○ acceptor protein
4. ○ DNA
5. ○ mRNA
6. ○ ribosome
7. ○ protein

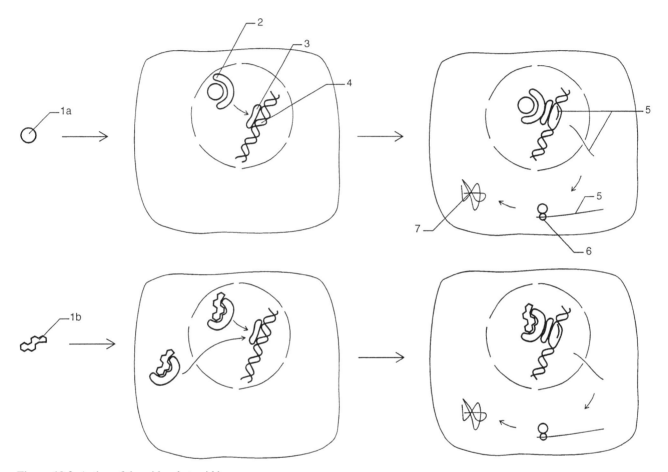

Figure 10.2. Action of thyroid and steroid hormones.

Exercise 10.2:

_____ 1. Thyroid hormones bind with receptors found in the _____ .

_____ 2. Steroid hormones bind with receptors in the _____ or _____ .

_____ 3. The hormone-receptor complex reacts with the _____ , which is attached to DNA.

_____ 4. Transcription of the activated gene produces _____ . (see protein synthesis, p. 39)

_____ 5. Translation of mRNA results in _____ .

_____ 6. Can changes in protein synthesis alter cell functioning?

Answers to Exercise 10.2: 1. nucleus; 2. cytoplasm, nucleus; 3. acceptor protein; 4. mRNA; 5. protein synthesis; 6. yes (the rates of biochemical reactions may change).

C. Plasma Membrane Mechanism of Action

Color and label:
1. ○ hormone (first messenger)
2. ○ receptor
3. ○ membrane
4. ○ mitochondrian

Color:
○ adenylate cyclase (or guanylate cyclase)
○ second messengers: cAMP (or cGMP)

IP$_3$ (inositol triphosphate)
DAG (diacylglycerol)
○ Ca^{++}
○ calmodulin
○ activated protein kinases

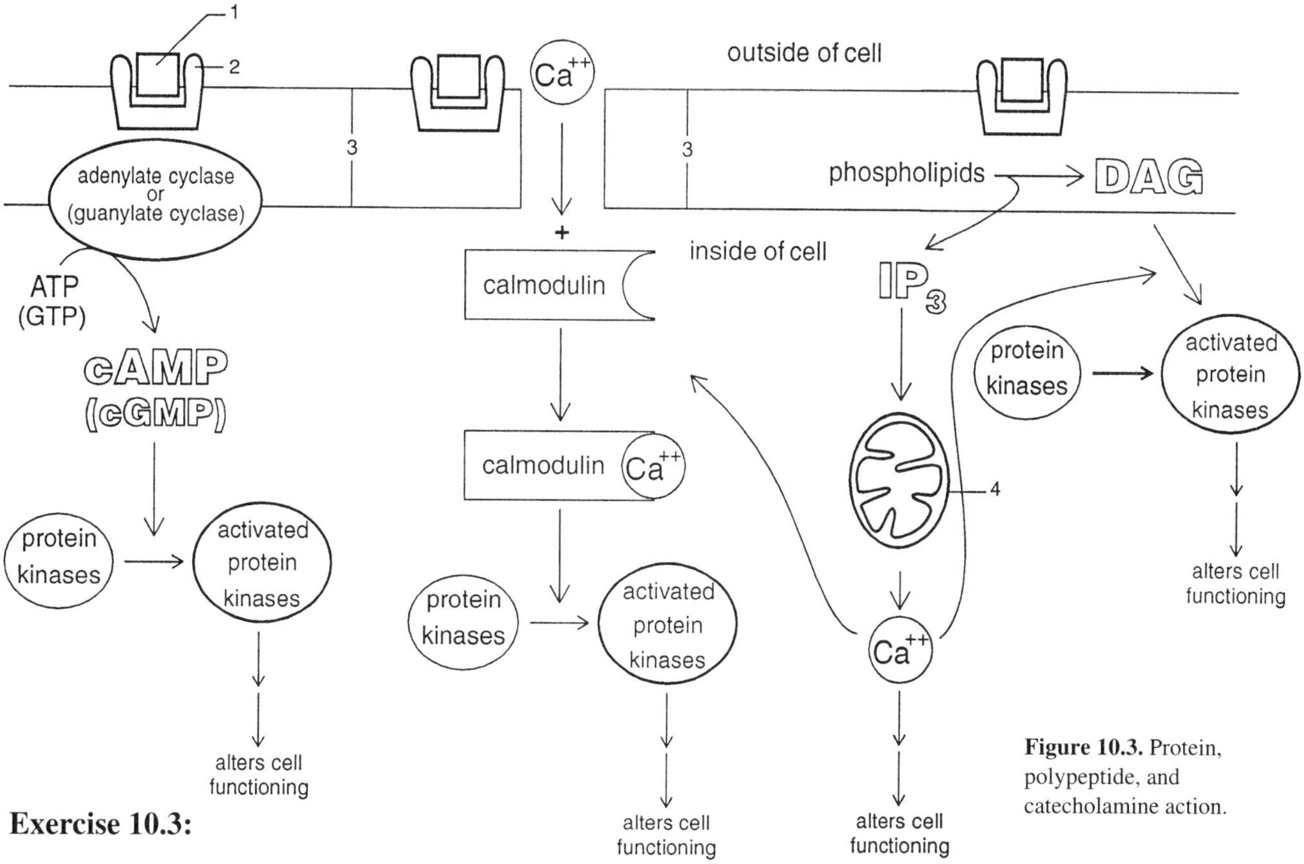

Figure 10.3. Protein, polypeptide, and catecholamine action.

Exercise 10.3:

1. Protein, polypeptide, and catecholamine hormones bind with receptors located _____ .
2. Depending upon the hormone-receptor interaction, three changes in the membrane are possible:
   a. the activation of adenylate cyclase (or guanylate cyclase), which catalyzes the breakdown of ATP (or GTP) to _____ ,
   b. the increased permeability of the membrane to extracellular _____ , or
   c. the breakdown of phospholipids to _____ and _____ .
3. Cyclic AMP activates enzymes called _____ , which alter cell functioning.
4. Ca^{++} binds with _____ , which then activates _____ , which alter cell functioning.
5. IP$_3$ triggers the release of organelle-stored _____ , which alters cell functioning, while DAG activates _____ , which alter cell functioning.
6. To summarize,
   a. hormones that trigger protein synthesis have a(n) _____ (plasma membrane, intracellular) mechanism of action.
   b. hormones that cause activation of enzymes already in the cell have a(n) _____ (plasma membrane, intracellular) mechanism of action.
7. Which mechanism of action would you expect to be faster?

Answers to Exercise 10.3: 1. on the membrane; 2a. cAMP (or cGMP); 2b. Ca^{++}; 2c. IP$_3$, DAG; 3. protein kinases; 4. calmodulin, protein kinases; 5. Ca^{++}, protein kinases; 6a. intracellular; 6b. plasma membrane; 7. plasma membrane.

D. Regulation of Hormones

Color and label:
1a. ○ hormone 1
1b. ○ hormone 2
2. ○ prohormone
3. ○ endocrine gland
4. ○ receptors
5. ○ target cell

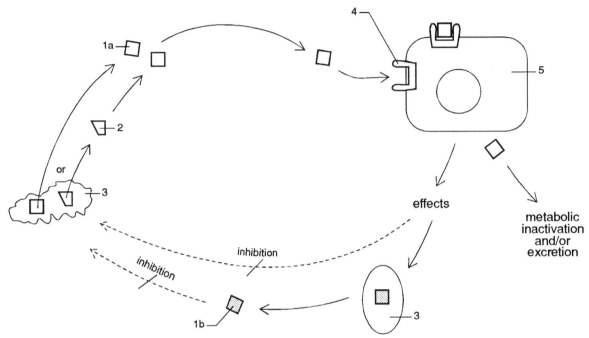

Figure 10.4a. Negative feedback.

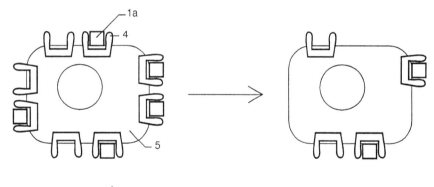

Figure 10.4b. Down regulation.

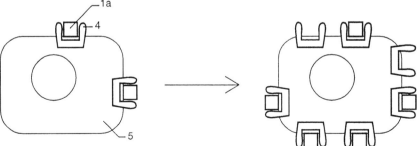

Figure 10.4c. Up regulation.

Exercise 10.4:

_____ 1. Hormones *may* be secreted from the endocrine gland as _____ , an inactive precursor.

_____ 2. Hormones travel to the target cells via the _____ . (see characteristics of hormones, p. 151)

_____ 3. As the hormone alters target cell functioning, a signal goes back to the endocrine gland that is *usually* _____ (inhibitory, stimulatory).

_____ 4. Inhibition of the endocrine gland means that the amount of hormone released _____ (increases, remains the same, decreases).

_____ 5. With less hormone, target cell effects _____ (increase, remain the same, decrease).

_____ 6. Inhibition (or stimulation) of the endocrine gland may occur indirectly, for example, via another _____ . (see figure 10.4a)

_____ 7. Hormone already present (remains in the system, is inactivated and/or excreted).

_____ 8. In summary, this regulation of target cell effects to a constant level is called _____ .

9. When a target cell is exposed to a hormone, the number of receptors *may* change. (see figure 10.4b and c)

_____ a. If the number of receptors decreases, target cell effects _____ (increase, remain the same, decrease).

_____ b. If the number of receptors increases, target cell effects _____ (increase, remain the same, decrease).

Answers to Exercise 10.4: 1. prohormone; 2. blood; 3. inhibitory; 4. decreases; 5. decrease; 6. hormone; 7. is inactivated and/or excreted; 8. homeostasis; 9a. decrease (down regulation); 9b. increase (up regulation).

E. Pituitary Gland

Color (trace) and label:
1. ◯ anterior pituitary (adenohypophysis)
2. ◯ posterior pituitary (neurohypophysis)
   a. infundibulum
3. ◯ hypothalamus
4. ◯ blood vessels (red)
5. ◯ nerves

Color:
◯ TSH (thyroid-stimulating hormone)
◯ GH (growth-hormone)
◯ ACTH (adrenocorticotropic hormone)
◯ PRL (prolactin)
◯ FSH (follicle-stimulating hormone)
◯ LH (luteinizing hormone)
◯ MSH (melanocyte-stimulating hormone)
◯ ADH (antidiuretic hormone)
◯ OCT (oxytocin)

Label:
6. thyroid gland
7. adrenal gland
8. gonads (ovaries)
9. skin (melanocytes)
10. kidneys
11. uterus

Exercise 10.5:

1. The pituitary gland extends from the _____ (part of brain) by a connecting stalk called the _____ .

2. The anterior pituitary makes and releases _____ , _____ , _____ , _____ , _____ , _____ , and _____ .

3. The hypothalamus makes _____ and _____ .

4. ADH and oxytocin travel via nerves to the _____ for storage.

5. Name each hormone with the metabolic effect listed.

   a. increases amino acid uptake and growth (increases bone length), increases blood sugar and fat breakdown

   b. increases adrenal gland activity

   c. increases thyroid activity

   d. increases water retention

6. Name each hormone with the reproductive function listed.

   a. stimulates follicle maturation and estrogen production in women, initiates sperm production in men

   b. causes ovulation and progesterone production in women, stimulates testosterone production in men

   c. stimulates milk production

   d. causes milk letdown

   e. causes uterine contractions

7. What hormone increases melanin production?

Answers to Exercise 10.5: 1. hypothalamus, infundibulum; 2. TSH, GH, ACTH, prolactin, FSH, LH, MSH; 3. ADH, oxytocin; 4. posterior pituitary; 5a. GH; 5b. ACTH; 5c. TSH; 5d. ADH; 6a. FSH; 6b. LH; 6c. prolactin; 6d. oxytocin; 6e. oxytocin, 7. MSH.

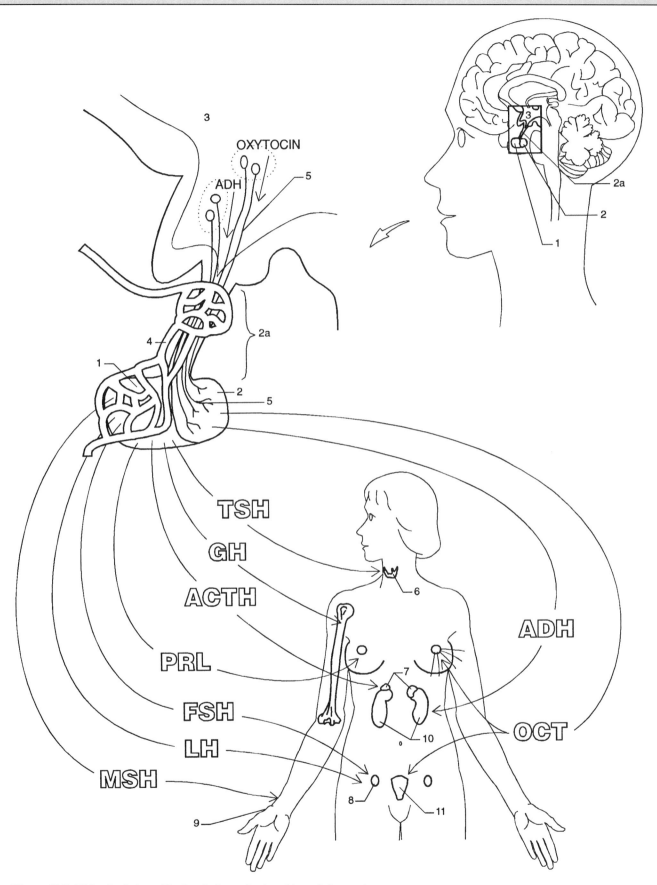

Figure 10.5. Midsagittal view of brain, pituitary gland, and hypothalamus; hormone targets.

Pituitary Gland

F. Pituitary Gland Regulation

Color and label:
1. ○ adenohypophysis (anterior pituitary)
2. ○ neurohypophysis (posterior pituitary)
    a. infundibulum
3. ○ hypothalamus
4. ○ portal system (red)
    a. first capillaries
    b. portal veins (hypothalamo-hypophyseal)
    c. second capillaries
5. ○ nerves (hypothalamo-hypophyseal tract)
6. ○ target cells

○ **Color all releasing hormones the same:**
  GHRH (growth hormone-releasing hormone)
  TRH (thyrotropin-releasing hormone)
  CRH (corticotropin-releasing hormone)
  GnRH (gonadotropin-releasing hormone)
  PRH (prolactin-releasing hormone)

○ **Color all inhibiting hormones the same:**
  GHIH (somatostatin, growth hormone-inhibiting hormone)
  PIH (prolactin-inhibiting hormone)

Exercise 10.6:

1. The release (or inhibition) of anterior pituitary hormones is controlled by hormones made in the _____ .
2. Capillaries at the base of the hypothalamus send blood to capillaries in the _____ via _____ veins.
3. Thus, the releasing (or inhibiting) hormones move from the hypothalamus to the anterior pituitary via _____ .
4. Which hypothalamic hormones stimulate the release of the following anterior pituitary hormones?
    a. GH
    b. TSH
    c. ACTH
    d. FSH and LH
    e. PRL
5. Which hypothalamic hormones inhibit the release of the following anterior pituitary hormones?
    a. GH
    b. PRL
6. For which anterior pituitary hormones have both releasing and inhibiting hormones been identified?
7. Since prolactin stimulates milk production, which would you expect to be the primary regulating hormone in non-nursing women—PRH or PIH?
8. Which hypothalamic hormone stimulates prolactin production?
    a. Suckling results in increased prolactin production by stimulating the release of more _____ .
    b. Is this an example of negative or positive feedback?
9. Can target cells inhibit anterior pituitary hormones
    a. by negative feedback on the anterior pituitary? (see negative feedback, p. 154)
    b. indirectly by altering the levels of hypothalamic releasing or inhibiting hormones?
10. The posterior pituitary stores and releases _____ and _____ .
11. ADH and oxytocin are made in the _____ and travel to the posterior pituitary via _____ .
12. The release of ADH and oxytocin from the posterior pituitary depends upon stimulation by nerves whose cell bodies are located in the _____ .
13. Osmoreceptors (receptors that detect solute concentrations in the anterior hypothalamus) stimulate hypothalamic nerves to trigger the release of _____ .
14. Uterine pressure and/or suckling stimulate hypothalamic nerves to trigger the release of _____ .

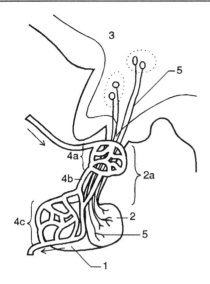

Figure 10.6a. Midsagittal view of pituitary gland and hypothalamus.

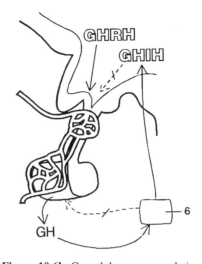

Figure 10.6b. Growth hormone regulation.

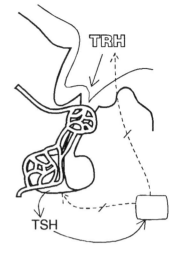

Figure 10.6c. TSH regulation.

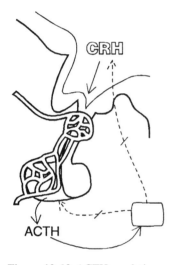

Figure 10.6d. ACTH regulation.

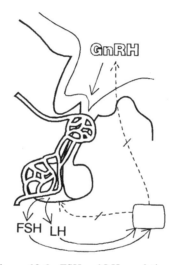

Figure 10.6e. FSH and LH regulation.

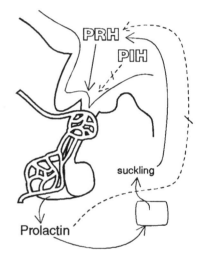

Figure 10.6f. PRL (prolactin) regulation.

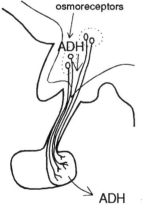

Figure 10.6g. ADH regulation.

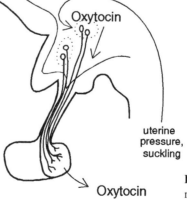

Figure 10.6h. OCT (oxytocin) regulation.

Answers to Exercise 10.6: 1. hypothalamus; 2. anterior pituitary, portal; 3. blood vessels; 4a. GHRH; 4b. TRH; 4c. CRH; 4d. GnRH; 4e. PRH; 5a. GHIH; 5b. PIH; 6. GH, prolactin; 7. PIH; 8. PRH; 8a. PRH; 8b. positive; 9a. yes; 9b. yes; 10. ADH, oxytocin; 11. hypothalamus, nerves; 12. hypothalamus; 13. ADH; 14. oxytocin.

Pituitary Gland Regulation

G. Thyroid Gland

Color and label:
1. ○ thyroid
2. ○ thyroid cartilage of larynx
3. ○ trachea
4. follicle
    a. ○ follicular cells
    b. ○ colloid
    c. ○ C cells (parafollicular cells)
5. ○ thyroid hormone
6. ○ calcitonin
7. ○ bone
8. ○ kidney

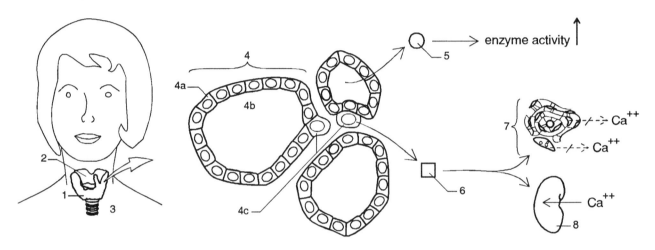

Figure 10.7. Anterior view of thyroid.

Exercise 10.7:

_____ 1. The thyroid gland is located anterior to the _____.

_____ 2. The thyroid is made of follicles that have a noncellular center called _____ surrounded by a single layer of _____ cells.

_____ 3. The follicles synthesize _____.

_____ 4. Between the follicles are _____ cells, which make _____.

_____ 5. Thyroid hormones _____ (increase, decrease, have no effect on) metabolic rate.

_____ 6. Calcitonin _____ (increases, decreases) bone resorption and _____ (increases, decreases) excretion of calcium.

_____ 7. Therefore, calcitonin can _____ (increase, decrease) calcium levels in the blood.

Answers to Exercise 10.7: 1. trachea; 2. colloid, follicular; 3. thyroid hormones; 4. C (parafollicular) cells, calcitonin; 5. increase; 6. decreases, increases; 7. decrease.

H. Parathyroid

Color and label:
1. ○ parathyroid glands
2. ○ thyroid gland
3. ○ pharynx
4. ○ esophagus
5. ○ trachea
6. ○ parathyroid hormone (PTH, parathormone)
7. ○ intestine
8. ○ bone
9. ○ kidney
10. ○ blood (red)

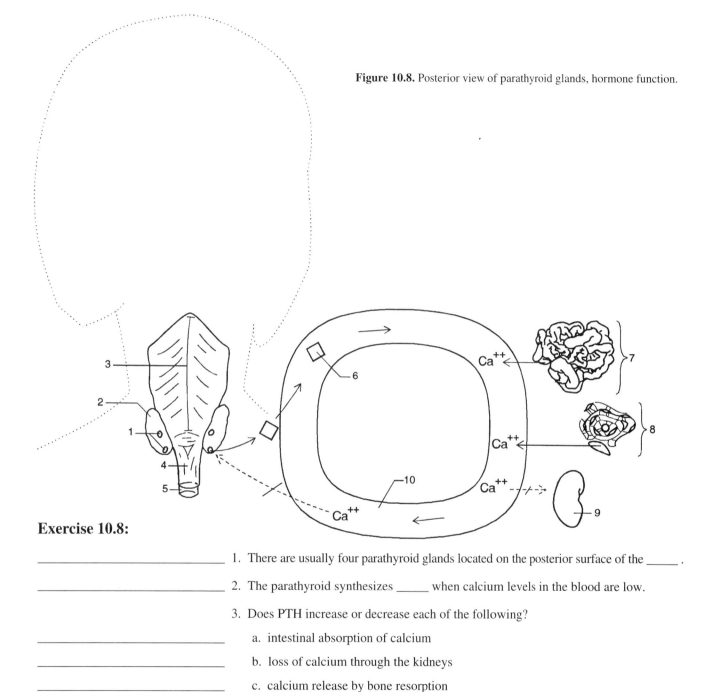

Figure 10.8. Posterior view of parathyroid glands, hormone function.

Exercise 10.8:

_____ 1. There are usually four parathyroid glands located on the posterior surface of the _____ .

_____ 2. The parathyroid synthesizes _____ when calcium levels in the blood are low.

3. Does PTH increase or decrease each of the following?

_____ a. intestinal absorption of calcium

_____ b. loss of calcium through the kidneys

_____ c. calcium release by bone resorption

_____ 4. Therefore, PTH _____ (increases, decreases) calcium levels in the blood.

_____ 5. When calcium levels in the blood are high, PTH levels _____ (increase, remain the same, decrease).

_____ 6. Do PTH and calcitonin have the same or antagonistic effects?

Answers to Exercise 10.8: 1. thyroid; 2. PTH (parathyroid hormone); 3a. increase; 3b. decrease; 3c. increase; 4. increases; 5. decrease; 6. antagonistic.

I. Adrenal Glands (Suprarenal Glands)

Color and label figure 10.9a:
1. ○ adrenal gland
2. ○ kidney

Color (trace) and label figure 10.9b:
3. adrenal cortex
   a. ○ zona glomerulosa
   b. ○ zona fasciculata
   c. ○ zona reticularis

4. ○ adrenal medulla
5. ○ catecholamines (epinephrine, norepinephrine)
6. ○ aldosterone (mineralocorticoids)
7. ○ cortisol (glucocorticoids)
8. ○ sex steroids
9. ○ spinal cord
10. ○ preganglionic neurons
11. ○ postganglionic neuron

Exercise 10.9:

_____ 1. The adrenal glands are on the superior surface of the _____ .

_____ 2. Catecholamines are made in the _____ , while steroid hormones are made in the _____ .

3. Stress causes

_____ a. the sympathetic nervous system to stimulate the adrenal medulla to release _____ .

_____ b. the adrenal cortex to release the anti-inflammatory hormone _____ .

_____ 4. Catecholamines trigger glucose production from _____ , while cortisol stimulates glucose production from _____ .

_____ 5. Which hormone controls glucose production during long-term stress?

_____ 6. Therefore, what would you expect would happen to nutrient reserves during long-term stress?

_____ 7. The production of cortisol is controlled by the anterior pituitary hormone _____ , which, in turn, is controlled by the hypothalamic hormone _____ .

_____ 8. Since ACTH acts on the entire adrenal cortex, what other hormones would increase in addition to cortisol?

_____ 9. What effect does aldosterone have on Na$^+$ retention?

_____ 10. As a secondary effect, what happens to water when Na$^+$ is retained? (see movement across a membrane, p. 28)

_____ 11. What function would water retention serve during stress?

Answers to Exercise 10.9: 1. kidneys; 2. adrenal medulla, adrenal cortex; 3a. catecholamines; 3b. cortisol; 4. glycogen, protein; 5. cortisol; 6. protein depletion; 7. ACTH, CRH; 8. aldosterone, sex steroids; 9. increases it; 10. water is also retained; 11. increase blood pressure.

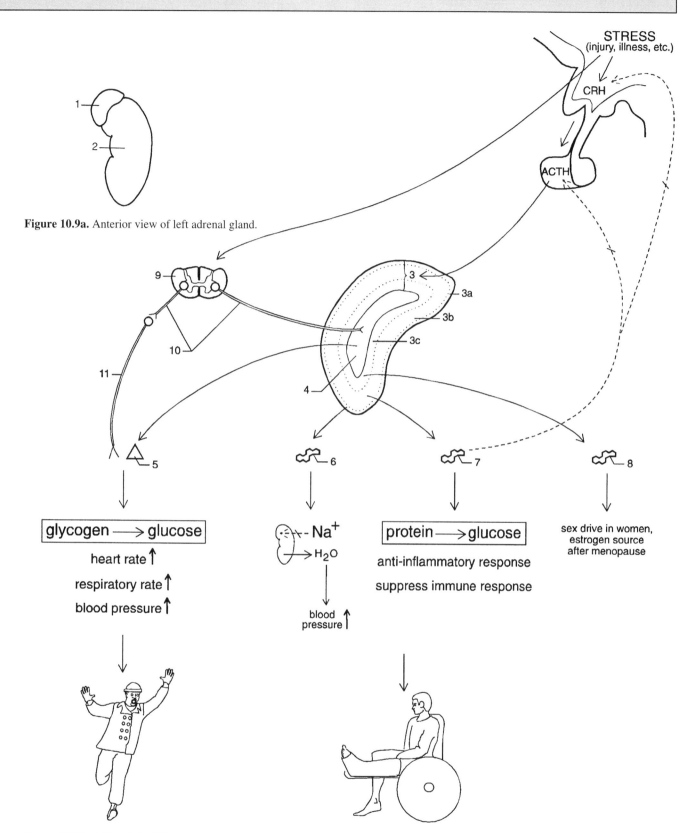

Figure 10.9a. Anterior view of left adrenal gland.

Figure 10.9b. Adrenal gland hormones.

Adrenal Glands (Suprarenal Glands) 163

J. Pancreas

Color and label figure 10.10a:
1. ○ pancreas
2. ○ stomach
3. ○ small intestine (duodenum)

Color and label figure 10.10b:
4. islets of Langerhans
   a. ○ A (alpha) cells
   b. ○ B (beta) cells
5. ○ exocrine cells

Color and label:
6. ○ glucagon
7. ○ insulin
8. ○ glucose
9. ○ blood (red)
10. ○ intestine
11. ○ liver

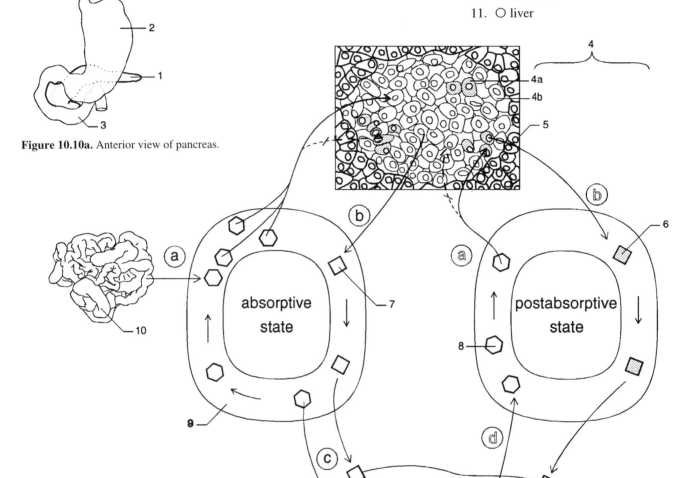

Figure 10.10a. Anterior view of pancreas.

Figure 10.10b. Histology of pancreas, glucose regulation.

164 Chapter 10 Endocrine System

Exercise 10.10:

_____ 1. The pancreas is on the posterior surface of the _____ .

_____ 2. Its exocrine cells make _____ while clusters of endocrine cells called _____ make hormones.

_____ 3. The islets of Langerhans contain alpha cells that make _____ and beta cells that make _____ .

4. Immediately after eating (absorptive state),

_____   a. blood glucose _____ (increases, decreases).

_____   b. This causes the amount of insulin to _____ (increase, decrease) and the amount of glucagon to _____ (increase, decrease).

_____   c. Therefore, liver cells (and most other tissues) _____ (take up, release) glucose, and blood glucose _____ (falls, rises).

_____   d. Insulin promotes the conversion of glucose to _____ .

5. Between meals (postabsorptive state),

_____   a. blood glucose _____ (increases, decreases).

_____   b. This causes the amount of insulin to _____ (increase, decrease) and the amount of glucagon to _____ (increase, decrease).

_____   c. Glucagon stimulates the _____ (formation, breakdown) of glycogen.

_____   d. Therefore, liver cells _____ (take up, release) glucose, and blood glucose _____ (falls, rises).

_____ 6. Do insulin and glucagon play primary roles in regulating the cyclical absorptive and postabsorptive variations in blood glucose levels?

_____ 7. Are the effects of insulin and glucagon the same or antagonistic?

8. Other hormones allow the body to meet specific needs for glucose. What hormones increase blood glucose levels for each of the following conditions?

_____   a. growth

_____   b. emergency situation

_____   c. illness, injury

Answers to Exercise 10.10: 1. stomach; 2. digestive enzymes, islets of Langerhans; 3. glucagon, insulin; 4a. increases; 4b. increase, decrease; 4c. take up, falls; 4d. glycogen, fats; 5a. decreases; 5b. decrease, increase; 5c. breakdown; 5d. release, rises; 6. yes; 7. antagonistic; 8a. growth hormone; 8b. catecholamines; 8c. cortisol; thyroid hormones.

K. Organs with Endocrine Function

Color and label:
1. ⭕ pineal gland
2. ⭕ thymus
3. ⭕ heart
4. ⭕ stomach
5. ⭕ intestine
6. ⭕ kidney
7. ⭕ placenta
8. ⭕ ovaries
9. ⭕ testes

Label:
10. hypothalamus
11. T cells
12. pancreas
13. uterus

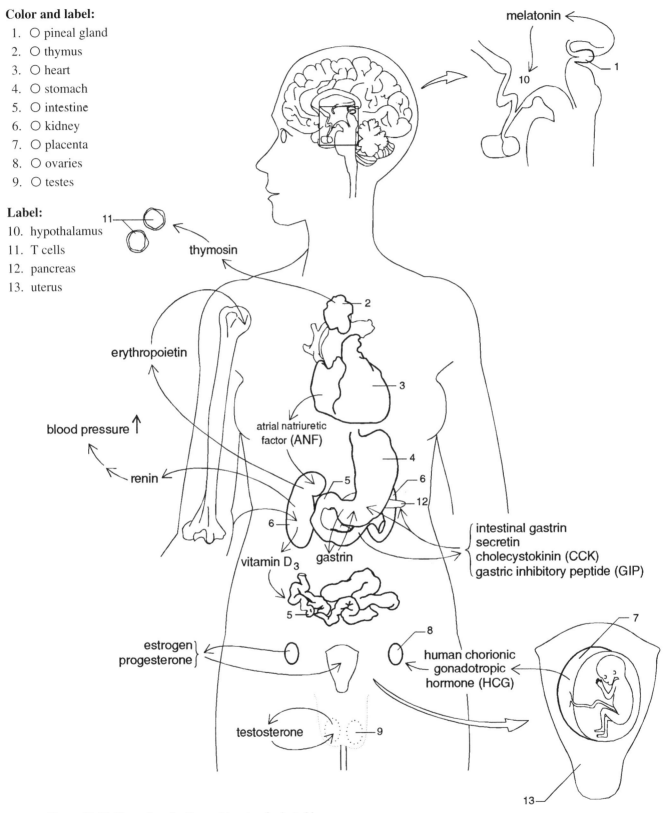

Figure 10.11. Sites of production and targets of selected hormones.

Exercise 10.11:

Use figure 10.11 to fill in the chart below.

Hormone(s) (a)	Site of Production (b)	Function
1		amounts vary during 24-hour period, inhibits GnRH release
2		stimulates ovary to make estrogen and progesterone during pregnancy
3		stimulates stomach to release HCl and pepsinogen
4		stimulates growth of uterus in preparation for implantation and pregnancy
5		stimulates small intestine to absorb calcium
6		stimulates bone marrow to make red blood cells
7		stimulates kidneys to excrete Na^+, relaxes blood vessels, results in decreased blood pressure
8		stimulates lymphocyte (T cell) development
9		controls pancreas function, inhibits stomach acidity
10		triggers series of events that lead to increased blood pressure
11		supports sperm production, male secondary sex characteristics

Answers to Exercise 10.11: 1a. melatonin; 1b. pineal gland; 2a. HCG; 2b. placenta; 3a. gastrin; 3b. stomach; 4a. estrogen and progesterone; 4b. ovaries; 5a. vitamin D_3; 5b. made by skin, activated by kidney; 6a. erythropoietin; 6b. kidney; 7a. atrial natriuretic factor; 7b. heart; 8a. thymosin; 8b. thymus; 9a. intestinal gastrin, secretin, cholecystokinin; 9b. intestine; 10a. renin; 10b. kidney; 11a. testosterone; 11b. testes.

L. Prostaglandins

Color (trace) and label figure 10.12a:
1. ○ phospholipids
2. ○ arachidonic acid
3. ○ prostaglandins
   ○ cAMP

Label figure 10.12a:
4. cell membrane
5. site of production
6. tissue fluid (or blood)
7. target cells

Color the prostaglandins in figure 10.12b:
○ I
○ E
○ F
○ G
○ H

Label:
TXA$_2$ thromboxane A$_2$

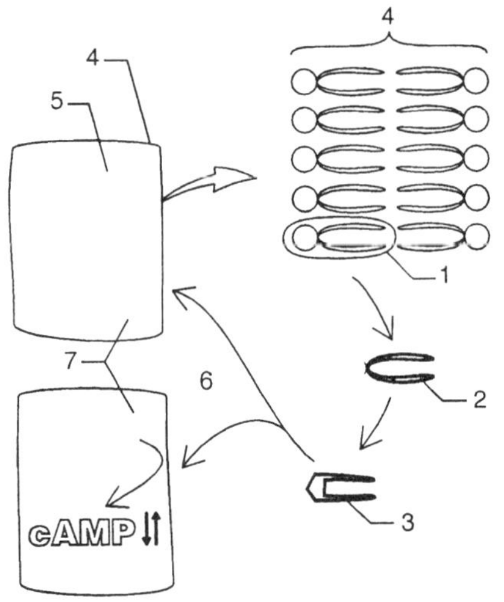

Figure 10.12a. Prostaglandin formation and action.

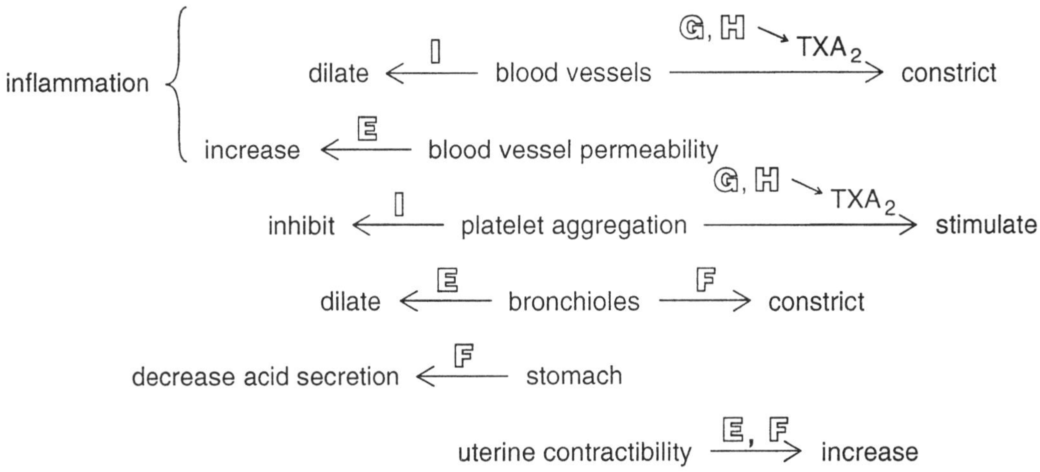

Figure 10.12b. Examples of prostaglandin functions.

Exercise 10.12:

_____ 1. Prostaglandins are made from the cell membrane components called _____ .

_____ 2. They travel through _____ to target cells.

_____ 3. Are the target cells the same or different from the site of production?

_____ 4. Prostaglandins may increase or decrease target cell levels of _____ .

_____ 5. Since many hormones use cAMP as a second messenger, would you expect prostaglandins to alter the effectiveness of these hormones?

_____ 6. Do different prostaglandins act on the same structures? (see figure 10.12b)

_____ 7. Are the effects of different prostaglandins the same or antagonistic?

Answers to Exercise 10.12: 1. phospholipids; 2. tissue fluid (or blood); 3. can be the same or different; 4. cAMP; 5. yes; 6. yes; 7. can be the same or antagonistic.

Chapter 11 Blood

A. Components of Blood

Color and label:
1. ○ plasma
2. ○ white blood cells, platelets
3. ○ red blood cells (red)
4. ○ nutrients
5. ○ gases
6. ○ ions (electrolytes)
7. ○ waste products
8. ○ hormones
9. ○ proteins

Label:
10. formed elements

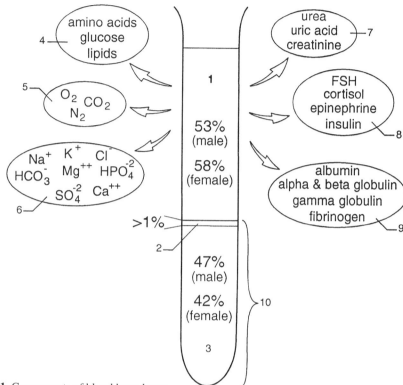

Figure 11.1. Components of blood by volume.

Exercise 11.1:

_____ 1. The formed elements of blood include _____ .

_____ 2. The average hematocrit (percentage of red blood cells) for males is _____ and for females is _____ .

_____ 3. The liquid portion of blood is called _____ .

4. Give examples of each of the components of plasma listed:

_____ a. nutrients   _____ d. waste products

_____ b. gases   _____ e. hormones

_____ c. ions   _____ f. proteins

_____ 5. Would you expect proteins found in plasma to be soluble or insoluble?

6. Match the major plasma proteins with their functions:

_____ a. antibodies made by lymphocytes

_____ b. involved in clotting

_____ c. provides osmotic pressure for blood vessels to absorb water from tissues

_____ d. transport lipids and fat-soluble vitamins.

Answers to Exercise 11.1: 1. red blood cells, white blood cells, platelets; 2. 47%, 42%; 3. plasma; 4a. glucose, amino acids, lipids; 4b. CO_2, O_2; 4c. Ca^{++}, Mg^{++}, Cl^-, SO_4^{-2}, HPO_4^{-2}, HCO_3^-; 4d. urea, creatinine, uric acid, bilirubin; 4e. any hormone (see endocrine chapter); 4f. alpha and beta globulin, gamma globulin, fibrinogen, albumin; 5. soluble (since plasma is liquid); 6a. gamma globulin; 6b. fibrinogen; 6c. albumin; 6d. alpha and beta globulins.

B. Red Blood Cells

Color and label:
1. ○ red blood cells (red)
2. ○ bone (marrow)
3. ○ erythropoietin
4. ○ kidney
5. ○ liver
6. ○ spleen
7. hemoglobin
   a. ○ globular polypeptide chain
   b. ○ heme (porphyrin ring + iron)
8. ○ bilirubin
9. ○ gallbladder

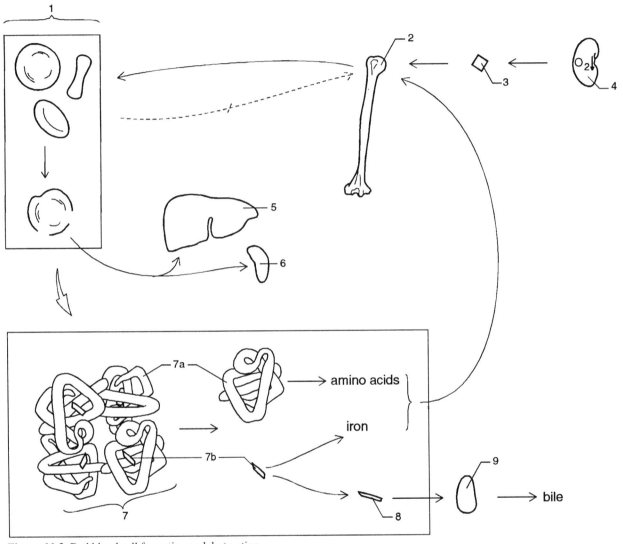

Figure 11.2. Red blood cell formation and destruction.

Exercise 11.2:

_____ 1. Red blood cells are made in the _____ when stimulated by the hormone _____ .

_____ 2. Erythropoietin forms when O_2 levels in the kidney are _____ (high, low).

_____ 3. A high red blood cell count _____ (stimulates, inhibits) further red blood cell production.

_____ 4. Old or damaged red blood cells are destroyed by the _____ and _____ .

_____ 5. Red blood cells contain the oxygen-carrying protein _____ .

_____ 6. Hemoglobin contains four _____ chains, each with a central _____ .

7. When hemoglobin breaks down, what happens to its components?

_____ a. globular polypeptide chains

_____ b. iron (from heme)

_____ c. porphyrin ring (from heme)

Answers to Exercise 11.2: 1. bone marrow, erythropoietin; 2. low; 3. inhibits; 4. liver, spleen; 5. hemoglobin; 6. globular polypeptide, heme; 7a. amino acid components recycled to make new protein; 7b. recycled to make new red blood cells; 7c. converted to bilirubin and excreted in bile.

C. White Blood Cells

Color cellular components as indicated (Wright's stain colors). Label cell type.

1. neutrophil
   - ○ nucleus (blue or blue-purple)
   - ○ granules (pink or light lilac)
2. eosinophil
   - ○ nucleus (light blue-purple)
   - ○ granules (red)
3. basophil
   - ○ nucleus (light blue-purple)
   - ○ granules (dark blue)
4. monocyte
   - ○ nucleus (blue-violet)
      a. macrophage
5. lymphocyte
   - ○ nucleus (blue-purple)
      a. T cell
      b. B cell
      c. natural killer (NK) cell

Color and label:

6. ○ colony stimulating factors
7. ○ bone (marrow)
8. ○ thymus
9. ○ secondary lymphatic tissue (lymph nodes, spleen, etc.)
10. ○ foreign antigen

Label the T cells:

11. CD8
12. CD4
13. T_C (cytotoxic or killer)
14. T_S (suppressor)
15. T_H (helper)
16. T_D (mediate delayed-type hypersensitivity)

Label the B cell:

17. plasma

Exercise 11.3:

_____ 1. Which cells are granulocytes (contain granules)?

_____ 2. Which cells are agranulocytes (lack granules)?

_____ 3. Granulocytes and monocytes are produced by the bone marrow when stimulated by _____ .

_____ 4. The lymphocytes that form in the bone marrow are _____ , _____ , and _____ .

Name the lymphocytes that:

_____ a. complete their development in the thymus.

_____ b. are genetically programmed for a specific immune reaction after exposure to a foreign antigen.

_____ c. function without previous exposure to foreign antigen.

_____ 5. T cells develop into _____ (number) functionally different cells. These cells look _____ (the same, different).

6. Complete the table below.

Cell Type	Description	Function
a.	multilobed nucleus, granules stain pale pink	phagocytic, acute response to infection
b.	bilobed nucleus, granules stain red	attack parasites, involved in allergic reactions
c.	lobed nucleus, granules stain blue	release histamines and heparin, inflammatory response, involved in allergic reaction
d.	largest white blood cells	phagocytic, become tissue macrophage and attack bacteria, make monokines (chemical messengers)
e.	large nucleus, little cytoplasm	provide _specific_ immune response, make lymphokines (chemical messengers)
f.	large lymphocyte, slightly indented nucleus	recognize and destroy tumor cells and viral infected cells

Answers to Exercise 11.3: 1. neutrophils, eosinophils, basophils; 2. monocytes, lymphocytes; 3. colony stimulating factors; 4. T cells, B cells, NK cells; 4a. NK cells; 4b. T cells, B cells; 4c. NK cells; 5. 4, the same; 6a. neutrophils; 6b. eosinophils; 6c. basophils; 6d. monocytes; 6e. lymphocytes (T and B cells); 6f. natural killer cells.

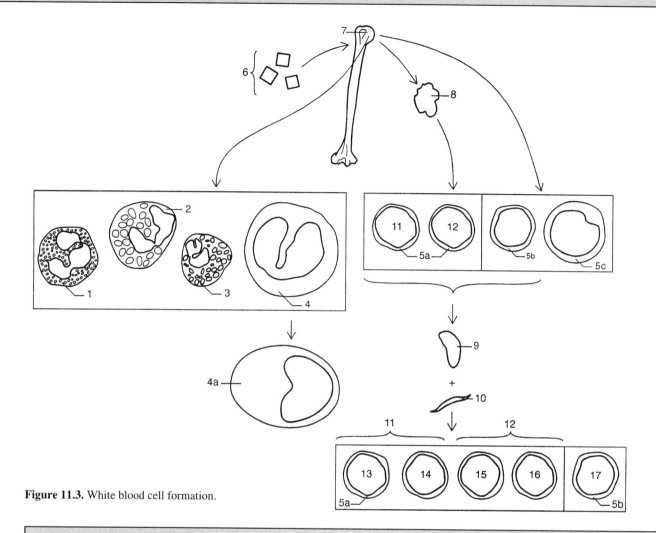

Figure 11.3. White blood cell formation.

D. Platelets

Color and label:
1. ○ megakaryocyte
2. ○ platelets (light blue)
3. ○ bone (marrow)

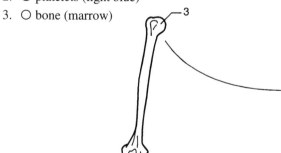

Figure 11.4. Platelet formation.

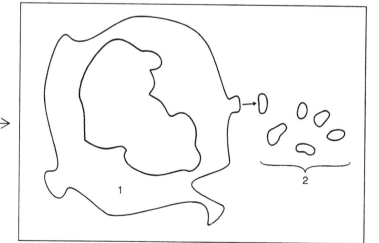

Exercise 11.4:

_____ 1. Platelets are cell fragments that form from _____ .

_____ 2. Platelet formation occurs in the _____ .

Answers to Exercise 11.4: 1. megakaryocytes; 2. bone marrow.

E. Clotting

Color and label:
1. blood vessel
   a. ○ endothelium (blood vessel lining)
   b. ○ collagen fibers
2. ○ platelets
3. ○ red blood cells (red)
4. ○ surrounding tissue
5. ○ fibrin

Outline:
○ intrinsic pathway
○ extrinsic pathway
○ common pathway

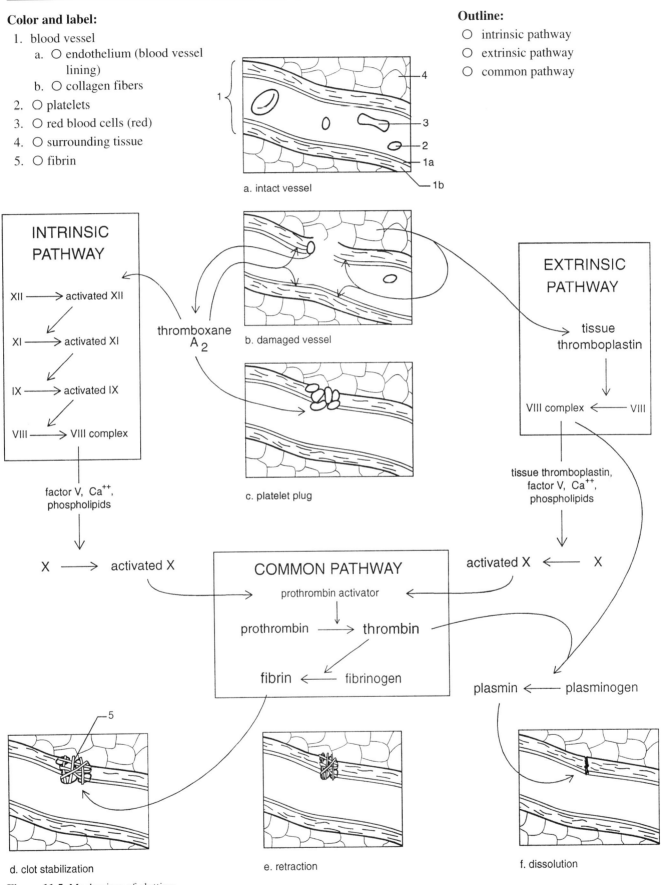

Figure 11.5. Mechanism of clotting.

Exercise 11.5:

_____ 1. Damage to a blood vessel exposes _____ within its wall.

_____ 2. This collagen attracts _____ .

3. Platelets release chemicals that

_____ a. cause blood vessel _____ (constriction, dilation) and therefore _____ (increase, decrease) blood loss,

_____ b. initiate the _____ pathway,

_____ c. increase platelet stickiness, resulting in _____ formation.

4. Damaged tissue surrounding the vessel

_____ a. triggers blood vessel _____ (constriction, dilation) and

_____ b. releases tissue thromboplastin, which begins the _____ pathway.

_____ 5. The blood clotting factors that are processed in the intrinsic and extrinsic pathways were originally present in the blood in _____ (active, inactive) form.

_____ 6. Does the cascading of activated factors in the intrinsic pathway serve to amplify the effects of the initial steps?

_____ 7. The intrinsic and extrinsic pathways lead to activation of factor _____ , which begins the _____ pathway.

_____ 8. The common pathway leads to the formation of insoluble _____ .

_____ 9. What effect does fibrin have on the platelet plug?

_____ 10. Are the precursors within the common pathway (factor X, prothrombin, and fibrinogen) in soluble or insoluble form in blood plasma?

_____ 11. Does the cascading of the common pathway serve to amplify these reactions?

_____ 12. After clot stabilization, actomyosin in platelets contracts. What effect does this have on the size of the clot and the wound?

_____ 13. Clot dissolution is triggered by _____ , which forms from _____ .

_____ 14. The chemicals that trigger plasmin formation form during the _____ and _____ pathways.

15. Substances present in the blood normally inhibit clotting when blood vessels are not damaged.

_____ a. Rostacyclin, which is secreted by endothelial cells, is an antagonist of thromboxane A2. What effect does prostacyclin have on clotting?

_____ b. Heparin, which is found on the surface of endothelial cells, inactivates thrombin. What effect does heparin have on clotting?

Answers to Exercise 11.5: 1. collagen; 2. platelets; 3a. constriction, decrease; 3b. intrinsic; 3c. platelet plug; 4a. constriction; 4b. extrinsic; 5. inactive; 6. yes; 7. X, common; 8. fibrin; 9. stabilizes it; 10. soluble; 11. yes; 12. clot retracts (brings wound edges closer); 13. plasmin, plasminogen; 14. extrinsic, common; 15a. inhibits it; 15b. inhibits it.

F. ABO Blood Group

Color and label:
1. ○ red blood cells (red)
2. ○ A antigen (agglutinogen)
3. ○ B antigen (agglutinogen)
4. ○ anti-A agglutinins
   (A antibodies)
5. ○ anti-B agglutinins
   (B antibodies)

Exercise 11.6:

_____ 1. The proteins normally present on the surface of cells are called _____ (antigens, antibodies).

_____ 2. The body makes _____ (antigens, antibodies) that react with foreign antigens.

3. Fill in the chart.

Blood Group	Antigen(s) Present	Antibodies Present
a. AB		
b. A		
c. B		
d. O		

4. In figure 11.6b, *small* amounts of type A blood are transfused into a person with type B blood.

_____ a. The quantity of donor antibodies is _____ (dilute, concentrated), while the quantity of recipient's antibodies is _____ (dilute, concentrated).

_____ b. Therefore, the antibodies of the _____ (donor, recipient) react with the red blood cells of the _____ (donor, recipient).

_____ c. Since a red blood cell has the same diameter as a capillary, what will happen when these cells agglutinate?

5. Predict the outcome for each of the following transfusions: (Assume small transfusion.)

_____ a. donor type B, recipient type A

_____ b. donor type B, recipient type AB

_____ c. donor type O, recipient type A

Answers to Exercise 11.6: 1. antigens; 2. antibodies; 3a. A and B antigens, no A or B antibodies; 3b. A antigens, B antibodies; 3c. B antigens, A antibodies; 3d. no A or B antigens, A and B antibodies; 4a. dilute, concentrated; 4b. recipient, donor; 4c. person may die (agglutinated cells clog blood vessels); 5a. agglutination (donor B cells agglutinate with recipient B antibodies); 5b. no agglutination; 5c. no agglutination.

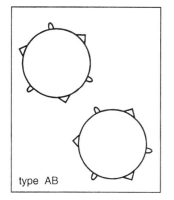

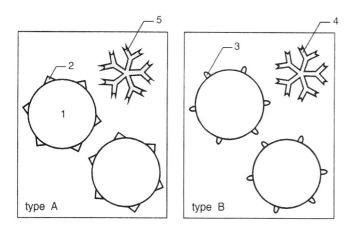

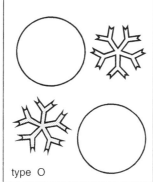

Figure 11.6a. Blood groups.

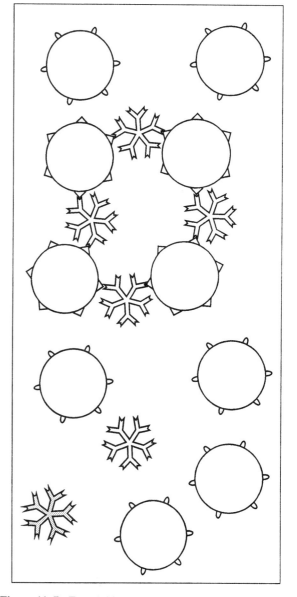

Figure 11.6b. Type A blood transfused into type B recipient.

ABO Blood Group

G. Rh Factor

Color and label:
1. ○ rh– red blood cells
2. ○ Rh+ (antigen D) red blood cells
3. ○ placenta
4. ○ Rh+ antibodies (made by mother)
5. ○ anti- RhD (injected Rh+ antibodies)

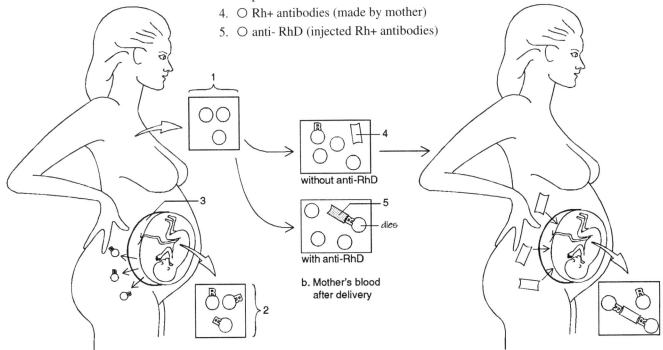

Figure 11.7. Rh factor in pregnancy.

Exercise 11.7:

1. The Rh factor called _____ is on the surface of red blood cells.
    a. In figure 11.7a, the fetus is _____ (Rh+, rh–),
    b. and the mother is _____ (Rh+, rh–).

2. If the placenta tears (most likely during delivery), _____ can cross into the mother.
    a. The Rh+ cells are considered _____ (foreign to, the same as) the mother's cells.
    b. Therefore, the mother's immune system makes _____ against these cells.

3. Since Rh+ antibody formation takes more than 72 hours, early destruction of the escaped Rh+ cells _____ (can, cannot) prevent antibody production.
    a. An injection of rh antibodies (RhoGAM) would cause the escaped Rh+ cells to _____ .
    b. RhoGAM _____ (does, does not) damage mother's rh– cells.

4. If RhoGAM is not given and Rh antibodies form, the mother's Rh+ antibodies _____ (can, cannot) cross the placenta in subsequent pregnancies.

5. Rh+ antibodies in an Rh+ fetus can cause _____ .

Answers to Exercise 11.7: 1. antigen D; 1a. Rh+; 1b. rh–; 2. Rh+ cells; 2a. foreign to; 2b. antibodies; 3. can; 3a. agglutinate; 3b. does not; 4. can; 5. agglutination of blood (possible death of fetus).

Chapter 12 Circulatory System

A. Overview of Circulation

Color and label:
1. ○ pulmonary arteries (blue)
2. ○ pulmonary veins (red)
3. ○ systemic arteries (red)
4. ○ systemic veins (blue)
5. ○ capillaries (purple)
6. ○ right atrium (blue)
7. ○ right ventricle (blue)
8. ○ left atrium (red)
9. ○ left ventricle (red)

Label:
10. tricuspid valve
11. mitral valve
12. lung

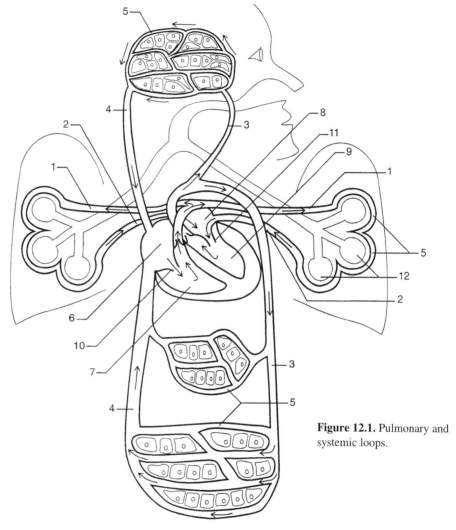

Figure 12.1. Pulmonary and systemic loops.

Exercise 12.1:

_____ 1. The vessels carrying blood to and from the lungs are called the _____ loop of the circulatory system.

_____ 2. The vessels carrying blood to and from tissues of the body (other than the lungs) are called the _____ loop of the circulatory system.

_____ 3. In the capillaries of the lungs the blood is exposed to_____ .

_____ 4. Usually blood draining from capillaries passes through _____ directly back to the_____ .

_____ 5. The chambers of the heart that receive blood from the veins are the right and left_____ .

_____ a. In the systemic loop, blood drains back to the heart from_____ into the_____ .

_____ b. In the pulmonary loop, blood drains back to the heart from the_____ into the_____ .

A. Overview of Circulation continued

_____ 6. Blood passes from the right atrium through the _____ valve into the _____ .

_____ 7. Blood passes from the left atrium through the _____ valve into the _____ .

_____ 8. The right ventricle pumps blood into the _____ loop.

_____ 9. The left ventricle pumps blood into the _____ loop.

Answers to Exercise 12.1: 1. pulmonary; 2. systemic; 3. air; 4. veins, heart, 5. atria; 5a. systemic veins, right atrium; 5b. pulmonary veins, left atrium; 6. tricuspid, right ventricle; 7. mitral, left ventricle; 8. pulmonary; 9. systemic.

B. Heart Anatomy

○ **Color the same. Label:**
1. aortic semilunar valve
2. pulmonary semilunar valve
   a. closed valve (seen from above)
   b. opened valve (seen from above)

○ **Color the same. Label:**
3. cusps of mitral valve (bicuspid)
4. cusps of tricuspid valve

Color and label:
5. ○ right coronary artery
6. ○ chordae tendineae

Color the same. Label:
7. papillary muscle
8. myocardium of right ventricle
9. myocardium of left ventricle
10. ventricular septum
11. cardiac muscle cells

Label:
12. apex
13. visceral pericardium
14. parietal pericardium

Exercise 12.2:

_____ 1. The semilunar valves are located at the beginning of the _____ and the _____ .

_____ 2. The arteries located behind the aortic semilunar valves are the _____ .

_____ 3. The valve between the right atrium and right ventricle is the _____ .

_____ 4. The valve between the left atrium and left ventricle is the _____ .

_____ 5. The atrioventricular valves have _____ at their free edges.

_____ 6. These chordae tendineae connect to the inside of the ventricles on _____ .

_____ 7. The myocardium is made up of _____ .

_____ 8. The chamber of the heart with the thickest myocardium is the _____ .

_____ 9. The outer surface of the heart is covered with _____ .

_____ 10. This membrane folds around the heart and is then known as the _____ .

Answers to Exercise 12.2: 1. pulmonary trunk, aorta; 2. coronary arteries; 3. tricuspid; 4. mitral (bicuspid); 5. chordae tendineae; 6. papillary muscles; 7. cardiac muscle; 8. left ventricle; 9. visceral pericardium; 10. parietal pericardium.

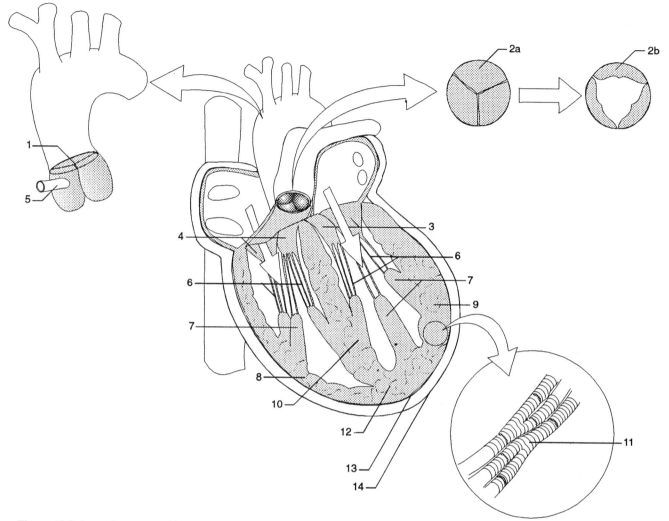

Figure 12.2. Internal anatomy of heart.

C. Cardiac Muscle Membranes

Color. (Color 1 and 3 the same.)
0. ○ phase 0
1. ○ phase 1
2. ○ phase 2
3. ○ phase 3
4. ○ phase 4
○ Na⁺
○ K⁺
○ Ca⁺⁺

Color and label:
5. ○ pacemaker cell
6. ○ contractile muscle cell

Label:
7. threshold potential

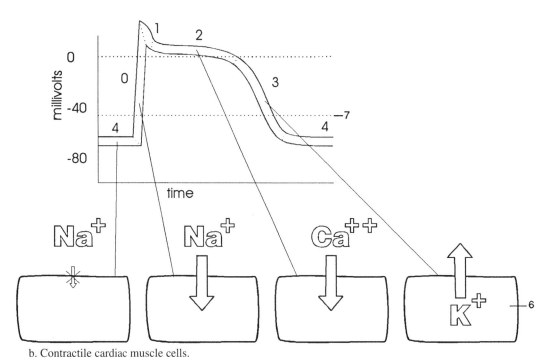

a. Pacemaker cells.

b. Contractile cardiac muscle cells.

Figure 12.3. Voltage-time graph for active membranes.

Chapter 12 Circulatory System

Exercise 12.3:

1. The myocardium (heart muscle) has two types of cells:_____ and_____ .

2. Compare phase 4 in figure 12.3a and b.

   a. In figure 12.3a the voltage_____ (becomes more negative, becomes more positive, does not change).

   b. In figure 12.3b the voltage_____ (becomes more negative, becomes more positive, does not change).

   c. When threshold voltage is reached, the rapid inflow of sodium ions (phase 0) causes the_____ .

   d. In which of the two figures will this occur spontaneously?

3. Phase 3 represents the outflow of_____ .

   a. This is called_____ .

   b. In figure b, the phases showing repolarization are_____ and_____ .

   c. What interrupts repolarization in contractile cardiac muscle cells?

4. During phase 2 the ions that flow through the membrane into the cell are_____ .

   a. Since calcium ions are positive, this prevents the membrane from_____ .

   b. This is called the_____ .

   c. The amount of time the contractile cardiac muscle cell spends in its absolute refractory period in relation to its time of depolarization is_____ (long, short).

   d. This is important because_____ .

Answers to Exercise 12.3: 1. pacemaker, contractile cardiac muscle cell; 2a. becomes more positive; 2b. does not change; 2c. action potential; 2d. figure a; 3. K^+; 3a. repolarization; 3b. 1, 3; 3c. phase 2; 4. Ca^{++}; 4a. repolarizing; 4b. absolute refractory period; 4c. long; 4d. it prevents the heart from tetanizing.

D. Regulation of Heartbeat

Color and label:
1. ○ cardiac conduction system
   a. sinoatrial node
   b. atrioventricular node
   c. atrioventricular bundle
   d. Purkinje fibers
2. ○ parasympathetic innervation
   a. medulla oblongata (brain)
   b. vagus nerve
   c. parasympathetic ganglion
   d. postganglionic fibers
3. ○ sympathetic innervation
   a. spinal cord
   b. preganglionic sympathetic nerve
   c. sympathetic ganglion
   d. postganglionic fibers

Label:
4. myocardium (heart muscle)
5. papillary muscles
6. ventricular septum
7. apex

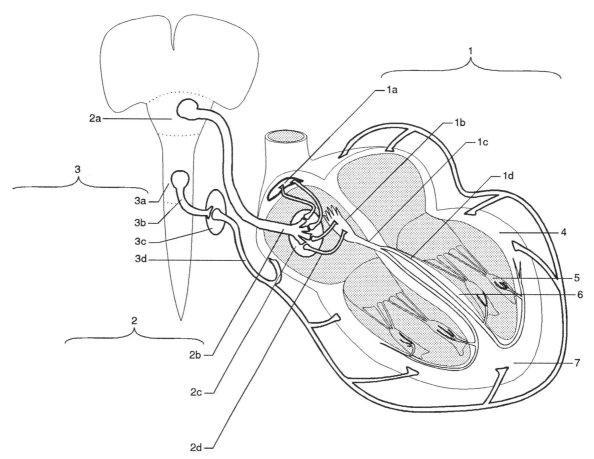

Figure 12.4. Coronal section of heart.

Chapter 12 Circulatory System

Exercise 12.4:

_____ 1. The sinoatrial node is called the pacemaker because it is normally the first part of the heart to_____ .

_____ a. Since some of its cells have unstable membrane potentials, the S-A node does not need outside stimulation to_____ .

_____ b. The depolarization of the S-A node causes the contraction of the_____ .

_____ 2. As the atria depolarize, the next part of the cardiac conduction system that is stimulated is the_____ .

_____ a. The A-V node is connected to the_____ .

_____ b. The A-V bundle leads the depolarization into the_____ of the heart along the_____ .

_____ 3. The first part of the ventricles to depolarize is the_____ , then the_____ .

_____ a. Then the depolarization continues back toward the_____ .

_____ b. This means that the ventricles contract from the_____ (apex up, valves down).

_____ 4. Stimulation from the vagus nerve (parasympathetic) decreases the activity of the_____ and_____ . This causes the heart rate to_____ (increase, decrease).

_____ 5. Stimulation from the sympathetic nerves increases both the force and rate of myocardial contraction. Therefore, this stimulation causes the heart to beat_____ (faster, slower) and_____ (more, less) vigorously.

Answers to Exercise 12.4: 1. depolarize; 1a. depolarize; 1b. atria; 2. atrioventricular node; 2a. atrioventricular bundle; 2b. ventricles, Purkinje fibers; 3. ventricular septum, apex; 3a. papillary muscles (of the valves); 3b. apex up; 4. S-A node, A-V node, decrease; 5. faster, more.

E. Cardiac Cycle

Color and label:
1. ○ myocardium
2. ○ right atrium (blue)
3. ○ left atrium (red)
4. ○ right ventricle (blue)
5. ○ left ventricle (red)

Label:
6. systemic veins
7. pulmonary veins
8. pulmonary arteries
9. aorta
10. atrioventricular valves

11. semilunar valves
    a. diastole
    b. atrial systole
    c. ventricular systole (isovolumetric)
    d. ventricular systole (ejection)

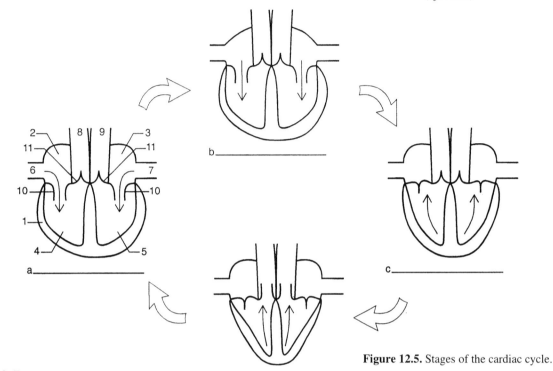

Figure 12.5. Stages of the cardiac cycle.

Exercise 12.5:

1. During diastole the myocardium is relaxed. (figure 12.5a)
    a. This allows blood to flow into the heart from the _____ and _____ veins.
    b. The semilunar valves are _____ (closed, open).
    c. This means the pressure in the aorta and pulmonary arteries must be _____ (greater, less) than that of the ventricles.

2. When the atria contract, more blood is forced into the _____ . (figure 12.5b)

3. When the ventricular myocardium begins to contract, the pressure in the ventricles _____ (increases, decreases). (figure 12.5c)
    a. The first effect of this rise in pressure is to cause the _____ valves to _____ .
    b. The semilunar valves do not open until the pressure in the ventricles exceeds the pressure in the _____ . (figure 12.5d)

4. The semilunar valves close when the pressure in the arteries becomes _____ (greater, less) than that of the ventricles. (figure 12.5a)

5. If the arterial blood pressure is elevated, it becomes (harder, easier) for the ventricles to eject blood.

Answers to Exercise 12.5: 1a. systemic, pulmonary; 1b. closed; 1c. greater; 2. ventricles; 3. increases; 3a. atrioventricular, close; 3b. arteries; 4. greater; 5. harder.

F. Blood Vessel Histology

Color and label:
1. ○ tunica adventitia
2. ○ tunica media
   a. elastic connective tissue
   b. smooth muscle
   c. vein valve
3. ○ tunica intima

Label:
4. precapillary sphincter
   a. aorta
   b. muscular artery
   c. capillary with arteriole
   d. vein

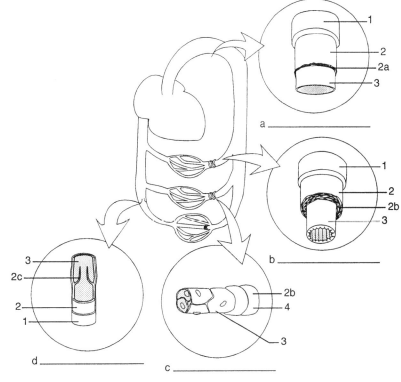

Figure 12.6. Blood vessel structure.

Exercise 12.6:

_____ 1. The layer (tunic) present in all the vessels shown is the_____ .

_____ 2. The tunica intima in the capillary is made of_____ (tissue type).

_____   a. Would this type of tissue allow interchange of materials between the blood and the surrounding interstitial fluid?

_____   b. If the precapillary sphincter in the arteriole constricts, the blood flow into the capillary will_____ (increase, decrease).

_____ 3. The tunica media in the aorta contains much_____ (tissue type). When the heart pumps blood into the aorta, this tissue allows the vessel wall to_____ .

_____ 4. The tunica media of a muscular artery is primarily composed of_____ . This process is called_____ (vasoconstriction, vasodilation).

_____   a. When the tone of smooth muscle increases, the vessel's lumen_____

_____   b. Therefore, the blood flow to different parts of the body can be controlled by altering the tone of_____ .

_____   c. If the tone in the smooth muscle in arteries and arterioles increases, then arterial blood pressure should _____ (increase, decrease).

_____ 5. The special feature of the vein is the_____ .

_____ 6. Valves in veins prevent blood from flowing back toward_____ .

Answers to Exercise 12.6: 1. tunica intima; 2. simple squamous epithelium; 2a. yes; 2b. decrease; 3. elastic connective tissue, stretch; 4. smooth muscle, vasoconstriction; 4a. decreases; 4b. smooth muscle; 4c. increase; 5. valve; 6. capillaries.

G. Blood Pressure

Color and label:
1. ○ myocardium
2. ○ arteriole

Color:
○ pressure arrows

Label:
3. aortic semilunar valve

4. aorta
5. large artery
6. capillary

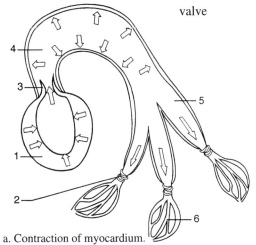
a. Contraction of myocardium.

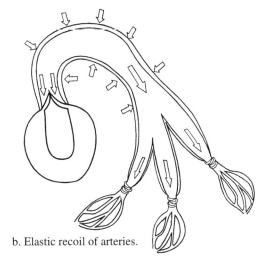
b. Elastic recoil of arteries.

Figure 12.7. Arterial blood pressure.

Exercise 12.7:

_____ 1. When the myocardium of the left ventricle contracts, the pressure in the ventricle_____ (increases, decreases). (figure 12.7a)

_____ a. When the pressure in the left ventricle exceeds the pressure in the aorta, the aortic semilunar valve_____ (opens, closes).

_____ b. This pressure forces blood into the aorta, causing its pressure to_____ (increase, decrease).

_____ 2. When the pressure in the aorta exceeds the pressure in the left ventricle, the aortic semilunar valve_____ (opens, closes). (figure 12.7b)

_____ 3. The pressure of the blood on the walls of the aorta causes its elastic connective tissue to_____ .

_____ 4. The pressure exerted by the heart muscle and the elastic rebound of the aorta and large arteries causes the blood to flow toward the_____ .

_____ 5. Before blood reaches the capillaries, it passes through_____ .

_____ a. The major structural feature of arterioles is_____ .

_____ b. When this smooth muscle contracts, the diameter of the arteriole becomes_____ (larger, smaller). This is called_____ (vasoconstriction, vasodilation).

6. Because arterioles restrict the flow of blood into capillaries, do they cause the pressure to be high or to be low in

_____ a. capillaries?

_____ b. arteries?

Answers to Exercise 12.7: 1. increases; 1a. opens; 1b. increase; 2. closes; 3. stretch; 4. arterioles and capillaries; 5. arterioles; 5a. smooth muscle; 5b. smaller, vasoconstriction; 6a. low; 6b. high.

H. Interstitial Fluid Balance

○ **Color the same. Label:**
1. arteriole
2. venule
3. vein

Color and label:
4. ○ lymph vessel
5. ○ tissue cells

6. ○ plasma flow out of capillary (arrows)
7. ○ flow of interstitial fluid (arrows)
8. ○ interstitial fluid return to capillary (arrows)
9. ○ interstitial fluid flow into lymph capillary (arrows)

Label:
10. fenestration

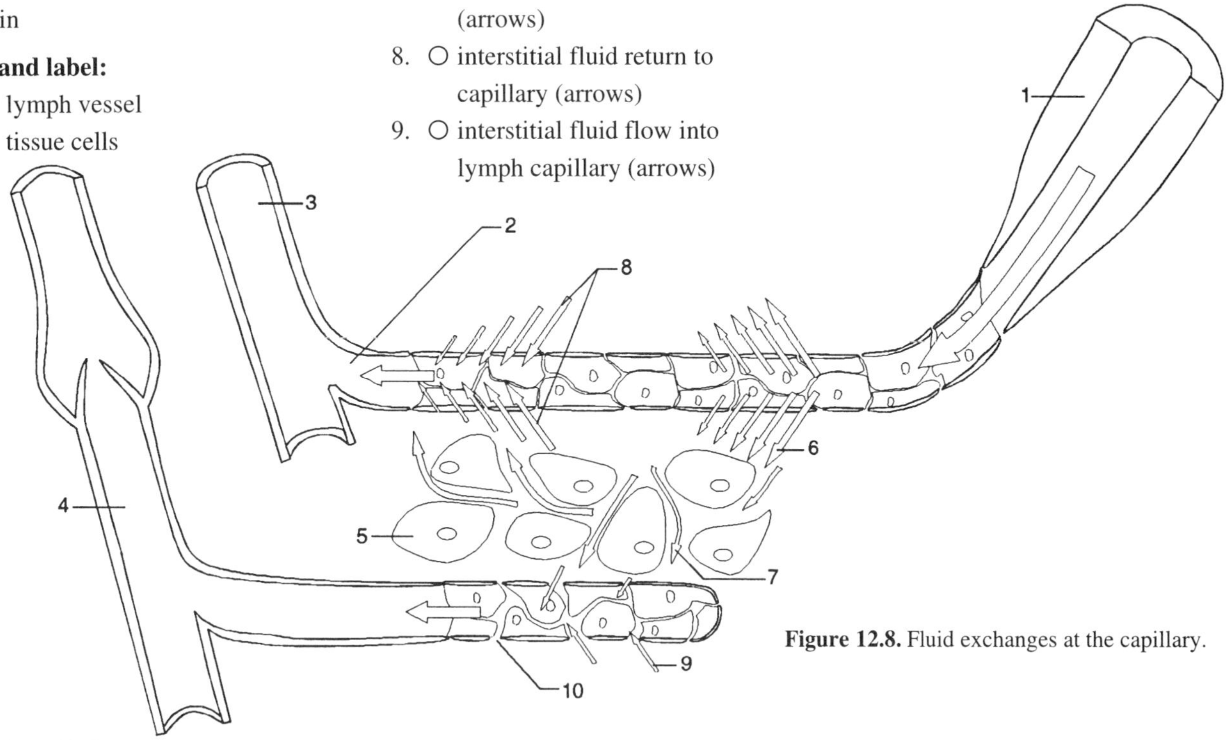

Figure 12.8. Fluid exchanges at the capillary.

Exercise 12.8:

1. Blood flows into capillaries from_____ .
2. Capillary walls are made of one layer of_____ .
   a. Water and dissolved materials are forced out of the capillaries by pressure from_____ .
   b. Materials produced by cells are released into_____ .
   c. Therefore, interstitial fluid is composed of_____ ,_____ , and_____ .
   d. Interstitial fluid is found_____ (inside, around) cells.
3. Plasma proteins are larger than the spaces between capillary wall cells.
   a. Are plasma proteins usually forced into interstitial fluid?
   b. The concentration of plasma proteins in the capillaries_____ (increases, decreases) as blood flows through capillaries.
   c. As blood moves toward the venous end of the capillaries, this increase in osmotic pressure (plasma oncotic pressure) causes fluid to flow_____ (into, out of) capillaries.
   d. Only about 90% of the interstitial fluid returns to capillaries. The rest returns through the_____ .
4. Small amounts of plasma proteins do leak out of capillaries.
   a. If proteins were to remain in interstitial fluid, the interstitial fluid would tend to_____ (gain, lose) water.
   b. What structure in lymph capillaries makes it possible to pick up larger molecules and other materials, such as microorganisms?

Answers to Exercise 12.8: 1. arterioles; 2. simple squamous endothelium; 2a. arteries; 2b. interstitial fluid; 2c. water, dissolved materials, cell products; 2d. around; 3a. no; 3b. increases; 3c. into; 3d. lymph vessels; 4a. gain; 4b. fenestration.

I. Vein Pump

○ **Color the same. Label:**
1. superficial vein
2. deep vein
3. inferior vena cava
4. heart

Color (trace) and label:
5. ○ skin
6. ○ skeletal muscle
7. ○ bone
8. ○ diaphragm
   a. relaxed
   b. contracted

Color:
9. ○ arrows outside of veins (pressure)
10. ○ arrows inside of veins (blood flow)

Label:
11. vein valve
    a. upper valve
    b. lower valve
12. thoracic cavity
13. abdominal cavity

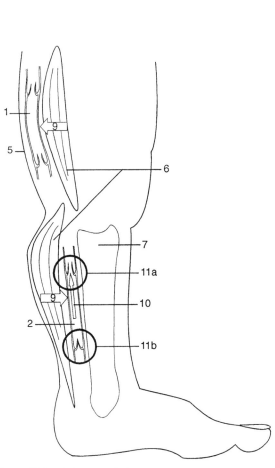

Figure 12.9a. Vein pump in leg.

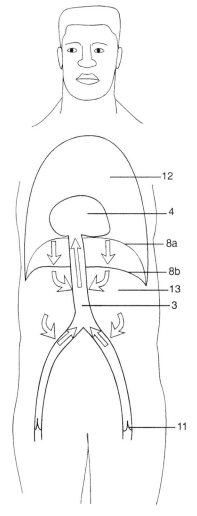

Figure 12.9b. Vein pump in trunk.

Exercise 12.9:

_____ 1. Blood drains from tissues into_____ .

_____   a. Since the hydrostatic (fluid) pressure in tissues is low, can this pressure push blood back toward the heart?

_____   b. To keep the blood from flowing back toward the capillaries, the veins in the limbs contain_____ .

_____ 2. In figure 12.9a, muscle contraction squeezes the deep vein against the_____ .

_____   a. This pressure forces the blood in the vein against the upper valve, causing it to_____ (open, close), and

_____   b. against the lower valve, causing it to_____ (open, close).

_____   c. Since the lower valve is closed, blood can only flow_____ .

_____ 3. In figure 12.9a, muscle contraction squeezes the superficial vein against the_____ .

_____ 4. Muscle contraction creates more pumping action against_____ (superficial, deep) veins.

_____ 5. Are valves present in abdominal veins?

_____ 6. When the diaphragm contracts, it pushes down on the_____ cavity.

_____   a. This causes the pressure in the abdominal cavity to_____ (increase, decrease) and the pressure in the thoracic cavity to_____ (increase, decrease).

_____   b. This causes the pressure in the inferior vena cava to_____ (increase, decrease).

_____   c. Therefore, blood in the inferior vena cava is forced toward the_____ .

_____ 7. Blood in the abdominal veins is prevented from flowing back into the legs by_____ in the veins of the legs.

Answers to Exercise 12.9: 1. veins; 1a. no; 1b. valves; 2. bone; 2a. open; 2b. close; 2c. upward; 3. skin; 4. deep; 5. no; 6. abdominal; 6a. increase, decrease; 6b. increase; 6c. heart; 7. valves.

J. Lymph Flow

○ **Color red and label:**
1. ○ right subclavian vein
2. ○ left subclavian vein

○ **Color (trace) green and label:**
3. thoracic duct
4. lymph nodes

5. cysterna chyli
6. lymph vessel
   a. valve
7. lymph capillary

○ **Lightly shade the parts of the body draining into the *right* branch of veins.**

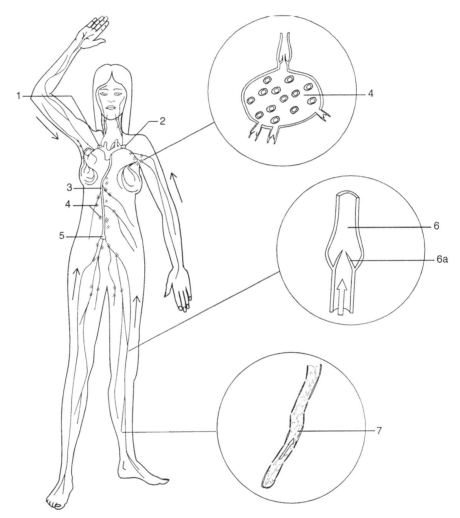

Exercise 12.10:   Figure 12.10. Anatomy of lymphatic system.

_____ 1. Lymph from the legs and most of the trunk drains into the _____ .

_____ 2. The thoracic duct drains into the _____ (right, left) subclavian vein.

_____ 3. Lymph is collected from interstitial fluid at the _____ .

_____ 4. Lymph does not normally flow back toward tissues because of _____ in lymphatic vessels.

_____ 5. Before draining into the veins, lymph usually passes through _____ .

_____ 6. Microorganisms or tumor cells which enter lymph capillaries are frequently caught in the _____ .

Answers to Exercise 12.10: 1. thoracic duct; 2. left; 3. lymph capillaries; 4. valves; 5. lymph nodes; 6. lymph nodes.

K. Arteries of the Systemic System

○ **Color (trace) red. Label:**
1. brachiocephalic artery
2. common carotid artery
3. subclavian artery
4. axillary artery
5. brachial artery
6. ulnar artery
7. radial artery
8. palmar arches
9. coronary arteries
10. descending aorta
11. common iliac artery
12. external iliac artery
13. internal iliac artery
14. femoral artery
15. posterior tibial artery
16. anterior tibial artery

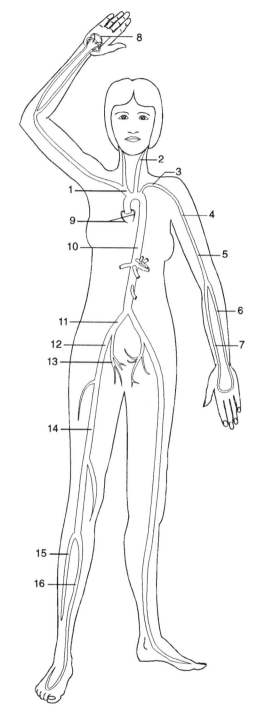

Figure 12.11. Anterior view of major arteries.

Arteries of the Systemic System

L. Veins of the Systemic System

○ **Color (trace) blue. Label:**
1. internal jugular vein
2. external jugular vein
3. subclavian vein
4. cephalic vein (superficial)
5. basilic vein (superficial)
6. brachial vein (deep)
7. intermediate antebrachial vein (superficial)
8. inferior vena cava
9. internal iliac vein
10. external iliac vein
11. femoral vein
12. great saphenous vein (superficial)
13. tibial vein (deep)
14. communicating veins

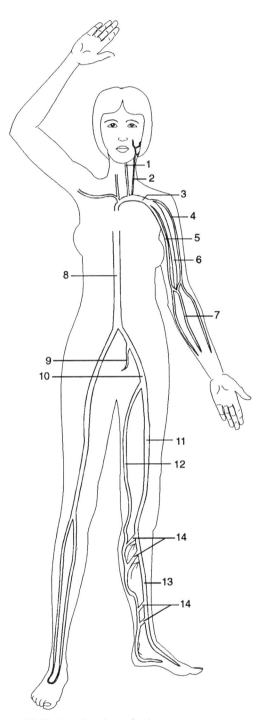

Figure 12.12. Anterior view of veins.

M. Blood Vessels of the Head and Neck

○ **Color blue. Label:**
1. internal jugular vein
2. external jugular vein
3. subclavian vein
4. superior vena cava
5. inferior vena cava
6. pulmonary veins

○ **Color red. Label:**
7. coronary arteries
8. pulmonary artery
9. aorta
10. subclavian artery
11. vertebral artery
12. common carotid artery
13. carotid sinus
14. external carotid artery
15. internal carotid artery

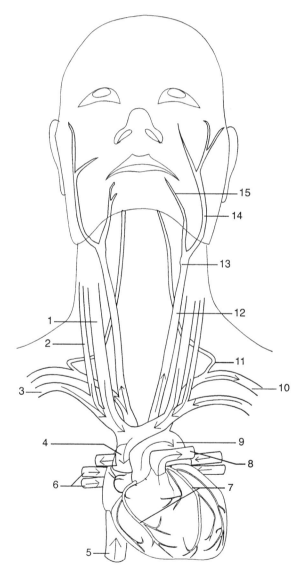

Figure 12.13. Arteries and veins of head and neck.

N. Abdominal Arteries

Color and label:
1. ○ kidney
2. ○ liver
3. ○ stomach
4. ○ spleen
5. ○ pancreas
6. ○ small intestine
7. ○ colon (large intestine)

Color (trace) red. Label:
8. renal arteries
9. celiac trunk
   a. hepatic artery
   b. left gastric artery
   c. splenic artery
10. abdominal aorta
11. superior mesenteric artery
12. inferior mesenteric artery
13. common iliac arteries

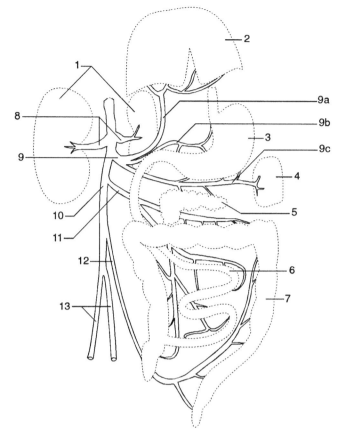

Figure 12.14. Three-quarter anterolateral view of abdomen.

Exercise 12.14:

Fill in the artery that supplies each of the listed organs.

Organ	Artery
1. spleen	
2. descending colon	
3. stomach	
4. liver	
5. small intestine	
6. ascending colon	
7. transverse colon	
8. pancreas	
9. kidney	

Answers to Exercise 12.14: 1. splenic artery; 2. inferior mesenteric artery; 3. left gastric artery; 4. hepatic artery; 5. superior mesenteric artery; 6. superior mesenteric artery; 7. superior mesenteric artery; 8. splenic artery; 9. renal artery.

O. Hepatic Portal System

Color and label:
1. ○ hepatic portal veins
2. ○ hepatic veins
3. ○ inferior vena cava

Label:
4. liver
5. stomach
6. spleen
7. small intestine
8. large intestine (colon)

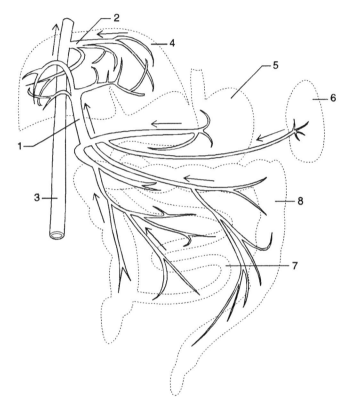

Figure 12.15. Three-quarter anterolateral view of hepatic portal system.

Exercise 12.15:

_____ 1. The hepatic portal veins drain the ____ , ____ , ____ , and ____ .

_____ 2. The blood in veins other than portal veins is carried directly back to the ____ .

_____ 3. Hepatic portal veins carry blood to the ____ .

_____ 4. The liver has two sources of blood, one from the hepatic arteries and the other from the ____ .

_____ 5. Blood leaves the liver through the ____ .

_____ 6. The hepatic portal system connects the capillaries of the ____ system with the ____ .

Answers to Exercise 12.15: 1. stomach, spleen, small intestine, large intestine; 2. heart; 3. liver; 4. portal system; 5. hepatic vein; 6. digestive, liver.

P. Fetal Circulation

Color and label:
1. ○ ○ aorta (light red-figure 12.16a, red-figure 12.16b)
2. ○ ductus arteriosus (pink)
3. ○ ○ pulmonary trunk (pink-figure 12.16a, blue-figure 12.16b)
   a. pulmonary artery
4. ○ ○ pulmonary veins (blue-figure 12.16a, red-figure 12.16b)
5. ○ inferior vena cava (blue)
6. ○ ductus venosus (red)
7. ○ hepatic portal veins (blue)
8. ○ umbilical vein (red)
9. ○ umbilical arteries (blue)

Label:
10. foramen ovale
11. lung
12. liver
13. umbilicus
14. placenta

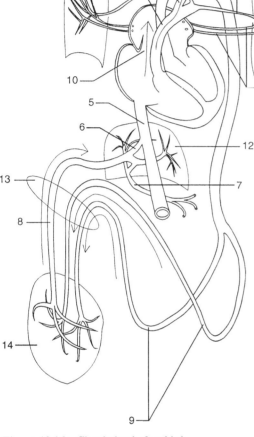

Figure 12.16a. Circulation before birth.

Figure 12.16b. Circulation after birth.

Exercise 12.16:

_____ 1. The fetus is joined to the mother at the_____ .

_____   a. Blood is carried from the fetus to the placenta in the_____ .

_____   b. Blood is carried back to the fetus from the placenta in the_____ .

_____   c. Since the fetus has no direct access to the external environment, it must obtain its oxygen and nutrients from the_____ .

_____ 2. The vessels that supply the lungs are the_____ .

_____   a. Since the fetal lungs contain no air, the pulmonary arteries deliver a relatively small amount of_____ .

_____   b. In the fetus, most of the blood pumped into the pulmonary trunk is shunted into the_____ by the _____ .

_____   c. Since a relatively small volume of blood is pumped to the fetal lungs, this means that the lungs return a small amount of blood to the_____ of the heart.

_____   d. The left atrium of the fetal heart receives most of its blood through the_____ .

_____ 3. The organ that directly receives much of the blood returning to the fetus in the umbilical vein is the_____ .

_____   a. The rest of the blood from the umbilical vein shunts through the_____

_____   b. directly to the_____ .

_____ 4. At birth the umbilical arteries and vein_____ (relax, constrict).

_____ 5. This causes blood flow to stop between the newborn and the_____ .

_____ 6. At birth the lungs fill with_____ and

_____   a. their resistance to blood flow_____ (increases, decreases)

_____   b. causing the flow of blood to_____ (increase, decrease).

_____ 7. When blood flow to the lungs increases, return of blood from the lungs_____ (increases, decreases).

_____ 8. This increase of flow into the left atrium causes the flap to close over the_____ .

_____ 9. The change to postnatal circulation is usually completed by the constriction of the_____ between the pulmonary trunk and the aorta.

Answers to Exercise 12.16: 1. placenta; 1a. umbilical arteries; 1b. umbilical vein; 1c. placenta; 2. pulmonary arteries; 2a. blood; 2b. aorta, ductus arteriosus; 2c. left atrium; 2d. foramen ovale; 3. liver; 3a. ductus venosus; 3b. inferior vena cava; 4. constrict; 5. placenta; 6. air; 6a. decreases; 6b. increase; 7. increases; 8. foramen ovale; 9. ductus arteriosus.

Fetal Circulation

Chapter 13: Lymphatic System and Immunity

A. Lymphatic Organs

Color and label:

1. ○ primary lymphatic tissue
   a. bone marrow
   b. thymus
2. ○ secondary lymphatic tissue
   a. tonsils
   b. lymph nodes
   c. spleen
   d. unencapsulated lymphatic tissue (Peyer's patch)

Outline each box and its corresponding circles the same color. Label:

3. ○ specific immunity
4. ○ nonspecific immunity

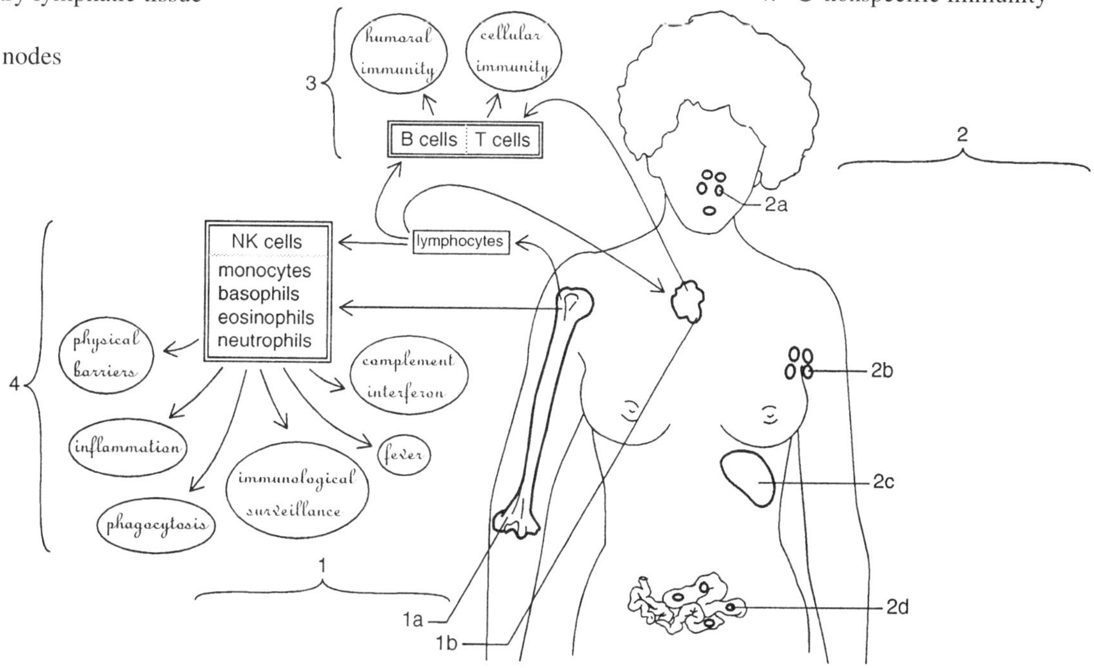

Figure 13.1. Types of lymphatic tissue.

Exercise 13.1:

1. The formation and maturation of white blood cells occurs in _____ (primary, secondary) lymphatic tissue.
2. These cells become activated and perform their functions in _____ (primary, secondary) lymphatic tissue, as well as body tissues in general.
3. The two types of immunity are _____ and _____ .
4. Nonspecific immunity involves the immediate, broad range of defenses against a variety of foreign materials (antigens).
   a. The white blood cells that participate in nonspecific immunity are: _____ , _____ , _____ , _____ , and _____ .
   b. One mechanism of nonspecific immunity is _____ .
5. Specific immunity is the recognition and destruction of specific antigens by specific cells or specific antibodies.
   a. The white blood cells that participate in specific immunity are _____ and _____ .
   b. One mechanism of specific immunity is _____ .

Answers to Exercise 13.1: 1. primary; 2. secondary; 3. nonspecific immunity, specific immunity; 4a. NK cells, monocytes, basophils, eosinophils, neutrophils; 4b. physical barriers (or inflammation, phagocytosis, immunological surveillance, fever, chemicals such as complement and interferon); 5a. B cells, T cells; 5b. humoral immunity or cellular immunity.

B. Secondary Lymphatic Tissue

Color and label:
1. ○ capsule
   a. trabeculae
2. ○ lymph nodule
   a. germinal center

For figure 13.2a, color and label:
3. ○ lymph
   a. afferent lymphatic vessels
   b. cortical sinuses
   c. medullary sinuses
   d. efferent lymphatic vessels
   e. valve
4. ○ cortex (except lymph nodules)
5. ○ medulla

For figure 13.2b, color and label:
6. ○ blood (red)
   a. artery
   b. venous sinuses
   c. vein
7. ○ red pulp (red)
8. ○ white pulp (same as lymph nodule)

For figure 13.2c and d, color and label:
9. ○ epithelial tissue
10. ○ connective tissue

Label:
11. crypt

Exercise 13.2:

_____ 1. Each type of secondary lymphatic tissue contains spherical masses of lymphocytes called _____ .

_____ a. Exposure to foreign materials stimulates division of lymphocytes. When B cells divide, a central concentration of B cells forms, called the _____ .

_____ b. Surrounding the lymph nodules are cells that engulf foreign materials and cellular debris called _____ . (see connective tissue, p. 44)

_____ 2. Lymph nodes filter foreign materials and cellular debris from _____ .

_____ a. The arrows in figure 13.2a show _____ .

_____ b. Lymph enters lymph nodes through vessels called _____ , travels through spaces called _____ , and leaves through vessels called _____ .

_____ c. When B cells divide in response to foreign materials, the size of the germinal centers _____ and, therefore, the size of the lymph nodes _____ .

_____ d. The more mobile T lymphocytes escape the lymph node via the _____ and travel to various parts of the body.

_____ 3. The spleen filters foreign materials and cellular debris from _____ .

_____ a. The arrows in figure 13.2b show _____ .

_____ b. Blood enters through arteries which are surrounded by lymph nodules. These lymph nodules are called the _____ .

_____ c. Exposure to foreign materials means the cells of the lymph nodules _____ .

_____ d. Then, blood flows from arteries into venous sinuses found in the _____ .

_____ e. Since large numbers of macrophages are found in red pulp, where are worn red blood cells and cellular breakdown products likely to be filtered?

_____ f. Since the venous sinuses can hold large volumes of blood, can they act as reservoirs for blood?

_____ 4. The tonsils and Peyer's patches respond to foreign materials found in the _____ and _____ respectively.

_____ a. Are these lymph nodules encapsulated?

_____ b. How does a lack of capsule help immune function?

_____ c. Deep indentations in the surface of the tonsil are called _____ .

_____ d. What effect do crypts have on the exposure of the tonsil to foreign antigen?

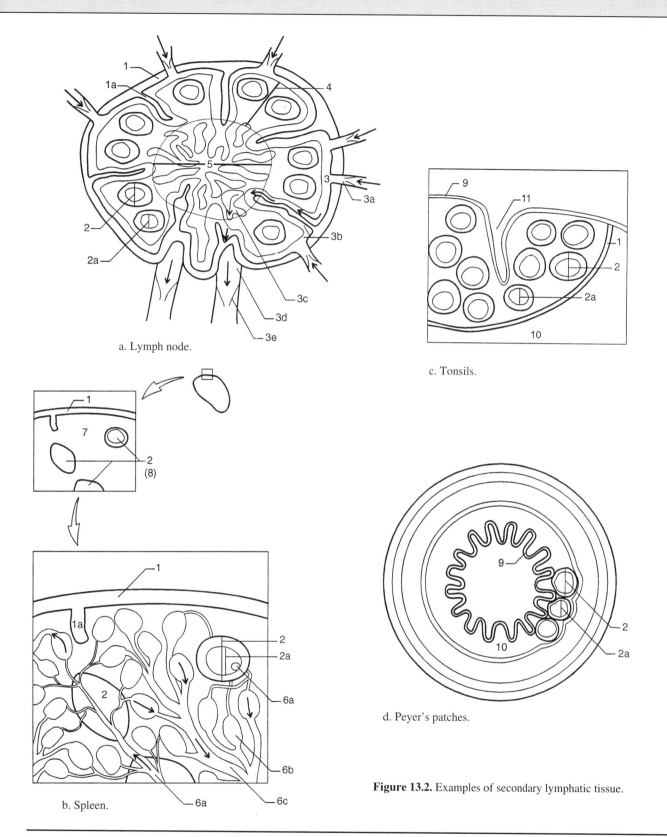

a. Lymph node.

b. Spleen.

c. Tonsils.

d. Peyer's patches.

Figure 13.2. Examples of secondary lymphatic tissue.

Answers to Exercise 13.2: 1. lymph nodules; 1a. germinal centers; 1b. macrophage; 2. lymph; 2a. direction of lymph flow; 2b. afferent lymphatic vessels, sinuses, efferent lymphatic vessels; 2c. increase, increases; 2d. efferent lymphatic vessels; 3. blood; 3a. direction of blood flow; 3b. white pulp; 3c. divide; 3d. red pulp; 3e. red pulp; 3f. yes; 4. mouth, pharynx; small intestines; 4a. tonsils—partial; Peyer's patches—no; 4b. easier for foreign materials to make contact with lymphocytes; 4c. crypts; 4d. increase it (can also harbor foreign organisms that can become a source of infection).

Secondary Lymphatic Tissue

C. Surface Barriers to Infection

Trace and label:
1. ○ integument

Outline and label the organs lined by:
2. ○ mucous membrane (solid lines)

Label the surfaces with the following chemical barriers:
3a. ○ oil
3b. ○ sweat
3c. ○ low pH
3d. ○ lysozymes

Label:
4. hairs
5. trachea
  a. cilia
6. stomach

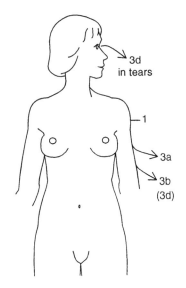

a. Integument.

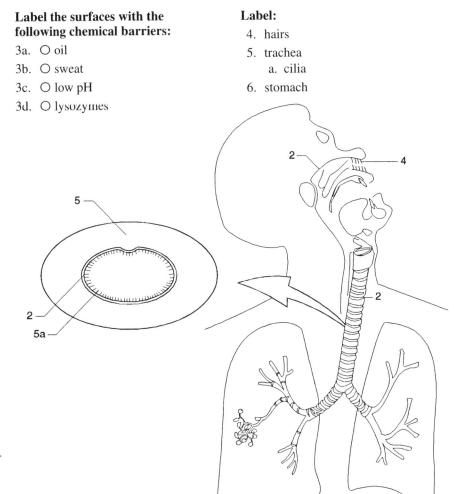

b. Respiratory tract.

Figure 13.3. Physical and chemical barriers.

Exercise 13.3:

_____ 1. Infection can enter the body across the following surfaces: _____ and _____ .

_____ 2. These surfaces, with potential or actual exposure to the external environment, are _____ (epithelial, connective, muscular, nervous) tissue. (see chapter 4, tissues)

3. Do each of the following epithelial barriers help prevent infection?

_____ a. tight junctions

_____ b. basement membrane

_____ c. stratification

_____ 4. Mucus made by the mucous membranes protects because _____ .

5. Match the following types of protection to their sites of action:

_____ a. oil (inhibits bacterial growth)

_____ b. sweat (inhibits bacterial growth)

Chapter 13   Lymphatic System and Immunity

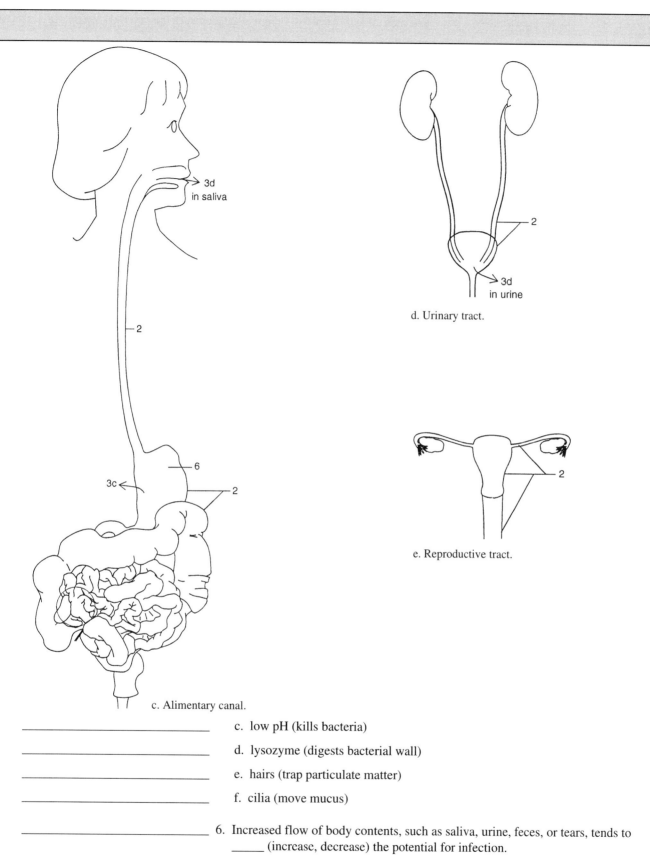

c. Alimentary canal.

d. Urinary tract.

e. Reproductive tract.

_____   c. low pH (kills bacteria)

_____   d. lysozyme (digests bacterial wall)

_____   e. hairs (trap particulate matter)

_____   f. cilia (move mucus)

_____ 6. Increased flow of body contents, such as saliva, urine, feces, or tears, tends to _____ (increase, decrease) the potential for infection.

Answers to Exercise 13.3: 1. integument, mucous membranes; 2. epithelial; 3a. yes; 3b. yes; 3c. yes; 4. it coats surfaces (prevents direct contact with noxious agents), traps bacteria (because it's sticky); 5a. integument; 5b. integument; 5c. stomach; 5d. saliva, sweat, tears, urine; 5e. nose; 5f. trachea; 6. decrease.

D. Inflammation Defined

Color (trace) and label:
1. ○ bacteria
2. ○ mast cell
3. ○ cells of capillary wall
4. ○ red blood cells (red)
5. ○ sensory neuron and receptor

Color and label the proteins:
- I ○ histamines
- F ○ fibrinogen
- H ○ heparin
- Y ○ antibodies
- K ○ kinins
- K* activated kinins
- C ○ complement proteins
- C* activated complement

Label:
6. epithelium
7. connective tissue
8. capillary diameter
9. space between cells

Exercise 13.4:

_____ 1. Tissue damaged by infection and/or trauma triggers the release of chemicals such as _____ and _____ from mast cells.

_____ 2. *Exposure* to bacteria can cause activation of _____ , which also stimulates histamine release.

_____ 3. Histamines cause blood vessels to _____ (dilate, constrict), thus reddening the damaged tissue and _____ (increasing, decreasing) its temperature.

_____ a. This increased blood flow _____ (increases, does not change, decreases) nutrient and oxygen deliver and waste removal by blood. (see capillaries, p. 189)

_____ b. An increase in temperature _____ (increases, does not change, decreases) the rates of chemical reactions. (see enzymes, p. 24)

_____ c. Therefore, increased blood flow and temperature _____ (increase, have no effect on, decrease) the rate of healing.

_____ 4. Histamines also _____ (increase, decrease) blood vessel permeability.

_____ a. Increased permeability means proteins such as _____ , _____ , _____ , and _____ leave the blood vessel.

_____ b. Increased permeability also means fluids _____ (enter, leave) the blood vessel.

_____ c. Increased fluids at the injury site cause _____ .

_____ 5. Chemicals, such as activated kinins, swelling, and/or increased temperature can stimulate sensory receptors, causing _____ .

_____ 6. Kinins cause dilation and increased capillary permeability. These effects are _____ (similar to, different from) the effects of histamine.

_____ 7. Therefore, kinins _____ (enhance, stop) inflammation.

Answers to Exercise 13.4: 1. histamine, heparin; 2. complement; 3. dilate, increasing; 3a. increases; 3b. increases; 3c. increase; 4. increase; 4a. fibrinogen, antibodies, complement, kinins, heparin; 4b. leave; 4c. swelling; 5. pain; 6. similar to; 7. enhance.

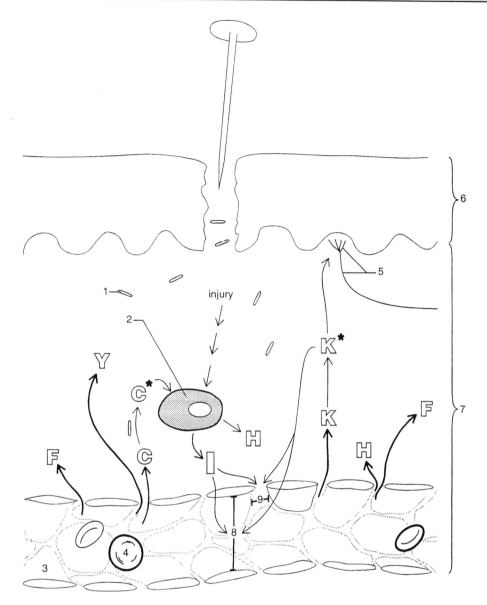

Figure 13.4. Inflammation: redness, warmth, swelling, pain.

E. Walling Off During Inflammation

Color and label:
1. ○ bacteria
   a. bacterial debris
2. ○ mast cell
3. ○ cells of capillary wall
4. ○ red blood cells (rcd)
5. ○ tissue macrophage
6. ○ cellular debris

Color and label:
F ○ fibrinogen and clot area
H ○ heparin
C* ○ complement
K* ○ kinins

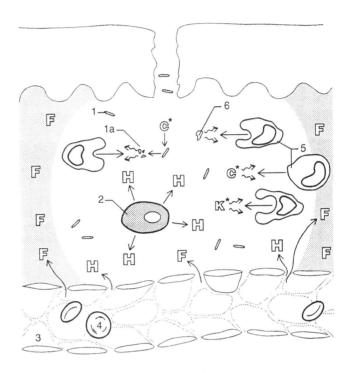

Figure 13.5. Clotting and tissue macrophage.

Exercise 13.5:

_____ 1. The clotting precursor that diffuses from blood vessels to the injury area is _____ .

_____ 2. Fibrinogen converts to fibrin and walls off the injured area. This serves to _____ .

_____ 3. Heparin is released from _____ and _____ .

_____ a. High concentrations of heparin in the immediate area of the injury _____ (prevent, enhance) clotting. (see clotting, p. 174)

_____ b. Thus, exchanges of waste products and nutrients between the injury site and blood _____ (can, cannot) occur.

4. Activated tissue macrophage are attracted to foreign materials and damaged tissue by chemotaxis.

_____ a. Chemotactic agents include _____ , _____ , _____ , and _____ .

_____ b. Bacterial debris forms when _____ acts on the invading organisms.

_____ 5. Since macrophage can stick to foreign materials for long periods of time, they _____ (can, cannot) aid in walling off.

Answers to Exercise 13.5: 1. fibrinogen; 2. prevent the spread of infection and toxins; 3. mast cells, blood vessels; 3a. prevent; 3b. can; 4a. bacterial debris, cellular debris, complement, kinins; 4b. complement; 5. can.

F. Cellular Response to Inflammation

Color and label:
1. ○ bacteria
    a. surface antigen
2. ○ blood vessel wall
3. ○ red blood cells (red)
4. ○ neutrophil
5. ○ monocyte
    a. macrophage (newly formed)
    b. tissue macrophage
6. ○ lymphocyte
7. ○ antibody

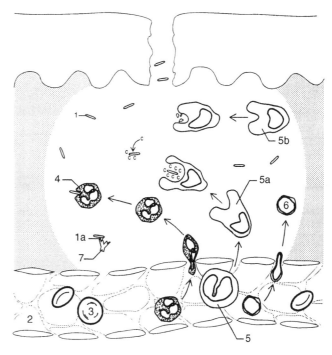

Figure 13.6. Processes: Diapedesis, chemotaxis, phagocytosis, opsonization.

Exercise 13.6:

_____ 1. Increased blood vessel permeability means that fluid and proteins leave the blood, as well as cells such as _____ , _____ , and _____ . This process is called _____ .

_____ 2. These cells are attracted to the injury site by _____ .

_____ 3. The monocytes develop into _____ in the tissues.

_____ 4. Foreign material and damaged tissue are engulfed by _____ and _____ (cell types).

_____ a. This process is called _____ . (see membrane transport, p. 28)

_____ b. Does phagocytosis involve recognition of specific bacterial antigens or is this a nonspecific response?

_____ c. Phagocytosis is easier after opsonization, when foreign material has been coated with antibodies or, as shown, with _____ .

_____ 5. A yellowish accumulation of phagocytes, infecting organisms, and damaged tissue, called _____ , may form.

_____ 6. Specific immunity begins with the time-consuming activation of _____ (cells) after exposure to foreign antigens.

_____ a. If there has been a *previous* exposure to a foreign material, its corresponding _____ may participate in a specific immune response fairly early in inflammation.

_____ b. Would this increase the effectiveness of the immune system?

Answers to Exercise 13.6: 1. neutrophils, monocytes, lymphocytes, diapedesis; 2. chemotaxis; 3. macrophage; 4. neutrophils, macrophage (both tissue macrophage and newly formed); 4a. phagocytosis; 4b. nonspecific response; 4c. complement; 5. pus; 6. lymphocytes; 6a. antibodies; 6b. yes.

G. Phagocytosis and Fever

Color (trace) and label:

1. phagocyte
   a. ○ nucleus
   b. ○ lysosome
   c. ○ membrane
2. ○ foreign material (or cellular debris)

P ○ pyrogens (interleukin-1)

Label:

3. phagosome
4. exocytosis
5. hypothalamus

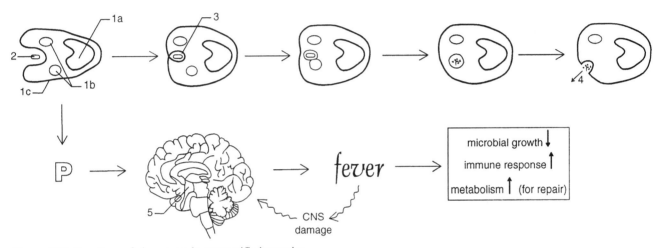

Figure 13.7. Functions of phagocytes in nonspecific immunity.

Exercise 13.7:

_____ 1. To review, the two types of phagocytes are _____ and _____ .

_____ 2. Phagocytosis of foreign material forms phagosomes, which are bounded by _____ .

_____ a. Phagosomes fuse with _____ , which contain _____ . (see cell structure, p. 32)

_____ b. These digestive enzymes _____ . (function)

_____ c. The breakdown products are released to the outside by _____ . (see movement across a membrane, p. 28)

_____ 3. Active macrophages and other agents release chemicals called _____ .

_____ a. Pyrogens can reset the temperature centers of the _____ , called _____ .

_____ b. The beneficial effects of fever include _____ , _____ , and _____ .

_____ c. When fever is too high, malfunctioning of the _____ can occur.

Answers to Exercise 13.7: 1. neutrophils, macrophage; 2. membrane; 2a. lysosomes, potent digestive enzymes; 2b. digest and destroy foreign material and cellular debris; 2c. exocytosis; 3. pyrogens; 3a. hypothalamus, fever; 3b. inhibited microbial growth, increased immune response, increased metabolism for repair; 3c. central nervous system.

H. Complement

Color (trace) and label:
1. ○ mast cell
2. ○ phagocyte
3. ○ target cell (bacterium)
   a. debris
   b. membrane
   c. pore
4. ○ complement
   C* activated complement
   Y ○ antibody
   I ○ histamine

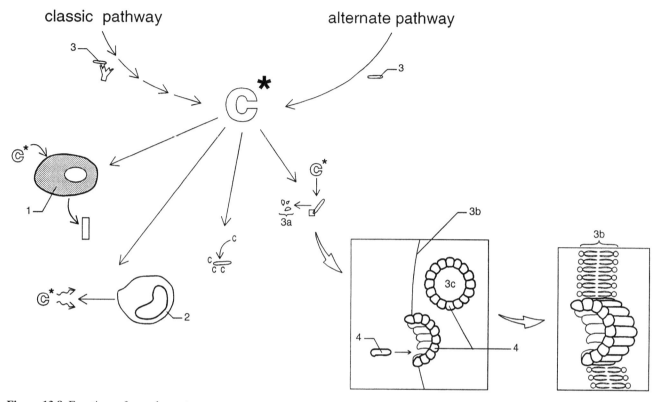

Figure 13.8. Functions of complement.

Exercise 13.8:

_____ 1. Complement is activated by the _____ and/or _____ pathways.

_____ a. The classic pathway's cascade of reactions is triggered by exposure to _____ , a specific immune reaction.

_____ b. The nonspecific alternate pathway is triggered by exposure to the _____ .

_____ c. The pathway most likely to activate complement during early inflammation is the _____ , because _____ .

_____ 2. During inflammation, the functions of complement proteins include _____ , _____ , _____ , and _____ . (see figures 13.3–13.5)

_____ 3. Target cell lysis occurs when complement proteins insert into the target cell's _____ , creating _____ .

_____ 4. How do you expect these pores to damage (and kill) target cells? (see movement across a membrane, p. 28)

Answers to Exercise 13.8: 1. classic, alternate; 1a. antibody-target cell complex; 1b. target cell; 1c. alternate, it is nonspecific (does not depend upon immune system programming); 2. stimulating histamine release, attracting phagocytes, opsonizing (coat bacteria for easier phagocytosis), lysing bacteria; 3. membrane, pores; 4. cellular contents leak out, water enters cell, cell lyses.

I. Immunological Surveillance

Color and label:
1. natural killer cells (NK cells)
   a. ○ nucleus
   b. ○ secretory granules
   c. ○ receptor
2. ○ tumor cell (or viral infected cell)
   a. dead cell
3. ○ tumor specific antigen (or viral antigen)
4. ○ perforin

Label:
5. pores
6. membrane

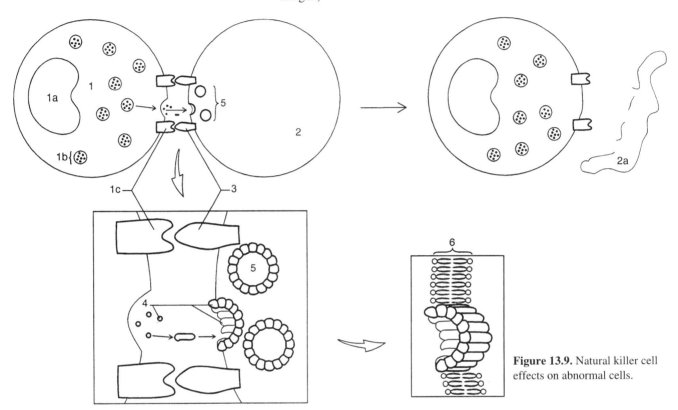

Figure 13.9. Natural killer cell effects on abnormal cells.

Exercise 13.9:

1. In addition to fighting bacteria and other foreign organisms, the immune system must recognize and eliminate abnormal body cells such as _____ and _____ cells.
   a. Tumor cells can be attacked by macrophage and nonspecific lymphocytes called _____ .
   b. Viral infected cells can be attacked by _____ .
   c. This recognition and destruction of abnormal cells is called _____ .
   d. Since NK cell action is nonspecific, would these cells attack viruses during inflammation?

2. Tumor cells and viral infected cells have surface proteins called _____ and _____ that are not present on normal body cells.
   a. NK cells recognize these antigens as _____ (self, nonself).
   b. These antigens react with the _____ on the NK cells.
   c. Do NK cells require *prior* exposure to tumor or viral antigen?

3. NK cells make a protein called _____ .
   a. Secretory granules release perforin into the intercellular space by the process of _____ .
   b. Perforin molecules insert into the _____ of the abnormal cells.
   c. These perforin molecules create _____ .
   d. How do these pores damage the target cells?
   e. This is similar to the action of _____ on the membrane.

Answers to Exercise 13.9: 1. tumor, viral infected; 1a. natural killer cells; 1b. natural killer cells; 1c. immunological surveillance; 1d. yes; 2. tumor specific antigens, viral antigens; 2a. nonself; 2b. receptors; 2c. no; 3. perforin; 3a. exocytosis; 3b. membrane; 3c. pores; 3d. cellular contents leak out, water enters cell, cell lyses; 3e. complement.

J. Interferon

Outline (trace) and label:
1. ○ cell 1
2. ○ cell 2
3. virus
   a. ○ viral genetic material

Color and label:
4. ○ interferon
5. ○ interferon receptor

Label:
6. nucleus

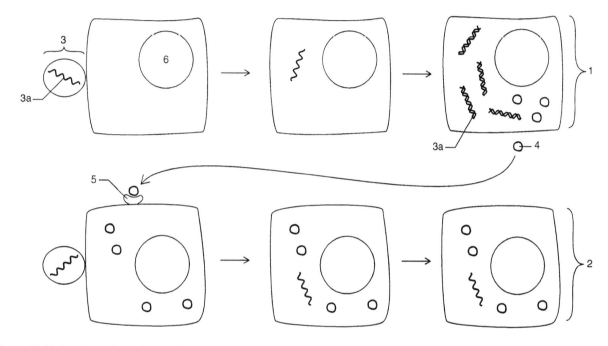

Figure 13.10. Interferon formation and function.

Exercise 13.10:

_____ 1. In order for a virus to replicate, its genetic material must enter a _____ .

_____ 2. A viral infected cell produces a substance called _____ .

_____ 3. Can interferon leave its cell of origin and enter other cells?

_____ 4. When viral genetic material enters cells containing interferon, the viral genetic material _____ (can, cannot) replicate.

_____ 5. Does interferon act by killing cells?

Answers to Exercise 13.10: 1. cell; 2. interferon; 3. yes; 4. cannot; 5. no.

K. Types of Specific Immunity

Color and label:
1. ○ antigen
2. ○ cell
   a. phagocyte
   b. abnormal cell (tumor cell, viral infected cell)
   c. MHC (major histocompatibility) proteins
3. ○ T cell
   a. TCR (T cell receptor)
   b. coreceptor
4. ○ activated B cell (plasma cell)
Y ○ antibody

Label:
a. humoral immunity
b. cellular immunity

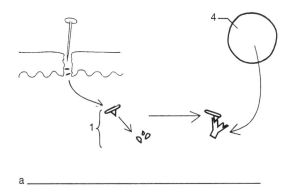

a _____

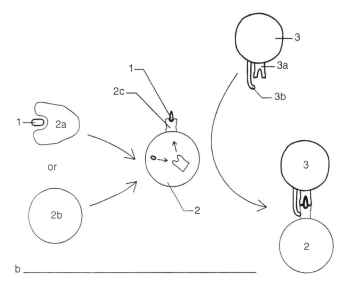

b _____

Figure 13.11. Antigens and specific immune response.

Exercise 13.11:

_____ 1. Antigens are molecules or portions of molecules that stimulate specific immune reactions. Are antigens found inside or outside cells, or both?

_____ 2. Free antigens, found in body fluids, react with specific _____ made by _____ cells.

_____   a. This type of immunity is called _____ .

_____   b. The antibodies found in blood are called _____ . (see components of blood, p. 169)

3. Within cells, ingested foreign organisms, tumor specific proteins, and viral proteins are constantly breaking down into molecular fragments.

_____   a. These molecular fragments (antigens) combine with _____ inside the cell.

_____   b. This antigen-MHC complex incorporates into the cell's _____ , and reacts with specific _____ cells.

_____   c. This type of immunity is called _____ .

_____ 4. The T cell reaction sites include the _____ and _____ (receptors).

Answers to Exercise 13.11: 1. both; 2. antibodies, activated B (plasma cell); 2a. humoral immunity; 2b. gamma globulin; 3a. MHC proteins; 3b. membrane, T; 3c. cellular immunity; 4. T cell receptor, coreceptor.

L. B Cell Differentiation

Outline and label:
1. ○ bone marrow
2. ○ secondary lymphatic tissue

Color and label:
3. ○ B cell
   a. plasma cell
   b. memory

4. ○ receptor
5. ○ self antigen
6. ○ foreign antigen
7. ○ antibodies
8. ○ helper T cell

a. Differentiation.

b. Activation and action.

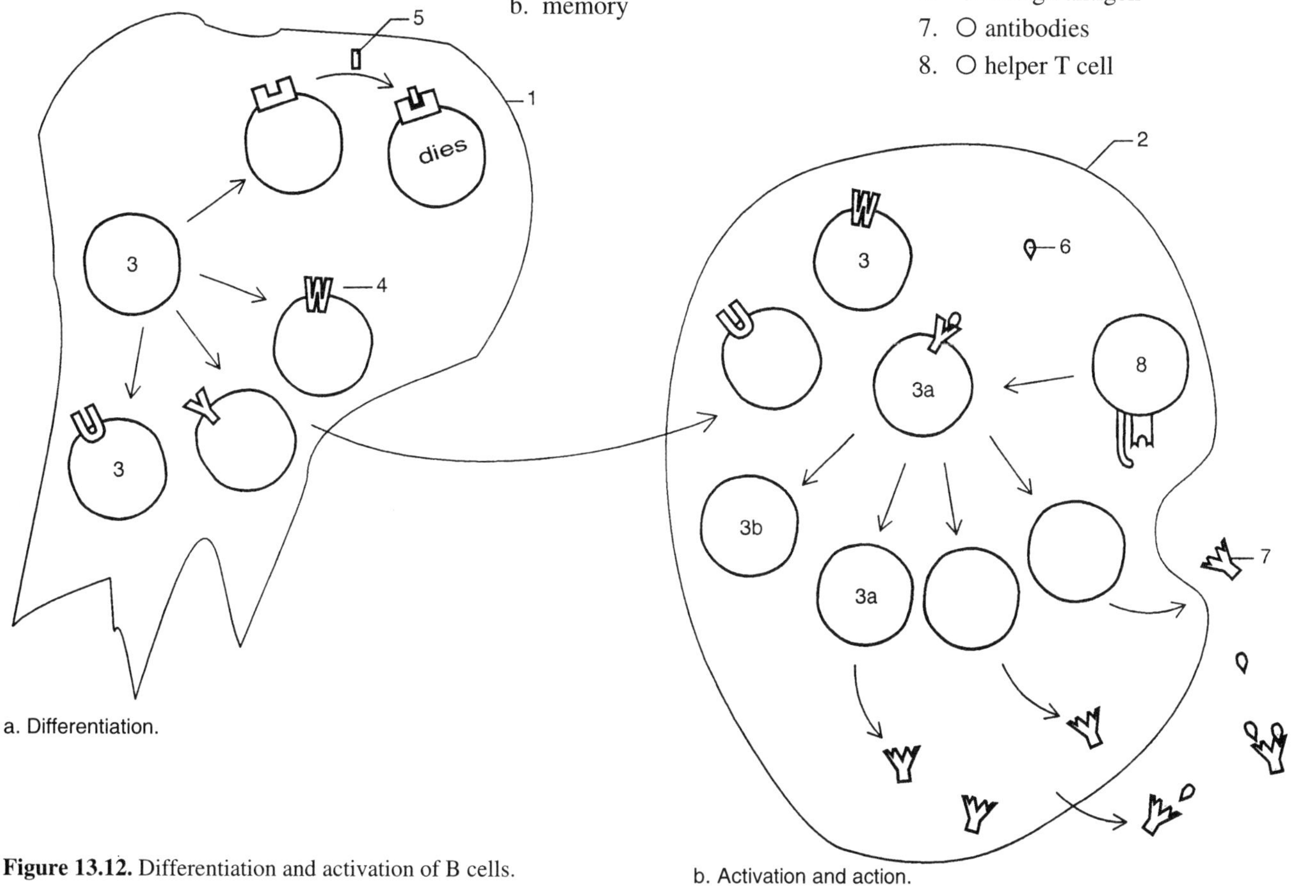

Figure 13.12. Differentiation and activation of B cells.

Exercise 13.12:

1. B cells form and mature in the _____ .
2. Many different B cells form through genetic rearrangement. Each type of B cell has its own unique _____ .
   a. How many different B cell receptors are shown in figure 13.12?
   b. Does B cell differentiation occur before or after exposure to foreign antigens?
   c. Do B cells that recognize self antigen survive?
3. Activation of specific B cells occurs in _____ .
   a. The specific B cell receptor reacts with its complementary _____ .
   b. For some antigens, this process is enhanced by the action of _____ .
4. Activated B cells divide and become _____ and _____ .
   a. Plasma cells make and secrete _____ .
   b. These antibodies react with antigens found in which—cells or body fluids?
   c. Specific memory cells become functional with future exposures to the same antigen. Why is this important?

Answers to Exercise 13.12: 1. bone marrow; 2. receptor; 2a. 4; 2b. before; 2c. no; 3. secondary lymphatic tissue; 3a. antigen; 3b. helper T cells; 4. plasma cells, memory cells; 4a. antibodies; 4b. body fluids; 4c. second immune response is faster (activation of B cells takes time).

M. Antibody Structure

Color and label:
1. ○ constant region
2. ○ variable region

Label:
3. light chain
4. heavy chain
5. binding site

6. IgG, IgD, or IgE
7. IgA
8. IgM

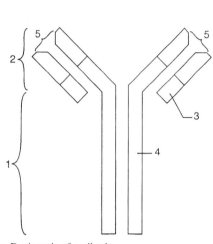

a. Basic unit of antibody.   b. Types of antibodies.

Figure 13.13. Structure and types of immunoglobulins.

Exercise 13.13:

_____ 1. All antibodies have a basic unit of four polypeptides: two _____ and two _____ chains.

_____ 2. Light and heavy chains have _____ and _____ regions.

_____ a. The antigen binding sites are part of the _____ regions of both chains.

_____ b. Each basic unit contains _____ (number) binding site(s).

_____ 3. Different types of antibodies contain _____ , _____ , or _____ (number) basic units.

Answers to Exercise 13.13: 1. light, heavy; 2. variable, constant; 2a. variable; 2b. 2; 3. 1, 2, 5.

N. Major Histocompatibility Complex Molecules

Color and label:
1. major histocompatibility complex
   a. ○ MHC class I genes (A, C, B genes)
   b. ○ MHC class II genes (D genes)
2. MHC proteins (also called HLA—human leukocyte antigens)
   a. ○ MHC class I
   b. ○ MHC class II
3. antigen (epitope)
   a. ○ proteins made inside cell
   b. ○ ingested proteins
4. ○ abnormal (nucleated) body cell
5. ○ antigen presenting cell (APC) (macrophage, mature B cells, helper T cells, dendritic cells)

Label:
6. chromosome 6

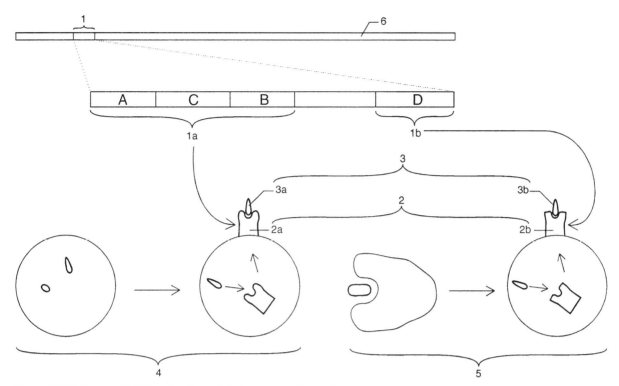

Figure 13.14. Types of MHC molecules and their corresponding cells.

Exercise 13.14:

_____ 1. Antigens may be present in or on cells such as _____ and _____ .

_____ a. Abnormal body cells, such as tumor or viral infected cells, contain antigens that form inside the cell. These antigens react with class _____ MHC proteins.

_____ b. Antigen presenting cells ingest antigens, such as bacterial toxins. These antigens react with class _____ MHC proteins.

_____ 2. These antigen-MHC complexes move to the _____ where they can be recognized by the T cells of the immune system.

_____ 3. MHC proteins are coded for by the _____ on chromosome 6.

_____ a. A, C, and B genes code for class _____ MHC proteins.

_____ b. D genes code for class _____ MHC proteins.

4. Since many different forms of these genes exist, individuals (except for identical twins) have unique sets of class I and II MHC proteins.

_____ a. Therefore, would antigen binding vary among individuals?

_____ b. Would different individuals vary in their ability to respond to disease?

5. Since MHC proteins are unique for each person, MHC proteins immunologically define self.

_____ a. For transplantations, donor tissue (not from an identical twin) contains cells with MHC molecules (HLA) that are _____ (the same as, different from) the recipient's MHC.

_____ b. Could the donor's tissue trigger an immune response?

Answers to Exercise 13.14: 1. abnormal body cells, antigen presenting cells; 1a. I; 1b. II; 2. membrane; 3. major histocompatibility complex; 3a. I; 3b. II; 4a. yes; 4b. yes; 5a. different from; 5b. yes.

Major Histocompatibility Complex Molecules

O. Primary Differentiation of T Cells

Outline and label:
1. ○ thymus

Color and label:
2. ○ T cells
   a. mature CD8
   b. mature CD4
3. ○ TCR (T cell receptor)
4. ○ CD4 coreceptor
5. ○ CD8 coreceptor
6. ○ thymocyte (thymus cell)
   a. ○ class I MHC molecule
   b. ○ class II MHC molecule
7. ○ self antigen

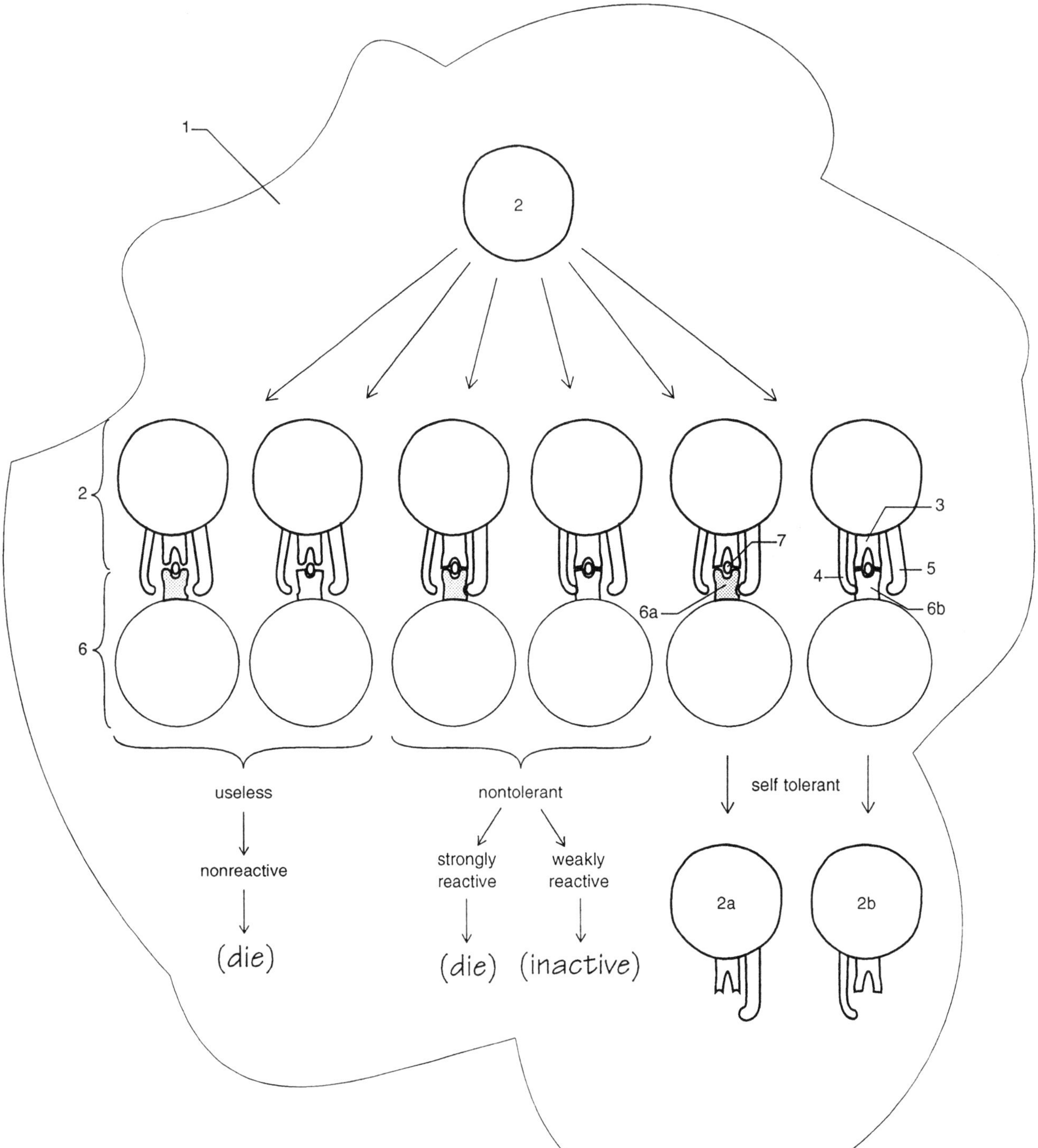

useless → nonreactive → (die)

nontolerant → strongly reactive → (die) / weakly reactive → (inactive)

self tolerant

Figure 13.15. T cell maturation in the thymus.

Exercise 13.15:

1. T cells form in the _____ and mature in the _____ . (see lymphatic organs, p. 201)
    a. Maturation involves the formation of _____ on the T cell membrane.
    b. The receptors that specifically react with different antigen-MHC complexes are the _____ .
    c. The receptors that stabilize these reactions are the _____ .
    d. T cell maturation occurs _____ (before, after) exposure to foreign antigens.

2. Through genetic rearrangement, more than one hundred million T cells form, each with its own unique _____ .
    a. How many different T cell receptors are shown in figure 13.15?
    b. In order for a T cell to function, its TCR must react with the _____ of the antigen presenting cell.
    c. Which T cells in figure 13.15 do not react with self MHC molecules?
    d. What happens to those T cells?

3. Which T cells in figure 13.15 react with self antigens?
    a. These self reactive T cells are _____ (nontolerant, self tolerant, useless).
    b. What happens to these T cells?
    c. If these T cells were to become functional, what could happen to body cells?

4. Which T cells in figure 13.15 have TCRs that do not react with self antigen, but do react with self MHC molecules?
    a. These T cells _____ (die, develop).
    b. These T cells are _____ (nontolerant, self tolerant, useless).
    c. Thus, the receptors that discriminate between self and nonself (self tolerance) are the _____ .

5. When self tolerant T cells react with cells in the thymus, these T cells differentiate into _____ and _____ cells.
    a. The T cells that react with MHC class I retain the _____ coreceptor, lose the _____ coreceptor, and become _____ .
    b. The T cells that react with MHC class II retain the _____ coreceptor, lose the _____ coreceptor, and become _____ .
    c. Therefore, the receptors that help determine the type of T cell are the _____ .

Answers to Exercise 13.15: 1. bone marrow, thymus; 1a. receptors; 1b. TCRs; 1c. coreceptors; 1d. before; 2. receptor; 2a. 3; 2b. MHC; 2c. two T cells on left; 2d. they die; 3. two middle T cells; 3a. nontolerant; 3b. they die (clonal deletion) or become inactive; 3c. they could be destroyed (autoimmune diseases); 4. two T cells on right; 4a. develop (positive selection); 4b. self tolerant; 4c. TCRs; 5. CD4, CD8, 5a. CD8, CD4, mature CD8 T cells; 5b. CD4, CD8, mature CD4 T cells; 5c. coreceptors.

P. Secondary Differentiation of T Cells

Color and label:
1. ○ secondary lymphatic tissue
2. ○ TCR
3. ○ CD8 T cell
   a. CD8 coreceptor
   b. T_c cytotoxic (killer) T cell
   c. T_s suppressor cell
4. ○ CD4 T cell
   a. CD4 coreceptor
   b. T_d mediate delayed type hypersensitivity cell
   c. T_H helper cell
5. ○ abnormal cell
   a. MHC class I
6. ○ APC (antigen presenting cells)
   a. MHC class II
7. ○ foreign antigen

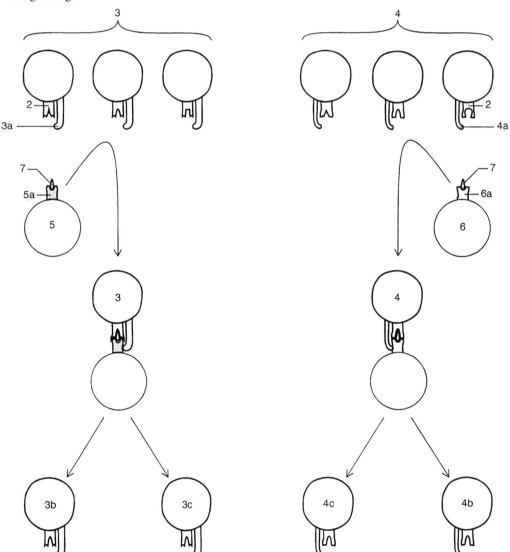

Figure 13.16. Activation of T cells in secondary lymphatic tissue.

Exercise 13.16:

_____ 1. Mature CD4 and CD8 T cells migrate from the _____ to _____ , where they may become activated.

_____ 2. Activation of a specific T cell depends upon reaction of its TCR with its complementary _____ .

_____   a. CD8 cells react with specific antigens displayed on _____ cells.

_____   b. CD4 cells react with specific antigens displayed on _____ cells.

   3. Activation triggers differentiation of

_____   a. CD8 cells into _____ and _____ cells, and

_____   b. CD4 cells into _____ and _____ cells.

Answers to Exercise 13.16: 1. thymus, secondary lymphatic tissue; 2. antigen; 2a. abnormal; 2b. antigen presenting cells; 3a. killer, suppressor; 3b. mediate delayed type hypersensitivity, helper.

Q. T Cell Function

Color and label:
1. ○ TCR
2. ○ CD8 T cell
   a. T_c cytotoxic (killer) T cell
   b. memory cell
   c. T_s suppressor cell
3. ○ CD4 T cell
   a. T_d mediate delayed type hypersensitivity cell
   b. memory cell
   c. T_4 helper cell
4. ○ abnormal cell
5. ○ APC (antigen presenting cells)
6. ○ foreign antigen
7. ○ perforin
8. ○ interleukins
9. ○ lymphokines

Q. T Cell Function Continued

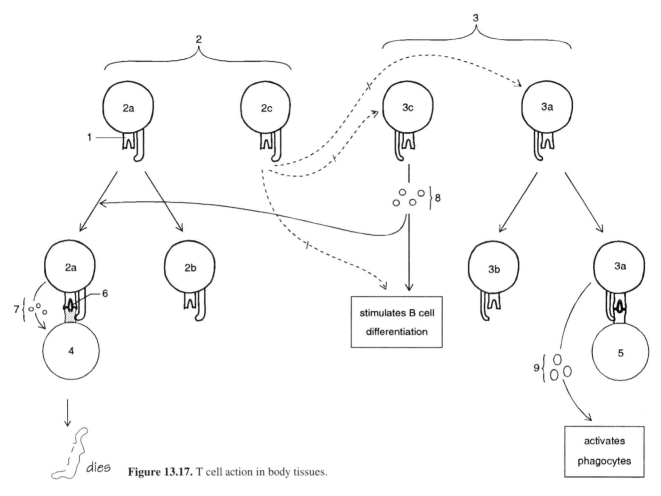

Figure 13.17. T cell action in body tissues.

Exercise 13.17:

_____ 1. Killer cells react with _____ cells and release the chemical _____ .

_____ a. Perforin causes abnormal cells to _____ . (see immunosurveillance, p. 212)

_____ b. Killer cells form memory cells that can become functional with future exposures to the same antigen. Why is this important?

_____ 2. T_D cells can react with _____ cells and release _____ .

_____ a. Lymphokines serve to _____ .

_____ b. Does T_D cell enhancement of nonspecific immunity occur immediately after antigen exposure?

_____ c. Will subsequent exposures to the same antigen result in a faster immune response? Why or why not?

_____ 3. The T cells that regulate the action of other lymphocytes are _____ and _____ T cells.

_____ a. Helper T cells release the chemical messengers _____ , which stimulate differentiation of _____ and _____ .

_____ b. Suppressor T cells block activation of _____ , _____ , and _____ .

_____ c. Suppressor cells develop after the other immune cells. Why is this timing important?

Answers to Exercise 13.17: 1. abnormal, perforin; 1a. die (by pore formation); 1b. second immune reaction occurs faster; 2. APC, lymphokines; 2a. activate phagocytes; 2b. no (differentiation of T_D cells takes time); 2c. yes, T_D cells can become memory cells; 3. helper, suppressor; 3a. interleukins, killer, B cells; 3b. T_H, T_D, B cells; 3c. way of turning off immune response.

Chapter 14 Respiratory System

A. Anatomy of the Respiratory System—Airways

Color and label:
1. ○ conchae
2. ○ palate
3. ○ epiglottis
4. ○ larynx
   a. glottis
5. ○ trachea
   a. cartilage
6. ○ primary bronchi
7. ○ lobar bronchus
8. ○ bronchiole
9. ○ alveolus

Label:
10. nasal cavity
11. oral cavity
12. pharynx
13. esophagus
14. pleurae

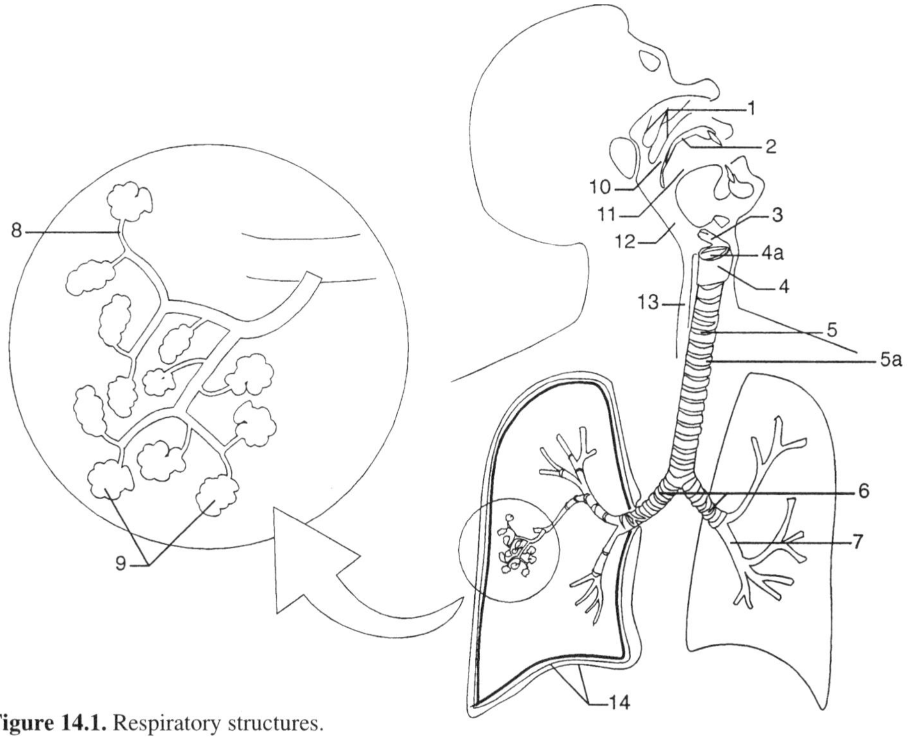

Figure 14.1. Respiratory structures.

Exercise 14.1:

_____ 1. In the nasal cavity, air passes over the _____ .

_____ 2. The nasal cavity is separated from the oral cavity by the _____ .

_____ 3. The nasal cavity and the oral cavity join to form the _____ .

_____ 4. The inferior openings of the pharynx open into the _____ and the _____ .

_____ a. The opening from the pharynx into the respiratory passages is the _____ .

_____ b. The glottis can be covered by the _____ .

_____ 5. The trachea is held open by rings of _____ .

_____ 6. The trachea divides into two _____ _____ .

_____ 7. After several branchings, the bronchi lead into _____ .

_____ 8. The bronchioles lead into the structures where gas is exchanged, called _____ .

Answers to Exercise 14.1: 1. conchae; 2. palate; 3. pharynx; 4. esophagus, larynx; 4a. glottis; 4b. epiglottis; 5. cartilage; 6. primary bronchi; 7. bronchioles; 8. alveoli.

B. Histology

Color and label:
1. ○ pseudostratified columnar epithelium
2. ○ goblet cell
3. ○ seromucus gland (cells)
4. ○ blood vessel
5. ○ ciliated columnar epithelium
6. ○ elastic connective tissue
7. ○ smooth muscle
8. ○ squamous (type I) cell
9. ○ type II cell
10. ○ red blood cell (red)
11. ○ white blood cell

Label:
12. cilia
13. capillary

Exercise 14.2:

_____ 1. Mucus and watery secretions are produced by _____ cells and _____ glands.

_____ 2. These fluids, along with foreign particles, are normally moved out of the air passages by the organized beating of the _____ .

_____ 3. The walls of bronchioles contain _____ muscle and _____ connective tissue.

_____ 4. The mouth of the alveolus is shaped by a ring of _____ tissue.

_____ 5. This smooth muscle and elastic connective tissue help the respiratory passages change shape with the ebb and flow of _____ .

_____ 6. The epithelial cells (type I) of the alveolus are _____ (tissue type).

_____ a. Type I cells are closely connected to the endothelium of _____ .

_____ b. Type I cells provide for _____ exchange with the blood.

_____ 7. Surfactant and fluid are produced by the cuboidal type _____ cells in the walls of the alveoli.

_____ a. The surfactant prevents the surface tension of the water from causing the collapse of the _____ .

_____ b. The fluid allows the diffusion of _____ across the respiratory membrane.

Answers to Exercise 14.2: 1. goblet, seromucus; 2. cilia; 3. smooth, elastic; 4. connective; 5. ventilation; 6. simple squamous; 6a. capillaries; 6b. gas; 7. II; 7a. alveoli; 7b. gases.

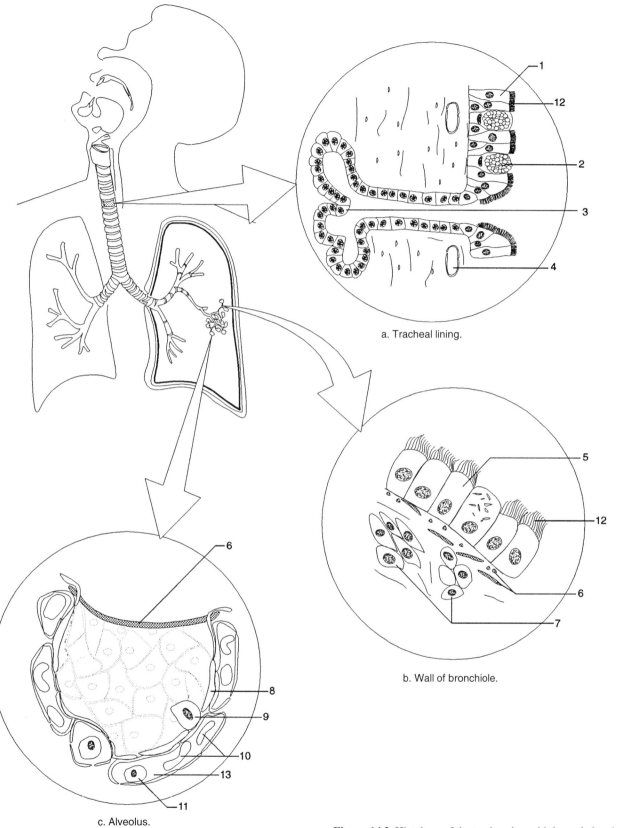

a. Tracheal lining.

b. Wall of bronchiole.

c. Alveolus.

Figure 14.2. Histology of the trachea, bronchiole, and alveolus.

C. Thoracic Cavity

Color (trace) and label:
1. ○ ribs
2. ○ sternum
3. ○ vertebrae
4. ○ intercostal muscles
5. ○ diaphragm
   a. central tendon
6. ○ heart
7. ○ lungs
8. ○ visceral pleura
9. ○ parietal pleura
10. ○ pleural space
11. ○ trachea
12. ○ bronchus

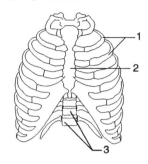

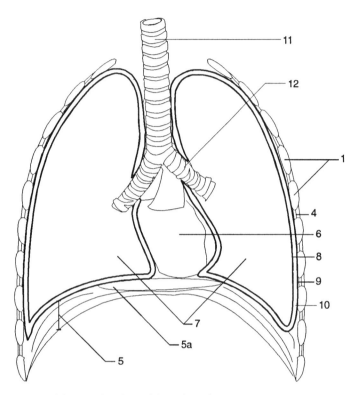

Figure 14.3. Anterior view of thoracic cavity.

Exercise 14.3:

_____ 1. The bones surrounding the thoracic cavity are the _____ , _____ , and _____ .

_____ 2. The thoracic cavity is separated from the abdominal cavity by a muscle called the _____ .

_____ 3. The major organs found in the thoracic cavity are the _____ and _____ .

_____ 4. The outer surface of the lungs is covered by a membrane called the _____ .

_____ 5. The inner surface of the thoracic cavity is covered by a similar membrane called the _____ .

_____ 6. These serous membranes produce a watery fluid that allows the lungs to _____ within the thoracic cavity.

_____ 7. The (potential) space between the pleura is the _____ .

_____ 8. Respiratory muscles found between the ribs are the _____ .

Answers to Exercise 14.3: 1. ribs, sternum, vertebrae, 2. diaphragm, 3. heart, lungs; 4. visceral pleura; 5. parietal pleura; 6. move; 7. pleural space; 8. intercostals.

D. Ventilation

Color (trace) and label:
1. ○ external intercostal muscles
2. ○ internal intercostal muscles
3. ○ diaphragm (muscle)
   a. central tendon
4. ○ sternum

Label:
a. inspiration
b. expiration

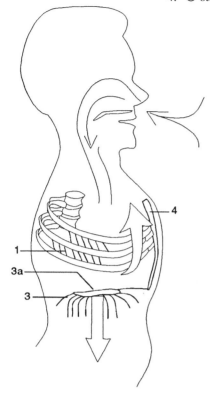
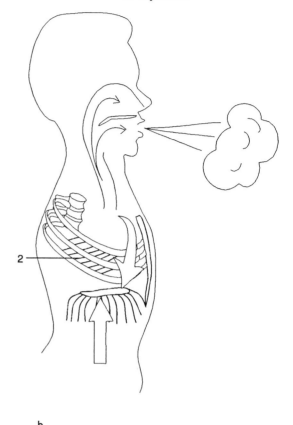

a _____   b _____

Figure 14.4. Changes in the thoracic cavity during ventilation.

Exercise 14.4:

_____ 1. When the diaphragm contracts, its central tendon is pulled inferiorly, causing the volume of the thoracic cavity to _____ (increase, decrease).

_____ 2. When the external intercostal muscles contract, the ribs pivot in a _____ (superior, inferior) direction. This causes the anterior-posterior dimension of the thorax to _____ (increase, decrease).

_____ 3. These two changes cause the total volume of the thoracic cavity to _____ (increase, decrease).

_____ 4. Since the volume of the thoracic cavity has increased, the pressure within it has _____ (increased, decreased).

_____ 5. If the pressure in the thorax falls below atmospheric pressure, then air moves _____ (into, out of) the lungs.

_____ 6. When the diaphragm relaxes, the abdominal pressure causes the central tendon to move back toward the thorax, and the thoracic pressure _____ (increases, decreases).

_____ 7. When the external intercostal muscles relax, the ribs fall back to their resting position, causing the anterior-posterior dimension of the thorax to _____ (increase, decrease).

_____ 8. These changes _____ (increase, decrease) the pressure on the air, causing it to move _____ (into, out of) the lungs.

Answers to Exercise 14.4: 1. increase; 2. superior, increase; 3. increase; 4. decreased; 5. into; 6. increases; 7. decrease; 8. increase, out of.

E. Lung Volumes

Color and label:
1. ○ tidal volume
2. ○ inspiratory reserve volume
3. ○ expiratory reserve volume
4. ○ vital capacity
5. ○ residual volume
○ FRC (functional residual capacity)

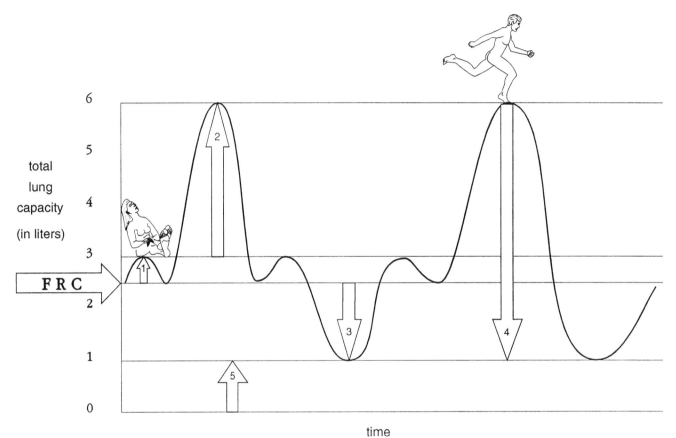

Figure 14.5. Lung volumes and capacities.

Exercise 14.5:

_____ 1. After exhaling, a person at rest has a lung volume called _____ .

_____ 2. Since this lung volume occurs from passive relaxation of muscles, does the body have to expend metabolic energy to arrive at it?

_____ 3. The volume of the resting inhalation described in exercise 14.4 is called _____ .

_____ 4. By using extra muscular force, the lungs can be enlarged to a greater volume than tidal. This reserve is called _____ volume.

_____ 5. When the lungs are at FRC, extra muscular force can cause lung volume to decrease expelling additional air. This reserve is called _____ volume.

_____ 6. The total volume change that the lung can accomplish is called the _____ .

_____ 7. When the thorax has forcefully expelled as much air as possible, the lungs are not completely empty. Their volume is called the _____ .

Answers to Exercise 14.5: 1. functional residual capacity (FRC), 2. no; 3. tidal volume; 4. inspiratory reserve; 5. expiratory reserve; 6. vital capacity; 7. residual volume.

F. Gas Pressures

Color and label:
1. partial pressures
   - ○ P O$_2$ (partial pressure of oxygen)
   - ○ P CO$_2$ (partial pressure of carbon dioxide)
   - ○ P N$_2$, H$_2$O (partial pressure of other gases)
2. ○ P = 760 (total atmospheric pressure)
3. ○ respiratory tract
4. ○ respiratory membranes
5. ○ pulmonary arteries (blue)
6. ○ capillaries (purple)
7. ○ pulmonary veins (red)
8. ○ systemic arteries (red)
9. ○ systemic tissues
10. ○ systemic veins (blue)

Label:
11. alveoli

Exercise 14.6:

1. The sum of the partial pressures of all gases in the atmosphere makes up the _____ . At sea level this is _____ mmHg.

2. Since inhaled air mixes with the air in the air passages,
   a. the oxygen content of inhaled air is _____ (higher, lower) than that of the atmosphere.
   b. the carbon dioxide content of inhaled air is _____ (higher, lower) than that of the atmosphere.

3. Since the air passes over moist surfaces in the respiratory passages, it usually becomes saturated with _____ .
   a. This water does not increase the total gas pressure, but does it contribute its own partial pressure?
   b. Therefore, it causes the partial pressure of oxygen to _____ (increase, decrease).

4. The partial pressure of oxygen in the air entering the alveoli is normally _____ mmHg.
   a. The partial pressure of oxygen in the pulmonary arteries is normally _____ mmHg.
   b. This means that the net flow of oxygen would be _____ (into, out of) the blood.
   c. The partial pressure of oxygen in the pulmonary veins is normally _____ mmHg.
   d. Therefore, would the net flow of oxygen continue if the blood remained in the lungs for a longer period of time?

5. In what part of the body does the blood normally lose oxygen?

6. The difference in the partial pressures of carbon dioxide in the systemic arteries and systemic veins indicates that carbon dioxide _____ (enters, leaves) the blood at the systemic tissues.

7. The difference in the partial pressures of carbon dioxide in the pulmonary arteries and pulmonary veins indicates that carbon dioxide _____ (enters, leaves) the blood at the respiratory membrane.

Answers to Exercise 14.6: 1. total atmospheric pressure, 760; 2a. lower; 2b. higher; 3. water; 3a. yes; 3b. decrease; 4. 104; 4a. 40; 4b. into; 4c. 104; 4d. no; 5. systemic tissues; 6. enters; 7. leaves.

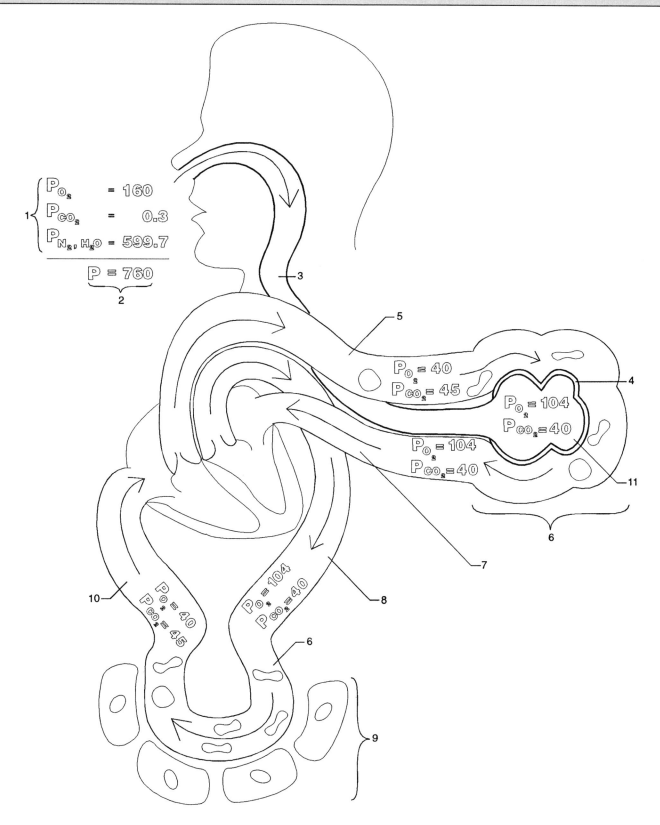

Figure 14.6. Partial pressures at sea level.

G. Oxygen Transport and Hemoglobin

Color and label:
- ○ Hb (hemoglobin)
- ○ O₂ (oxygen)
1. ○ systemic tissue
2. ○ lungs

Trace:
3. ○ oxygen dissociation curve
4. ○ oxygen dissociation curve in metabolically active tissue (dotted line)

Exercise 14.7:

_____ 1. Hemoglobin is located in the _____ .

_____ 2. A single hemoglobin molecule can carry a maximum of _____ oxygen molecules.

_____   a. Oxygen enters the blood in the _____ of the lungs.

_____   b. Oxygen is released from hemoglobin in the _____ .

_____ 3. Oxygen partial pressures between 60 and 100 mmHg occur in the _____ .

_____   a. If the oxygen partial pressure is over 100 mmHg, what is the percent oxygen saturation of hemoglobin?

_____   b. If the oxygen partial pressure falls to 60 mmHg, what is the percent oxygen saturation of hemoglobin?

_____ 4. The oxygen partial pressure of systemic tissues is usually _____ (the same as, lower than) that of the lungs.

_____   a. This lower oxygen partial pressure causes hemoglobin to _____ (release, take up) oxygen.

_____   b. At partial pressures typical of tissues, the slope of the curve is _____ (steeper, shallower) than in the lungs.

_____   c. The percent oxygen saturation of hemoglobin ranges from _____ to _____ .

_____   d. This means that, compared to the lungs, the oxygen saturation of hemoglobin in typical tissues is _____ (more, less) sensitive to partial pressure changes.

_____ 5. Heat and acidic metabolic by-products are produced by _____ (inactive, highly active) tissues.

_____   a. The physiological changes in active tissues shift the curve (dotted line) to the _____ (right, left).

_____   b. The percent oxygen saturation of hemoglobin ranges from _____ to _____ .

_____   c. This shows that hemoglobin gives up _____ (more, less) oxygen to active tissue than to inactive tissues.

Answers to Exercise 14.7: 1. blood (erythrocytes); 2. four; 2a. alveoli; 2b. systemic tissues; 3. lungs; 3a. 100%; 3b. 90%; 4. lower than; 4a. release; 4b. steeper; 4c. 82%, 54%; 4d. more; 5. highly active; 5a. right; 5b. 78%, 33%; 5c. more.

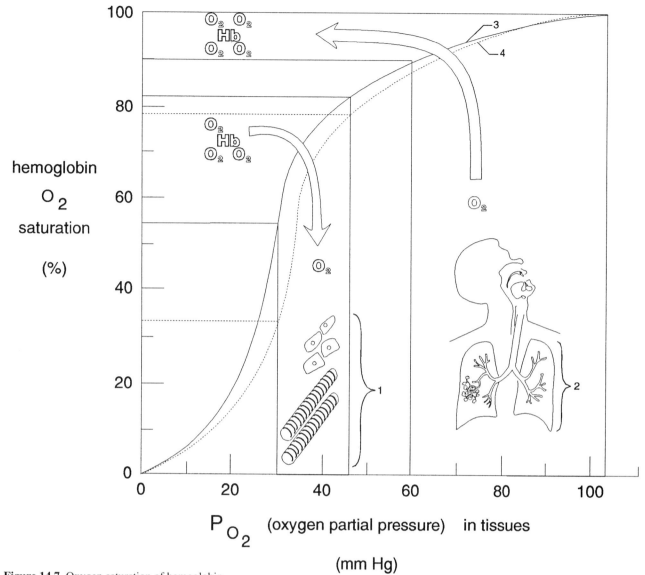

Figure 14.7. Oxygen saturation of hemoglobin.

H. Carbon Dioxide Transport

Color:
- ○ CO_2
- ○ H_2O
- ○ H_2CO_3
- ○ H^+
- ○ HCO_3^-
- ○ Hb (hemoglobin)
- ○ Cl^-
- ○ plasma protein
- ○ AIR

Color and label:
1. ○ capillary wall
   a. plasma
2. ○ cell
3. ○ alveolar wall

Label:
4. red blood cell
5. interstitial fluid

Exercise 14.8:

_____ 1. Cells produce carbon dioxide as a waste product of _____ .

_____ 2. Carbon dioxide diffuses into the _____ fluid, and then into the _____ .

_____ 3. From the plasma, a large amount of carbon dioxide diffuses into the _____ .

_____   a. Most of the carbon dioxide in the red blood cells is converted into _____ by the enzyme _____ .

_____   b. This immediately splits into _____ and _____ .

_____   c. Some of the carbon dioxide in the red blood cells attaches to the amino acids of _____ .

_____ 4. Hemoglobin also absorbs the _____ formed along with bicarbonate.

_____ 5. Most of the bicarbonate formed in the red blood cell diffuses into the _____ .

_____ 6. Since bicarbonate is a cation, the electrical balance is maintained in the system by the diffusion (shift) of _____ into the red blood cell.

_____ 7. In the lung, _____ diffuses out of the plasma, causing a change in chemical equilibrium.

_____   a. This results in bicarbonate ions being converted back to _____ .

_____   b. This equilibrium change also causes hemoglobin to give up _____ and _____ .

Answers to Exercise 14.8: 1. metabolism; 2. interstitial, plasma; 3. red blood cells; 3a. H_2CO_3 (carbonic acid), carbonic anhydrase; 3b. H^+ (proton), HCO_3^- (bicarbonate); 3c. hemoglobin; 4. H^+; 5. plasma; 6. chloride; 7. carbon dioxide; 7a. carbon dioxide; 7b. carbon dioxide, H^+.

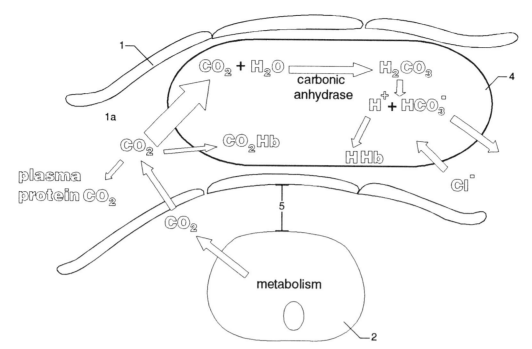

Figure 14.8a. CO_2 exchange at the systemic tissue.

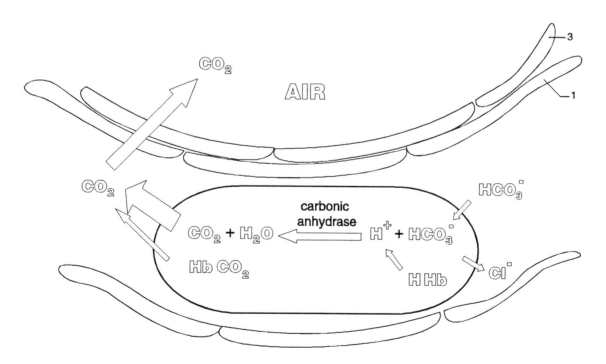

Figure 14.8b. CO_2 exchange at the respiratory membrane.

Carbon Dioxide Transport

I. Respiratory Feedback

Color and label:
1. ○ respiratory centers in pons
2. ○ respiratory centers in medulla

Color:
○ arrows to respiratory centers

Label:
3. cerebral cortex (voluntary control)
4. central chemoreceptors (CO_2, H^+)
5. carotid chemoreceptors (CO_2, H^+, O_2)
6. aortic chemoreceptors (CO_2, H^+, O_2)
7. irritant and stretch receptors (airways and alveoli)
8. muscle and tendon stretch and proprioceptors

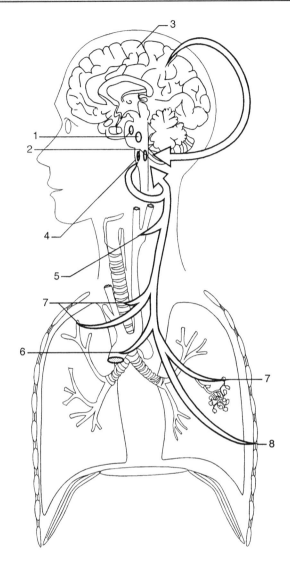

Figure 14.9. Input into respiratory centers.

Exercise 14.9:

_____ 1. The central nervous system centers for control of respiration are in the _____ and _____ .

_____ 2. Behavioral actions under voluntary control are initiated in the _____ of the brain.

_____ 3. Inhalation of smoke or ammonia would stimulate _____ receptors in the _____ .

_____ 4. Respiratory centers react to stretch receptors found in _____ , _____ , _____ , and _____ .

_____ 5. Chemoreceptors would respond primarily to _____ and _____ .

_____ 6. This array of feedback would primarily affect the respiratory centers in the _____ .

Answers to Exercise 14.9: 1. pons, medulla; 2. cerebral cortex; 3. irritant, air passages; 4. muscle, tendon, airways, lungs; 5. carbon dioxide, pH; 6. medulla.

J. Regulation of Respiratory Output

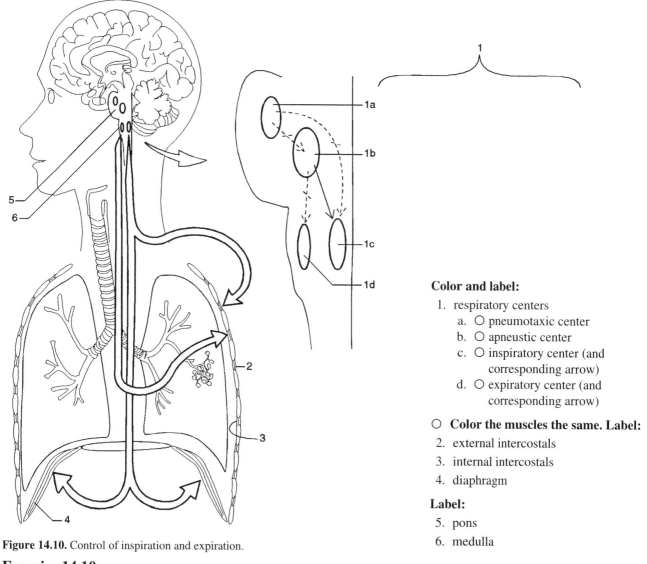

Color and label:
1. respiratory centers
   a. ○ pneumotaxic center
   b. ○ apneustic center
   c. ○ inspiratory center (and corresponding arrow)
   d. ○ expiratory center (and corresponding arrow)

○ **Color the muscles the same. Label:**
2. external intercostals
3. internal intercostals
4. diaphragm

Label:
5. pons
6. medulla

Figure 14.10. Control of inspiration and expiration.

Exercise 14.10:

1. The inspiratory and expiratory centers are located in the _____ .
   a. The inspiratory center stimulates the _____ and _____ to contract.
   b. This causes the volume of the thorax to _____ (increase, decrease). (see ventilation, p. 227)

2. At the end of a tidal inspiration the chest wall and lungs passively return to the lung volume known as _____ . (see lung volumes, p. 228)

3. During more forceful expiration, the expiratory center stimulates the _____ intercostals.

4. The pneumotaxic and apneustic centers are located in the _____ .
   a. The apneustic center stimulates the _____ center.
   b. Therefore, the apneustic center tends to promote _____ .
   c. The pneumotaxic center inhibits the _____ and _____ centers.
   d. Therefore, the pneumotaxic center tends to terminate _____ .

Answers to Exercise 14.10: 1. medulla; 1a. diaphragm, external intercostal muscles; 1b. increase; 2. functional residual capacity (FRC); 3. internal; 4. pons; 4a. inspiratory; 4b. inspiration; 4c. apneustic, inspiratory; 4d. inspiration.

Chapter 15 Digestive System

A. Function of the Digestive System

Color and label:
1. ○ carbohydrates
2. ○ proteins
3. ○ lipids
4. ○ blood (red)
5. ○ body tissues

Label:
6. mouth
7. esophagus
8. stomach
9. small intestine
10. large intestine
11. liver
12. pancreas

Label the processes:
13. digestion
14. absorption

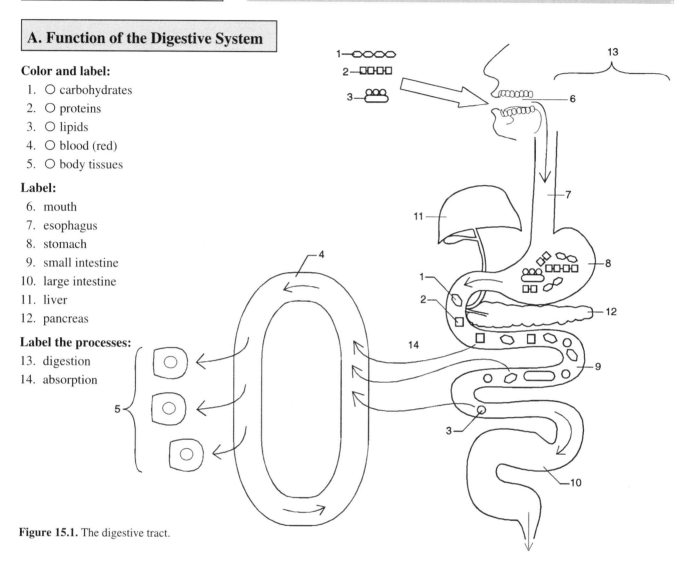

Figure 15.1. The digestive tract.

Exercise 15.1:

_____ 1. The process of breaking large organic molecules into small, soluble components is called _____ .

_____ 2. Does this process occur entirely within the passages of the digestive tract?

_____ 3. These soluble components then pass through the walls of the gut in the process of _____ .

_____ 4. After being absorbed, the nutrients usually are carried by the _____ to the _____ of the body.

_____ 5. If a substance that has been eaten cannot be digested and/or absorbed, will it enter the blood and be carried to body tissues? This substance will instead be _____ .

Answers to Exercise 15.1: 1. digestion; 2. yes; 3. absorption; 4. blood, tissues; 5. no, eliminated.

B. The Mouth

Color:
- ○ water
- ○ mucus
- ○ salivary amylase
- ○ electrolytes

Color and label:
1. ○ salivary glands
    a. parotid gland
    b. sublingual gland
    c. submandibular gland
    d. salivary gland duct
2. ○ tongue
3. ○ tooth
4. ○ enamel
5. ○ dentin
6. ○ dental pulp
7. ○ cementum
8. ○ periodontal membrane

Label:
9. root canal
10. nerves
11. blood and lymph vessels
12. alveolus (socket)
13. gingiva (gum)
14. crown

Exercise 15.2:

_____ 1. When food enters the mouth, it is made wet and slippery by the _____ and _____ in saliva.

_____ 2. Its initial chemical digestion begins with the enzyme _____ .

_____ 3. The food is mechanically divided and mixed by the _____ and _____ .

_____ 4. Each tooth is a living structure with a hard outer surface of _____ and a pulp cavity containing _____ and _____ .

Answers to Exercise 15.2: 1. water, mucus; 2. salivary amylase; 3. teeth, tongue; 4. enamel, nerves, blood vessels.

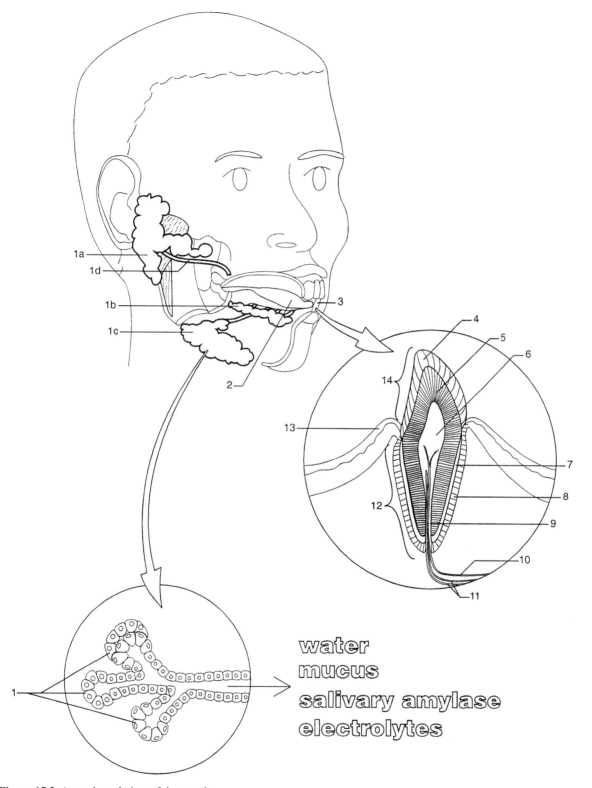

Figure 15.2. Anterolateral view of the mouth.

The Mouth

C. Microscopic Anatomy of the Alimentary Canal

Color and label:
1. ○ mucus cell
2. ○ chief cell
3. ○ parietal cell
4. ○ mucosa
5. ○ submucosa
6. ○ circular muscle
7. ○ longitudinal muscle
8. ○ serosa
9. ○ epithelial cell
10. ○ goblet cell
11. ○ capillary (red)
12. ○ lacteal
13. ○ lamina propria
14. ○ muscularis mucosa

Label:
15. villus
16. crypt of Lieberkühn
17. brush border
18. gastric pit

Exercise 15.3:

_____ 1. The gastric glands are in openings called _____ in the wall of the stomach.

_____ a. Cells in the necks of gastric glands produce _____ .

_____ b. Chief cells in gastric glands produce the inactive form of the enzyme _____ .

_____ c. This enzyme is activated by the _____ produced by the parietal (oxyntic) cells.

_____ d. The gastric epithelium protects the stomach lining by producing large amounts of _____ .

_____ 2. The entire gut wall, as shown here in the small intestine, contains at least two layers of _____ muscle.

_____ 3. The lining of the small intestine contains many projections called _____ .

_____ a. These villi are covered with _____ epithelial cells.

_____ b. The interior of each villus contains tufts of _____ and a _____ .

_____ c. The mucosal layer also contains _____ tissue and _____ muscle.

_____ 4. A watery fluid is produced at the bases of the villi by intestinal glands called _____ .

_____ 5. Lysozyme (an antibacterial enzyme) is produced by _____ cells at the bases of these crypts.

_____ 6. The epithelium of the villi have _____ on their apical surfaces.

_____ 7. These microvilli greatly increase the _____ of the surface.

Answers to Exercise 15.3: 1. gastric pits; 1a. mucus; 1b. pepsin; 1c. HCl; 1d. mucus; 2. smooth; 3. villi; 3a. columnar; 3b. capillaries, lacteal; 3c. connective, smooth; 4. crypts of Lieberkühn; 5. Paneth's; 6. brush border; 7. area.

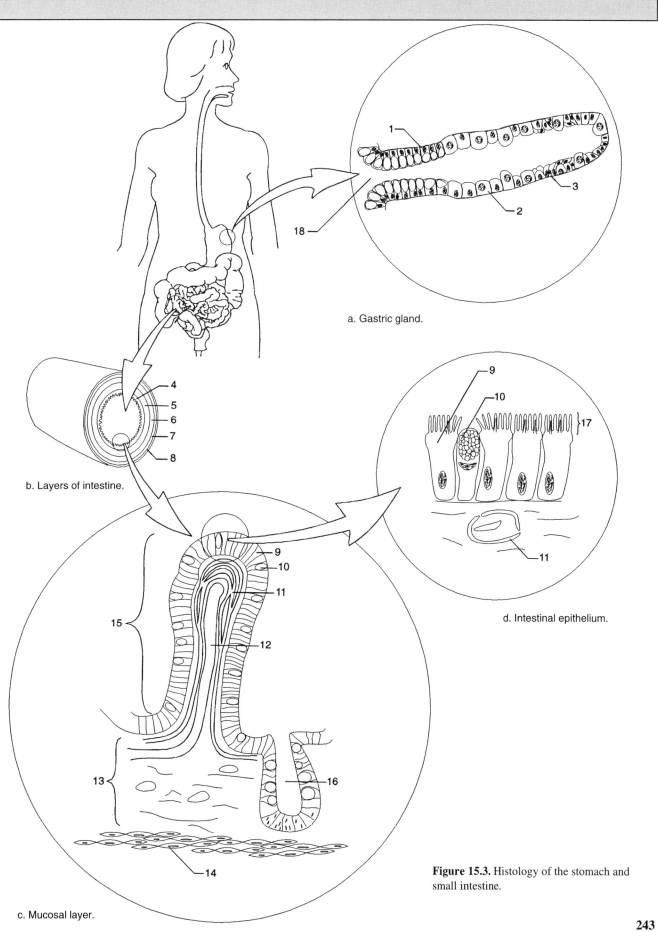

a. Gastric gland.

b. Layers of intestine.

c. Mucosal layer.

d. Intestinal epithelium.

Figure 15.3. Histology of the stomach and small intestine.

243

D. Movements of the Digestive System

Color and label:

1. ○ food bolus
2. ○ epiglottis
3. ○ muscle
   a. longitudinal
   b. circular

Label:

4. tongue
5. larynx
6. hard palate
7. soft palate
8. pharynx
9. esophagus
10. intestine
11. contraction
12. relaxation

Exercise 15.4:

_____ 1. Are the breathing passages open during the oral stage of food consumption? Are they open during the pharyngeal phase?

_____ 2. When food is in the pharynx, the epiglottis covers the _____ .

_____ 3. During the late pharyngeal stage, the muscles of the _____ relax to receive the bolus.

_____ 4. The bolus in the smooth muscle segment of the esophagus causes peristalsis, a reflex in which the muscle around the bolus _____ (contracts, relaxes) and the muscle in front of it _____ (contracts, relaxes).

_____ 5. Segmentation in the intestine, a mixing action, is the result of alternating parts of the circular muscle _____ and then _____ .

Answers to Exercise 15.4: 1. yes, no; 2. larynx; 3. esophagus; 4. contracts, relaxes; 5. contracting, relaxing.

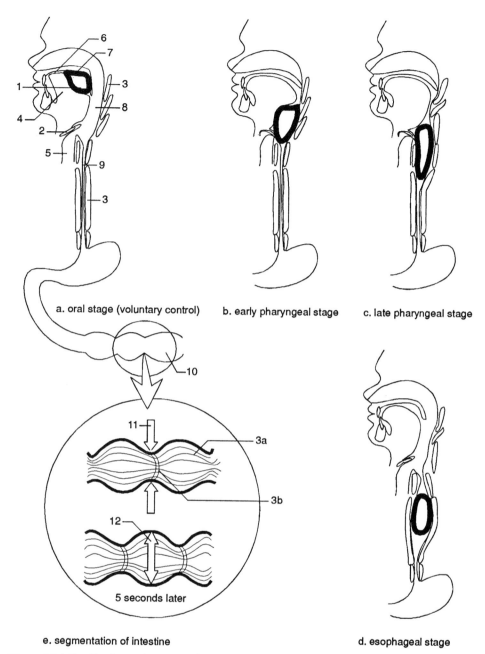

Figure 15.4. Swallowing and segmentation movements.

E. Liver and Pancreas

Color and label:
1. ○ liver
   a. ○ hepatic ducts (green)
2. ○ gallbladder (green)
3. ○ cystic duct (green)
4. ○ common bile duct (green)
5. ○ pancreas
   a. ○ pancreatic duct
6. ○ portal vein (blue)
7. ○ hepatic vein (blue)
8. ○ inferior vena cava (blue)
9. ○ aorta (red)
10. ○ hepatic artery (red)

Label:
11. Oddi's sphincter
12. duodenum

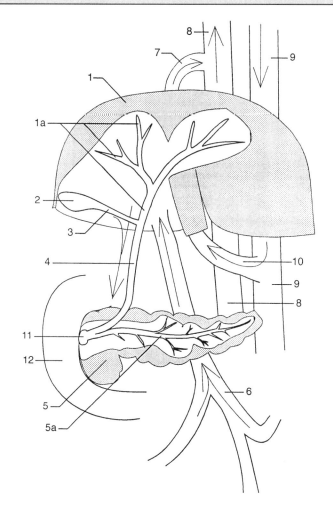

Figure 15.5. Liver, pancreas, and vessels.

Exercise 15.5:

1. The liver receives blood from two sources, the _____ and _____ .
   a. The hepatic artery carries blood from the _____ .
   b. The portal vein carries blood from the _____ . (see hepatic portal system, p. 197)
2. Blood leaves the liver through the _____ and enters the _____ .
3. Bile leaves the liver through the _____ .
4. This bile may be carried directly to the duodenum through the _____ or it may be stored in the _____ .
5. The pancreatic duct joins the common bile duct at _____ .

Answers to Exercise 15.5: 1. hepatic artery, hepatic portal vein; 1a. aorta; 1b. digestive tract; 2. hepatic vein, inferior vena cava; 3. hepatic duct; 4. common bile duct, gallbladder; 5. Oddi's sphincter.

F. Carbohydrates

Color and label:
1. ○ glucose
2. ○ fructose
3. ○ galactose

Label:
4. maltose
5. sucrose
6. lactose
7. amylopectin
8. amylose
   a. monosaccharides
   b. disaccharides
   c. polysaccharides

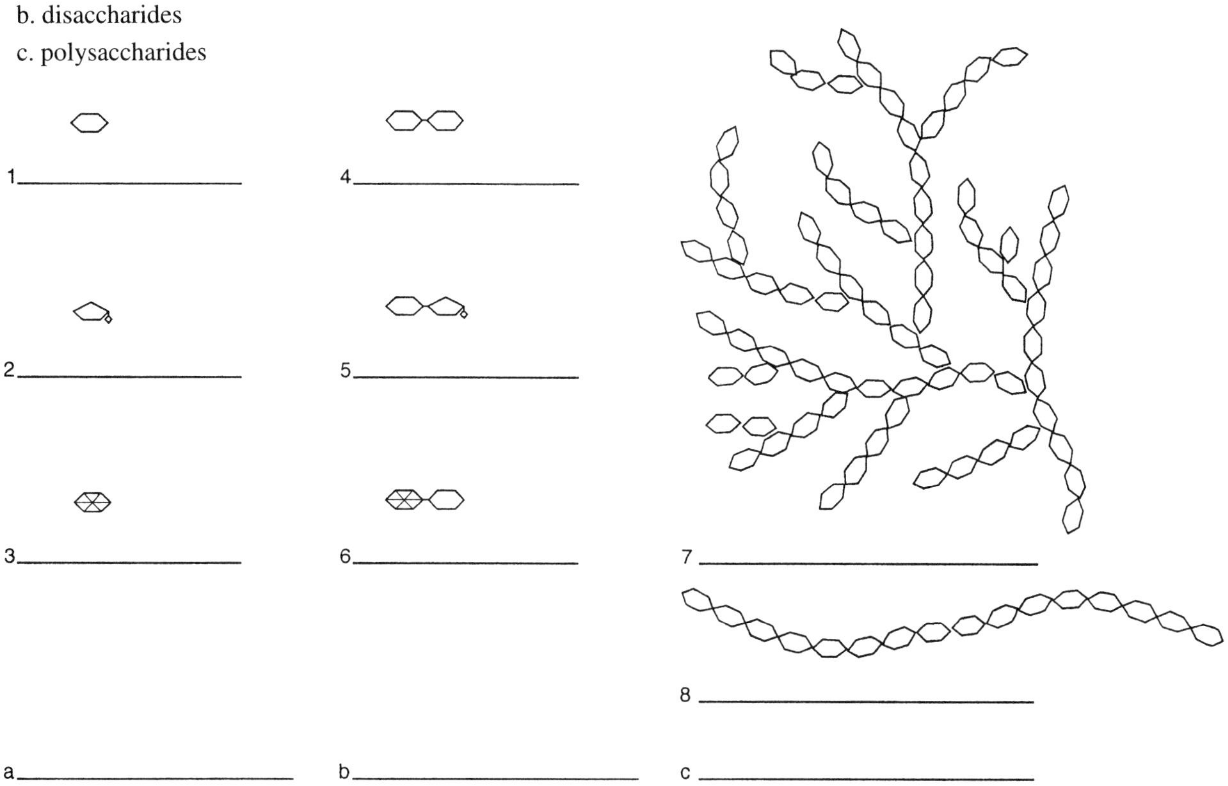

1 _____   4 _____

2 _____   5 _____

3 _____   6 _____   7 _____

8 _____

a _____   b _____   c _____

Figure 15.6. Major nutritional carbohydrates.

Exercise 15.6:

_____ 1. Disaccharides are made of two _____ molecules.

_____   a. Maltose is made of two _____ molecules.

_____   b. Sucrose is made of a _____ and _____ molecule.

_____   c. Lactose is made of a _____ and _____ molecule.

_____ 2. Polysaccharides are made of many molecules of _____ .

Answers to Exercise 15.6: 1. monosaccharide; 1a. glucose; 1b. glucose, fructose; 1c. galactose, glucose; 2. glucose.

G. Digestion of Carbohydrates

Color and label:
1. ○ glucose
2. ○ fructose
3. ○ galactose
4. ○ amylase (salivary amylase, pancreatic amylase)
   ○ lactase
   ○ maltase
   ○ sucrase

Label:
5. maltose
6. sucrose
7. lactose
8. amylopectin
9. cellulose
10. salivary gland
11. chyme
12. pancreas

Exercise 15.7:

_____ 1. Carbohydrate digestion begins in the _____ .

_____   a. The enzyme that begins this digestion is _____ and is made in the _____ .

_____   b. When a mass of food reaches the stomach, is it *immediately* mixed with stomach acid?

    c. Therefore, does salivary amylase continue to digest carbohydrates for a short time in the stomach?

_____ 2. The mass of food and gastric juices in the stomach is called _____ .

_____ 3. As the chyme leaks into the duodenum of the small intestine, the pancreas is stimulated to release _____ .

_____ 4. Disaccharides and other small fragments produced by amylase digestion are further digested at the _____ .

_____   a. The enzymes of the brush border primarily break down the class of carbohydrates called _____ .

_____   b. If one of the specific enzymes at the brush border is missing, will its substrate (disaccharide) be absorbed?

_____   c. If that unabsorbed disaccharide continues into the large intestine, can it serve as a food source for the bacteria living there?

_____   d. The likely result of this bacterial metabolism would be _____ .

_____ 5. Since there are no enzymes for cellulose and certain other plant fibers, would they be digested and absorbed?

_____ 6. These undigested carbohydrates would pass through the large intestine and be _____ .

Answers to Exercise 15.7: 1. mouth; 1a. salivary amylase, salivary glands; 1b. no; 1c. yes; 2. chyme; 3. pancreatic amylase; 4. brush border; 4a. disaccharides; 4b. no; 4c. yes; 4d. gas production; 5. no; 6. eliminated.

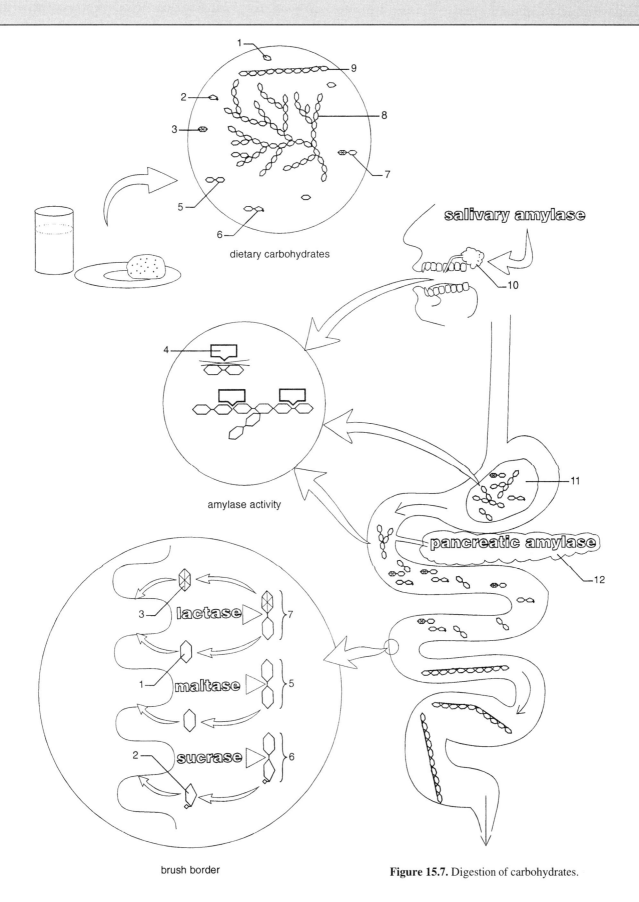

Figure 15.7. Digestion of carbohydrates.

Digestion of Carbohydrates

H. Digestion of Proteins

Color and label:
1. ◯ amino acids
2. ◯ pepsin
   ◯ pepsinogen
   ◯ HCl
   ◯ HCO_3^-
   ◯ aminopeptidases
   ◯ trypsin
   ◯ chymotrypsin
   ◯ elastase
   ◯ carboxypeptidase

Label:
3. peptide bond
4. stomach
5. pancreas

Exercise 15.8:

_____ 1. Proteins are long chains of _____ held together by _____ bonds.

_____ 2. Protein digestion begins in the _____.

_____ 3. Chief cells in gastric glands produce the inactive enzyme _____.

_____ 4. Parietal cells in the gastric glands produce _____, which activates pepsinogen into _____.

_____ 5. When chyme from the stomach leaks into the duodenum, the pancreas is stimulated to release the electrolyte _____ and a set of _____.

_____ 6. At the brush border the polypeptide fragments (proteoses, peptones, dipeptides) are hydrolyzed into _____ by _____.

_____ 7. The final products that are absorbed are _____.

Answers to Exercise 15.8: 1. amino acids, peptide; 2. stomach; 3. pepsinogen; 4. HCl, pepsin; 5. HCO_3^-, enzymes; 6. amino acids, aminopeptidases; 7. amino acids.

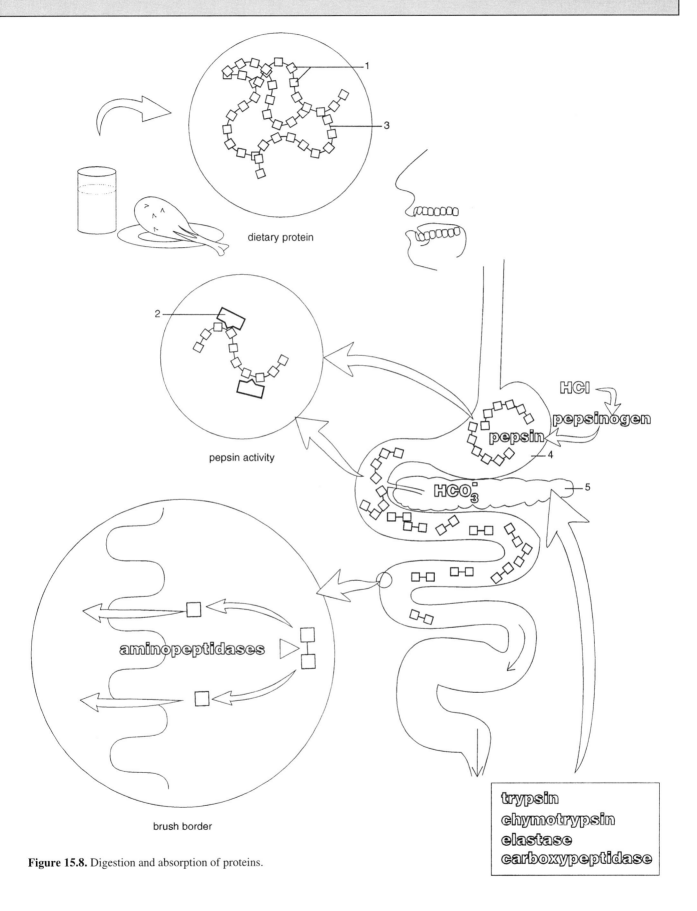

Figure 15.8. Digestion and absorption of proteins.

Digestion of Proteins

I. Digestion of Lipids

Color and label:
1. ○ phospholipid
2. ○ triglyceride
3. ○ sterol
4. ○ lipase (gastric lipase, pancreatic lipase)
5. ○ lacteal
6. ○ portal vein
   ○ bile acids
   ○ HCO_3^-

Label:
7. stomach
8. pancreas
9. micelle
10. chylomicron

Exercise 15.9:

_____ 1. The enzymes that digest lipids are produced in the _____ and _____ .

_____ 2. Bile is produced in the _____ and is concentrated and stored between meals in the _____ .

_____ a. Bile acids emulsify lipids, which means that they cause lipids to become _____ (more, less) soluble in water.

_____ b. Thus, digestion in the _____ (stomach, small intestine) is more efficient.

_____ c. Bile acids bind the digested lipids into _____ , which can enter the epithelial cells at the brush border.

_____ d. From the intestinal epithelium the bile acids return to the liver through the _____ .

_____ 3. In the intestinal epithelium, triglycerides are reassembled into _____ containing lipids and lipoproteins.

_____ 4. These chylomicrons are then released by exocytosis into _____ .

Answers to Exercise 15.9: 1. stomach, pancreas; 2. liver, gallbladder; 2a. more; 2b. small intestine; 2c. micelles; 2d. portal veins; 3. chylomicrons; 4. lacteals.

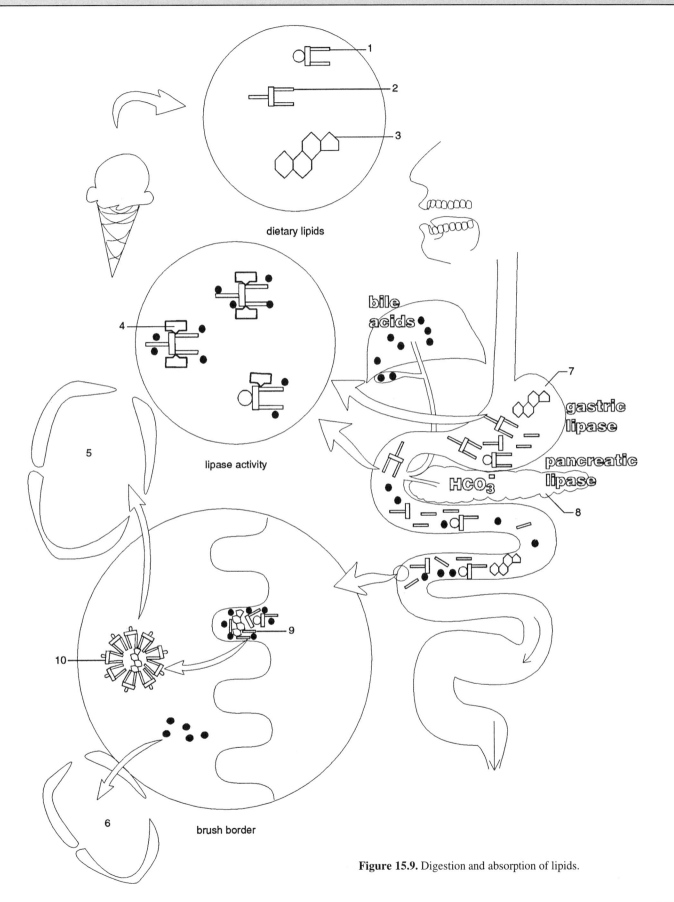

Figure 15.9. Digestion and absorption of lipids.

Digestion of Lipids

J. Absorption by Carriers

Color:
- ○ vit B_{12} (vitamin B_{12})
- ○ Ca^{++}
- ○ Fe^{++}
- ○ vitamin D

Color and label:
1. ○ intrinsic factor
2. ○ calcium binding factor
3. ○ transferrin
   a. plasma transferrin
4. ○ portal vein

Label:
5. gastric gland
6. brush border

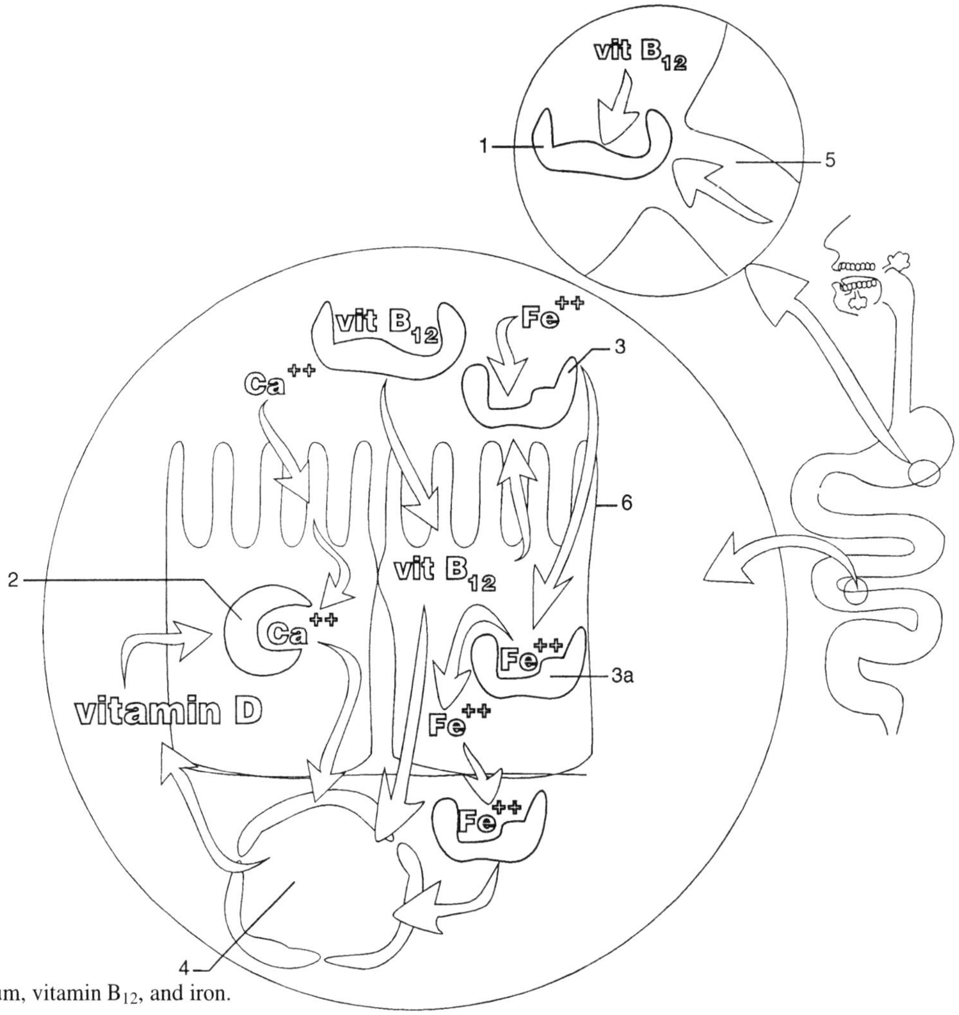

Figure 15.10. Absorption of calcium, vitamin B_{12}, and iron.

Exercise 15.10:

_____ 1. Vitamin B_{12} attaches to _____ in the stomach.

_____ a. Intrinsic factor is produced by parietal cells in _____ .

_____ b. Intrinsic factor permits vitamin B_{12} to be absorbed at the _____ .

_____ 2. The efficient transport of calcium through the intestinal epithelium depends upon _____ .

_____ 3. The production of calcium binding protein is stimulated by _____ .

_____ 4. The iron carrying protein transferrin is produced at the _____ .

_____ 5. Iron moves through the intestinal epithelium and is picked up by _____ .

_____ 6. Each of the above nutrients enters the _____ after absorption.

Answers to Exercise 15.10: 1. intrinsic factor; 1a. gastric glands; 1b. brush border; 2. calcium binding protein; 3. vitamin D; 4. brush border; 5. plasma transferrin; 6. portal veins.

K. Water Balance in the Digestive System

Color the arrows. Label:
1. ○ ingestion
2. ○ secretions
3. ○ absorption
4. ○ waste

Label:
5. salivary glands
6. liver
7. stomach
8. pancreas
9. small intestine
10. colon

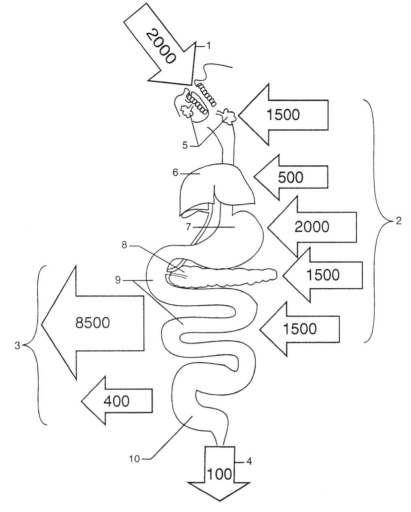

Figure 15.11. Average daily water movements (ml).

Exercise 15.11:

_____ 1. Most of the water in the digestive tract comes directly from _____, not from ingested water.

_____ 2. The daily average volume of water secreted into the digestive tract is _____.

_____ 3. Most of the water in the digestive tract is absorbed by the _____.

_____ 4. Most of the nutrients in the digestive tract are absorbed by the _____.

_____ 5. If the contents of the digestive tract were allowed to become dry, would the chemical reactions of digestion and absorption continue?

_____ 6. If the small intestine is irritated, it will usually secrete more water and absorb less. This increase in intestinal water will cause the condition called _____.

Answers to Exercise 15.11: 1. secretions; 2. 7000 ml; 3. small intestine; 4. small intestine; 5. no; 6. diarrhea.

L. Nervous Control of the Digestive System

Color (trace) and label:

1. ○ enteric nervous system
    a. myenteric plexus
    b. submucosal plexus
2. ○ autonomic nerves
    a. ○ parasympathetic nervous system
    b. ○ sympathetic nerves
3. ○ gastrocolic reflex (long reflex)
4. ○ peristaltic reflex (short reflex)

Label:

5. mesentery
6. serosa
7. longitudinal muscle
8. circular muscle
9. submucosa
10. mucosa
11. vagus nerve
12. sacral nerves
13. sympathetic ganglion

Exercise 15.12:

_____ 1. The enteric nervous system is made of nerves located in the walls of the gut. Its major networks are the _____ plexus and the _____ plexus.

_____ 2. The myenteric plexus is located between the _____ layers.

_____ 3. The submucosal plexus is located between the _____ and _____ layers of the gut.

_____ 4. These networks are entirely within the walls of the gut, but they communicate with the central nervous system through _____ nerves.

_____ 5. The autonomic control coordinates the activity of the digestive system with the rest of the body. For example, if a person is exercising, the activity and blood supply of the entire gut will _____ (increase, decrease).

_____ 6. Since the myenteric plexus is located between the circular and longitudinal muscle layer, it coordinates the actions of _____ .

7. An example of a long reflex controlled by the enteric nervous system is the gastrocolic reflex.

_____ a. This reflex coordinates the activities of the _____ with the _____ .

_____ b. When the stomach is distended (stretched), the activity of the colon _____ (increases, decreases).

_____ 8. In short reflexes, such as peristalsis, when the circular muscle of one segment of the gut contracts, the area immediately distal (in front of) _____ (contracts, relaxes).

Answers to Exercise 15.12: 1. myenteric, submucosal; 2. muscle; 3. submucosal, muscular; 4. autonomic; 5. decrease; 6. smooth muscles; 7a. stomach, colon; 7b. increases; 8. relaxes.

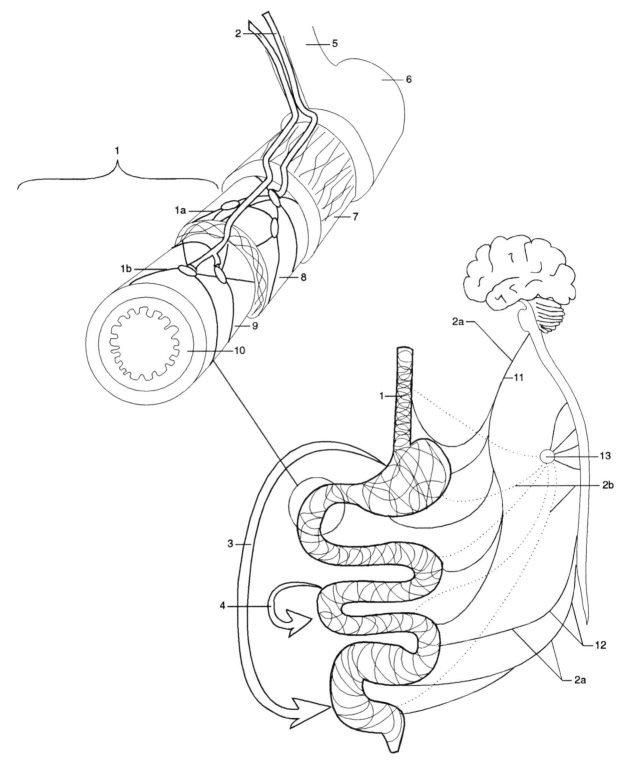

Figure 15.12. Enteric and autonomic nerves of the gut.

M. Chemical Control

Color:
- ○ bile acids
- ○ acid
- ○ fats
- ○ cholecystokinin (CCK)
- ○ secretin
- ○ HCO_3^-
- ○ enzymes

Label:
1. liver
2. gallbladder
3. stomach
4. chyme
5. pancreas
6. duodenum

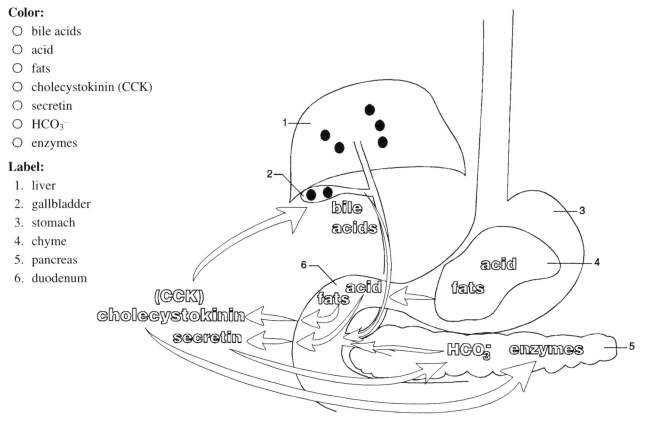

Figure 15.13. Stomach, duodenum, liver, pancreas.

Exercise 15.13:

_____ 1. The chyme that leaks through the pyloric valve is usually _____ (acidic, basic).

_____ a. The acidity stimulates the release of _____ by the duodenal mucosa.

_____ b. Secretin stimulates the pancreas to release watery secretions rich in _____ .

_____ c. This _____ the acid from the stomach.

_____ 2. The fats in the chyme stimulate the duodenal mucosa to release _____ .

_____ a. CCK stimulates both the _____ and the _____ .

_____ b. When the smooth muscle of the gallbladder contracts, _____ is released into the _____ .

_____ c. The bile acids emulsify the _____ in the small intestine.

_____ d. CCK also stimulates the pancreas to release _____ .

Answers to Exercise 15.13: 1. acidic; 1a. secretin; 1b. bicarbonate (HCO_3^-); 1c. neutralizes; 2. cholecystokinin (CCK); 2a. gallbladder, pancreas; 2b. bile, duodenum; 2c. fats; 2d. enzymes.

Chapter 16 Metabolism

A. Introduction to Metabolism

Color the words and their corresponding arrows:
- food
- energy
- materials
- work
- structure

Label:
1. mechanical work
2. membrane work
3. growth
4. repair, replace

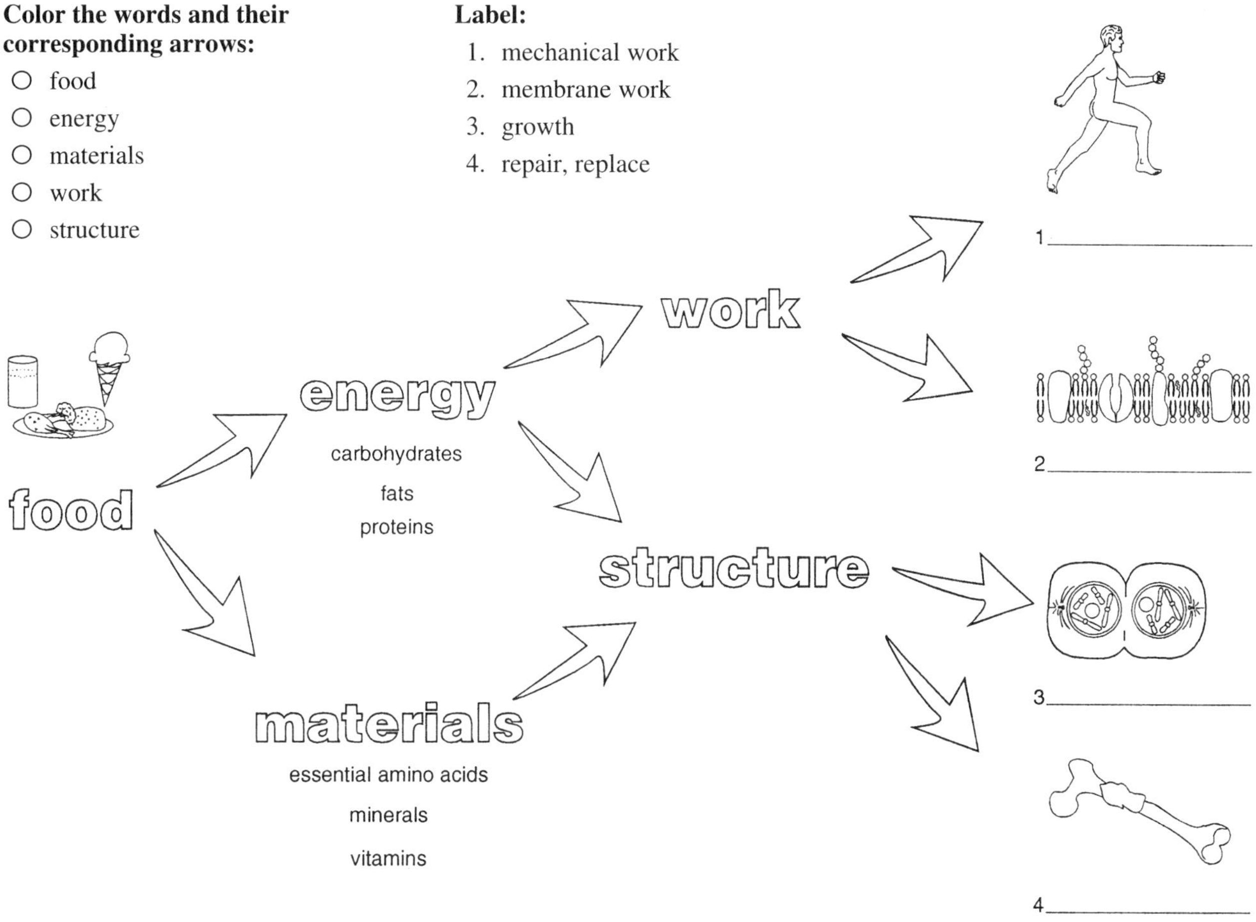

food → energy (carbohydrates, fats, proteins)
food → materials (essential amino acids, minerals, vitamins)
energy → work
energy → structure
materials → structure

1. _____
2. _____
3. _____
4. _____

Figure 16.1. Uses of food.

Exercise 16.1:

_____ 1. Food is used for both _____ and _____ .

_____ 2. The major sources of energy in food are _____ , _____ , and _____ .

_____ 3. To build the structures of the body both _____ and _____ are needed.

_____ 4. Work is done during muscle activity and during molecule transport across _____ .

_____ 5. Amino acids can be used for both _____ and _____ .

Answers to Exercise 16.1: 1. energy, materials; 2. carbohydrates, fats, proteins; 3. energy, materials; 4. membranes; 5. energy, materials.

B. Glucose Metabolism

Color and label:
1. ○ glucose
2. ○ glycogen

Label:
3. general circulation
4. liver
5. portal vein
6. small intestine

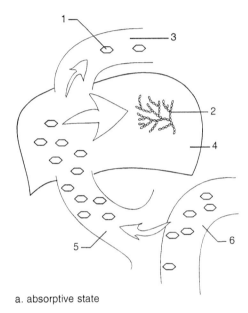
a. absorptive state

b. postabsorptive state

Figure 16.2. Absorptive and postabsorptive states.

Exercise 16.2:

_____ 1. The time period when carbohydrates are digested in the intestine is called the _____ .

_____ a. During this time, the portal veins transport large amounts of _____ to the liver.

_____ b. When large amounts of glucose are present, the liver converts much of it to _____ .

_____ c. What hormone signals the liver to do this? (see p. 164)

_____ 2. The time period when no carbohydrates remain in the intestines is called the _____ .

_____ a. In response to the decrease in circulating glucose, the liver breaks down _____ .

_____ b. What hormone signals the liver to do this? (see p. 164)

_____ 3. When comparing the absorptive and postabsorptive states, the glucose levels in general circulation are _____ (the same, different).

Answers to Exercise 16.2: 1. absorptive state; 1a. glucose; 1b. glycogen; 1c. insulin; 2. postabsorptive state; 2a. glycogen; 2b. glucagon; 3. the same.

C. Cellular Energy Conversion

Color and label:
1. ○ glucose
2. ○ other hexoses (6 carbon)
3. ○ fructose 1, 6-diphosphate
4. ○ pyruvic acid
5. ○ lactic acid
6. ○ ADP
7. ○ ATP
8. ○ mitochondrion
○ NAD
○ NADHH

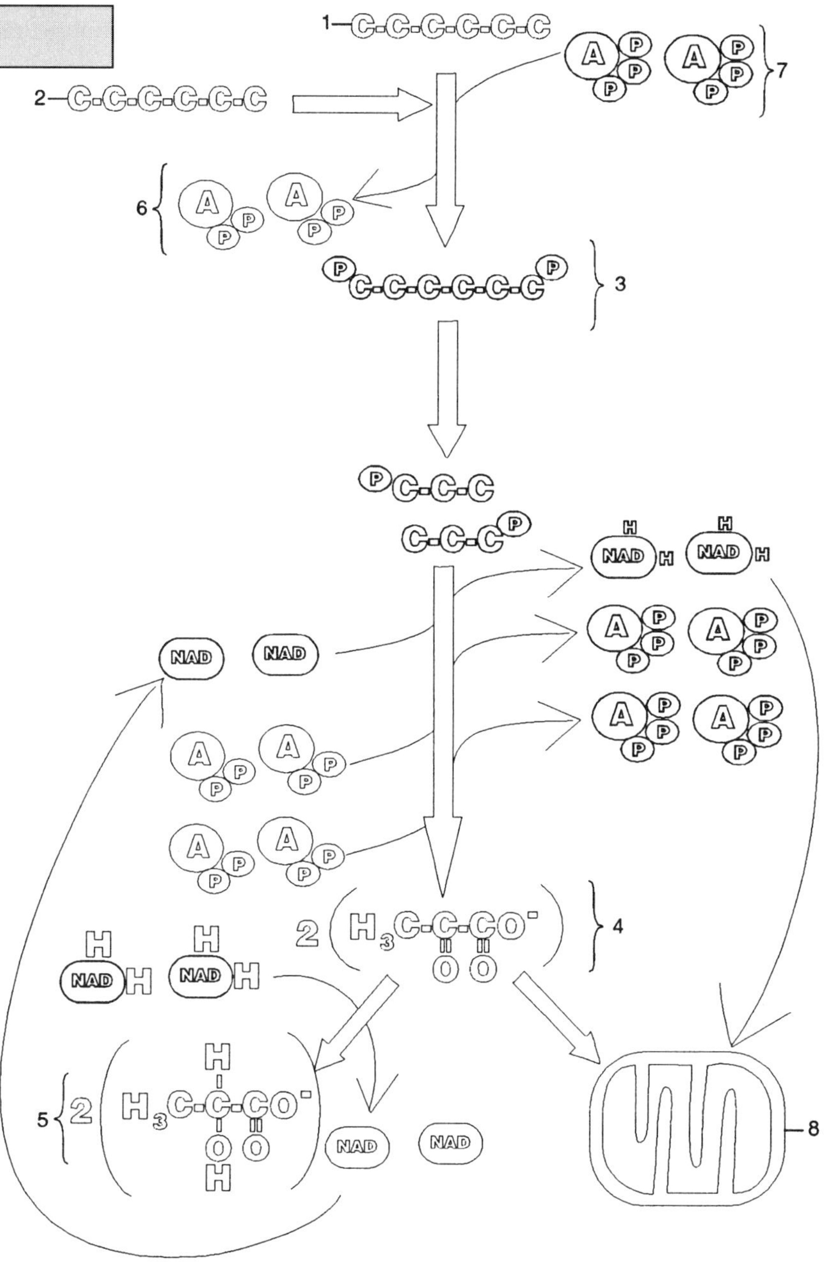

Figure 16.3. Glycolysis.

Exercise 16.3:

_____ 1. The metabolic pathway of glycolysis refers to the breakdown of the molecule _____ .

_____ a. This set of chemical reactions takes place in the _____ of the cell.

_____ b. If other sugars or intermediate molecules are present, can they enter along the pathway?

_____ 2. When ATP transfers a phosphate to another molecule, the ATP _____ (gains, loses) energy.

_____ 3. The molecule that receives the phosphate _____ (gains, loses) energy.

_____ 4. When NAD becomes reduced (gains H), it _____ (gains, loses) energy.

_____ a. The organelle that accepts the energy of this molecule is the _____ .

_____ b. If the mitochondrion cannot accept the H, NAD reacts with _____ in order to recycle.

_____ c. If NAD is oxidized (gives up its H) by pyruvic acid, _____ forms.

_____ 5. At the end of glycolysis, which molecules have gained energy?

_____ 6. If pyruvic acid is not converted to lactic acid by NAD, it enters the _____ .

Answers to Exercise 16.3: 1. glucose; 1a. cytoplasm; 1b. yes; 2. loses; 3. gains; 4. gains; 4a. mitochondrion; 4b. pyruvic acid; 4c. lactic acid; 5. ATP and NAD; 6. mitochondrion.

D. Krebs Cycle and Oxidative Phosphorylation

Color:
- ○ NAD
- ○ H (hydrogen)
- ○ glycolysis
- ○ Krebs cycle
- ○ respiratory chain

Color and label:
1. ○ pyruvic acid
2. ○ acetyl CoA
3. ○ ATP
4. ○ oxygen
5. ○ water
6. ○ carbon dioxide

Label:
7. mitochondrion

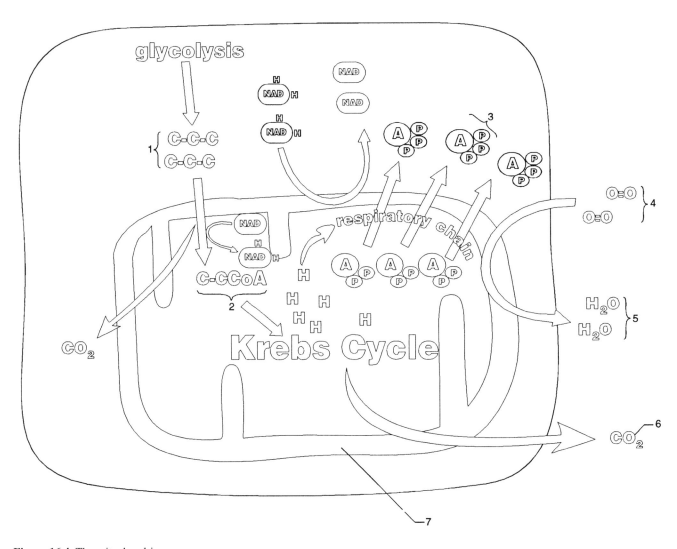

Figure 16.4. The mitochondrion.

Exercise 16.4:

_____ 1. Two products of glycolysis are provided to the mitochondrion. They are _____ and _____ .

_____ 2. In the mitochondrion, pyruvic acid loses a _____ and two _____ .

_____ 3. It then joins coenzyme A to result in the molecule called _____ .

_____ 4. This acetyl group now enters the _____ .

_____ 5. In the Krebs cycle, additional energy is extracted, most of which is used in the _____ to make _____ .

_____ 6. The fragments from the original molecules are released by the Krebs cycle in the form of _____ .

_____ 7. Oxygen is needed to carry off the _____ at the end of the respiratory chain. The product of this reaction is _____ .

_____ 8. The total yield of ATP from the mitochondrion is 36. The net yield of ATP from the metabolism of one molecule of glucose is _____ . (See exercise 16.3)

_____ 9. If all of the pyruvic acid produced by glycolysis cannot be processed in the mitochondrion, the excess is reduced to _____ .

Answers to Exercise 16.4: 1. pyruvic acid, H from NAD; 2. CO_2, H; 3. acetyl CoA; 4. Krebs cycle; 5. respiratory chain, ATP; 6. CO_2; 7. H, water; 8. 38; 9. lactic acid.

Krebs Cycle and Oxidative Phosphorylation

E. Synthesis of Triglycerides

Color and label:
1. ○ glucose
2. ○ pyruvic acid
3. ○ glycerol
4. acetyl CoA
   a. ○ acetyl (C-C)
   b. ○ CoA
5. ○ fatty acid
6. ○ triglyceride

Label:
7. glycolysis

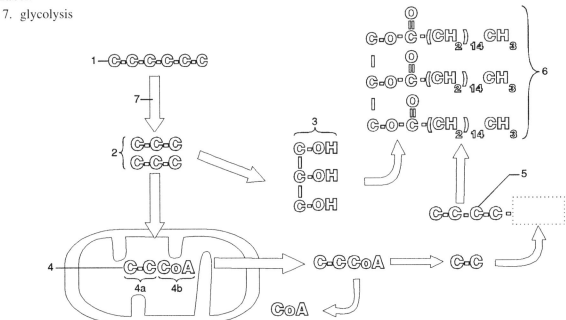

Figure 16.5. Triglyceride synthesis in liver, adipose, and active mammary gland cells.

Exercise 16.5:

_____ 1. Glycerol is synthesized from intermediates of _____ .

_____ 2. The two-carbon acetyl group becomes part of a(n) _____ acid.

_____ 3. Glycerol and three fatty acids join to form a(n) _____ .

4. In the previous exercises,

_____ a. we saw that glucose could be broken down to supply _____ .

_____ b. we saw that glucose not immediately needed for energy could be stored as _____ .

_____ 5. Since the cell's ability to store glycogen is limited, the body also stores energy in _____ .

Answers to Exercise 16.5: 1. glycolysis; 2. fatty; 3. triglyceride; 4a. energy; 4b. glycogen; 5. triglycerides.

F. Lipid Transport

Color and label:
1. ○ HDL
2. ○ LDL
3. ○ VLDL
4. ○ liver
5. ○ muscle cells
6. ○ adipose cell

Label:
7. FFA (free fatty acids)
8. triglyceride
9. lipoprotein
10. cholesterol
11. circulatory system

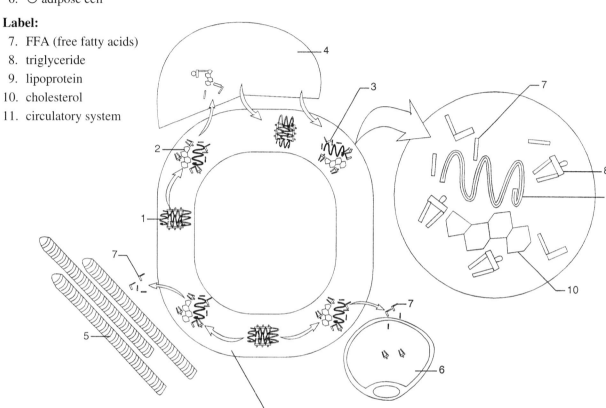

Figure 16.6. Lipid particles in circulation.

Exercise 16.6:

_____ 1. High density lipoproteins (HDL), low density lipoproteins (LDL), and very low density lipoproteins (VLDL) are produced in the _____ .

_____ 2. LDLs and VLDLs contain _____ , _____ , and _____ .

_____ 3. Tissues such as adipose and muscle break down low density particles. This process is facilitated by _____ .

_____ 4. Fatty acids are taken into adipose tissue for _____ and into muscle cell for _____ .

_____ 5. Since LDLs may damage the lining of arteries, how could HDLs have a protective effect?

Answers to Exercise 16.6: 1. liver; 2. cholesterol, fatty acids, lipoproteins; 3. HDL; 4. storage, energy; 5. remove LDLs.

G. Proteins

Color and label:
1. ○ essential amino acids
2. ○ nonessential amino acids
3. ○ metabolism (arrows)
4. ○ protein synthesis (arrow)

Label:
5. capillary
6. amino acid pool
7. polypeptide (protein)
8. cell

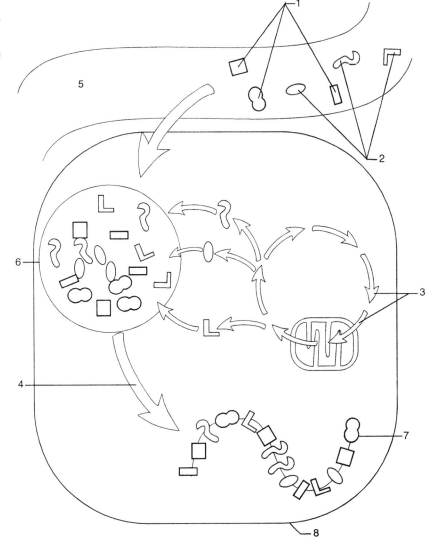

Figure 16.7. Amino acid metabolism in cells.

Exercise 16.7:

_____ 1. Are nonessential amino acids necessary for protein synthesis?

_____ 2. If they are not present in the diet, nonessential amino acids are produced by _____ .

_____ 3. Can the cell synthesize essential amino acids?

_____ 4. The cell gets essential amino acids from _____ .

_____ 5. If the amino acid pool in a cell does not contain sufficient amounts of all twenty amino acids, will protein synthesis be completed?

Answers to Exercise 16.7: 1. yes; 2. the cell; 3. no; 4. the diet; 5. no.

Chapter 17 Urinary System

A. Organs of the Urinary System

Color and label:
1. ○ right kidney
2. ○ left kidney
3. ○ ureter
4. ○ bladder
5. ○ urethra

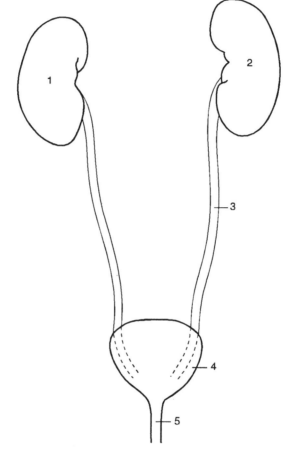

Figure 17.1. Anterior view of urinary system.

Exercise 17.1:

_____ 1. Urine is produced in the _____ .

_____ 2. Urine is carried from the kidney to the bladder by the _____ .

_____ 3. Urine is stored in the _____ .

_____ 4. Urine is carried from the bladder out of the body by the _____ .

Answers to Exercise 17.1: 1. kidneys; 2. ureters; 3. bladder; 4. urethra.

B. Abdominal Relationships

- ○ Color and label the urinary structures (yellow):
1. right kidney
2. left kidney
3. ureter
4. bladder
5. urethra

Color and label:
6. ○ dorsal aorta (red)
7. ○ renal artery (red)
8. ○ renal vein (blue)
9. ○ inferior vena cava (blue)

For figure 17.2a, color and label:
10. ○ diaphragm (muscle)
11. ○ adrenal gland
12. ○ rectum
13. ○ uterus

For figure 17.2b, color (outline) and label:
14. ○ vertebra
15. ○ ribs
16. ○ fat capsule
17. ○ parietal peritoneum
18. ○ peritoneal cavity

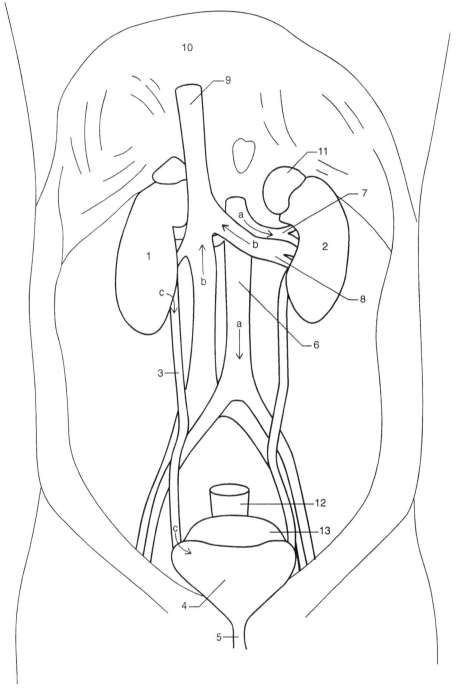

Figure 17.2a. Anterior view of abdomen.

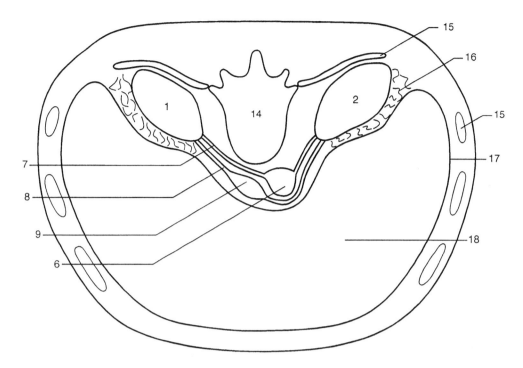

Figure 17.2b. Transverse section of abdomen.

Exercise 17.2:

_____ 1. The hilus (notch) of the kidney is on its _____ (anterior, posterior, lateral, medial, superior, inferior) surface.

_____ 2. The ureter and renal blood vessels enter the kidney at the _____ .

_____   a. Arrows labeled *a* show the movement of _____ . This fluid is coming from the _____ .

_____   b. Arrows labeled *b* show the movement of _____ . This fluid is going to the _____ .

_____   c. Arrows labeled *c* show the movement of _____ .

_____ 3. The dorsal aorta is _____ (right, left) of the inferior vena cava.

_____ 4. The adrenal glands, which belong to the _____ system, are on the _____ surface of the kidneys.

_____ 5. The kidneys are _____ (inside, outside) the peritoneal cavity.

_____   a. The kidneys are surrounded by a _____ (tissue type).

_____   b. This tissue serves to _____ (function).

_____ 6. The bladder is _____ (anterior, posterior) to the uterus, which is _____ (anterior, posterior) to the rectum.

Answers to Exercise 17.2: 1. medial; 2. hilus; 2a. arterial blood, heart; 2b. venous blood, heart; 2c. urine; 3. left; 4. endocrine, superior; 5. outside; 5a. fat capsule; 5b. cushion kidneys; 6. anterior, anterior.

Abdominal Relationships

C. Internal Anatomy of the Kidney

Color (trace) and label:
1. ○ nephron
   a. collecting tubule
2. ○ capsule
3. ○ cortex
4. medulla
   a. ○ renal pyramids
   b. ○ renal columns
5. ○ renal sinus (yellow)
   a. minor calyces
   b. major calyces
   c. pelvis
6. ○ ureter (yellow)
7. ○ **Color arteries red. Label:**
   a. renal artery
   b. segmental artery
   c. interlobar artery
   d. arcuate artery
   e. interlobular artery
8. ○ **Color veins blue. Label:**
   a. interlobular vein
   b. arcuate vein
   c. interlobar vein
   d. segmental vein
   e. renal vein

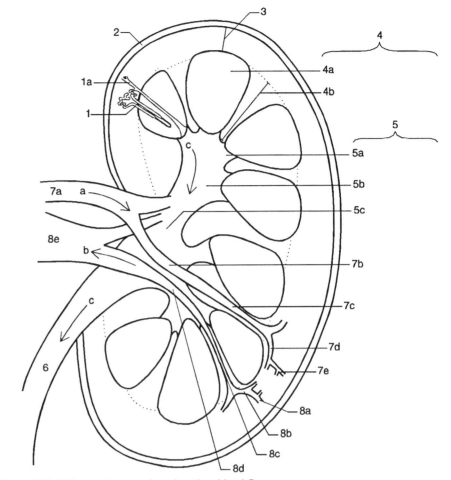

Figure 17.3. Kidney regions, nephron location, blood flow.

Exercise 17.3:

_____ 1. The functional unit of the kidney is the _____ .

_____ 2. Nephrons are located in the _____ and _____ (regions).

_____ 3. Blood enters the kidney through the _____ , which branches into smaller arteries that travel through the renal _____ (pyramids, columns) to the nephron.

_____ 4. Blood leaves the vicinity of the nephron through veins that travel through the renal _____ (pyramids, columns) and then leaves the kidney via the _____ .

5. The following arrows represent the direction of flow of what fluids?

_____ a. a arrows

_____ b. b arrows

_____ c. c arrows

_____ 6. Urine collects in the _____ (region) before leaving the kidney.

Answers to Exercise 17.3: 1. nephron; 2. cortex, renal pyramids; 3. renal artery, columns; 4. columns, renal veins; 5a. arterial blood; 5b. venous blood; 5c. urine; 6. renal sinus.

D. Nephron Structure and Location

Color and label:
1. ◯ Bowman's capsule (glomerular capsule)
2. ◯ proximal convoluted tubule
3. loop of Henle
   a. ◯ descending limb
   b. ◯ ascending limb
4. ◯ distal convoluted tubule
5. ◯ collecting tubule
6. ◯ papillary duct

Label:
7. cortex
8. renal pyramid (medulla)
9. minor calyx (renal sinus)
10. juxtamedullary nephron
11. cortical nephron

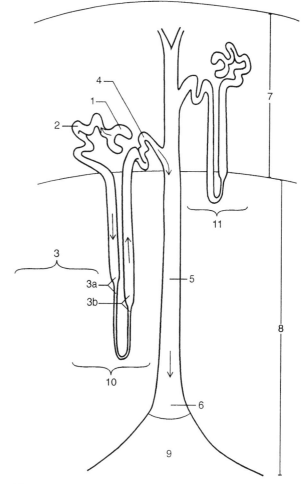

Figure 17.4. Structure of nephron and regional location.

Exercise 17.4:

_____ 1. The cup-shaped portion of the nephron is _____ .

_____ 2. The convoluted portions of the nephron are the _____ and _____ .

_____ 3. Bowman's capsule faces the _____ convoluted tubule.

_____ 4. The straight portions of the nephron are the _____ and _____ .

_____ 5. The nephron empties its fluid, filtrate, into the _____ .

_____ 6. The collecting tubule fluid, urine, empties into the _____ .

_____ 7. The two types of nephrons are _____ and _____ .

_____ a. The nephron with a loop of Henle that extends deep into the medulla is the _____ .

_____ b. The nephron at the cortical-medullary junction is the _____ .

Answers to Exercise 17.4: 1. Bowman's capsule; 2. proximal, distal convoluted tubules; 3. distal; 4. descending, ascending limbs of the loop of Henle; 5. collecting tubule; 6. minor calyx (renal sinus); 7. juxtamedullary, cortical; 7a. juxtamedullary; 7b. cortical.

E. Nephron Circulation

○ **Color the nephron parts yellow. Label:**
1. Bowman's capsule
2. proximal convoluted tubule
3. loop of Henle
   a. descending limb
   b. ascending limb
4. distal convoluted tubule
5. collecting tubule

Color and label:
6. ○ arcuate artery (red)
7. ○ interlobular artery (red)
8. ○ afferent arteriole (red)
9. ○ glomerulus (purple)
10. ○ efferent arteriole (red)
11. ○ peritubular capillary (purple)
12. ○ vasa recta (purple)
13. ○ interlobular vein (blue)
14. ○ arcuate vein (blue)

Exercise 17.5:

1. Bowman's capsule surrounds the _____ .
   a. Blood enters the glomerulus from the _____ , and
   b. leaves through the _____ .
2. Blood flows from the efferent arterioles into the _____ and _____ .
   a. The peritubular capillaries are next to the _____ , _____ , and _____ tubules.
   b. The vasa recta are next to the _____ and _____ of the juxtamedullary nephrons.
3. What type of blood vessels are the glomerulus and vasa recta?
4. The following arrows represent the direction of flow of what fluids?
   a. a arrows
   b. b arrows
   c. c arrows

Answers to Exercise 17.5: 1. glomerulus; 1a. afferent arteriole; 1b. efferent arteriole; 2. peritubular capillaries, vasa recta; 2a. proximal convoluted, distal convoluted, collecting; 2b. descending limb, ascending limb of the loop of Henle; 3. capillaries; 4a. arterial blood; 4b. venous blood; 4c. urine.

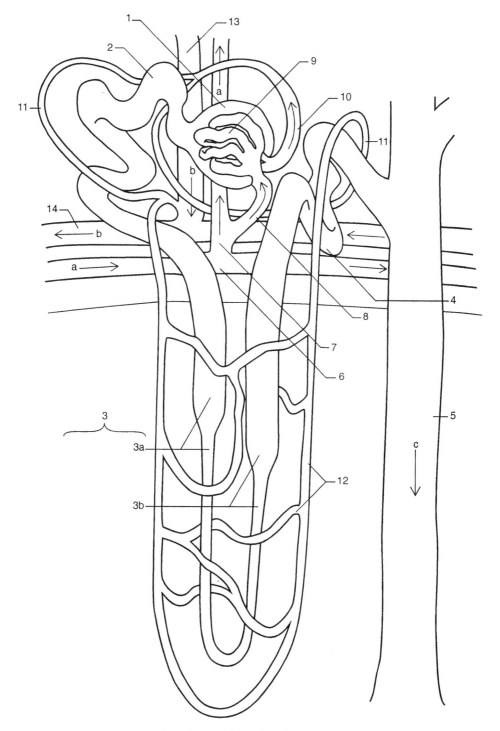

Figure 17.5. Relationship of blood flow to nephron.

Nephron Circulation

F. Filtration, Reabsorption, Secretion—Defined

Color and label:
1. ○ blood (red)
2. ○ filtrate (yellow)
3. ○ urine (yellow)

Color and label arrows:
4. ○ filtration
5. ○ reabsorption
6. ○ secretion

Label:
7. glomerulus
8. Bowman's capsule
9. tubule

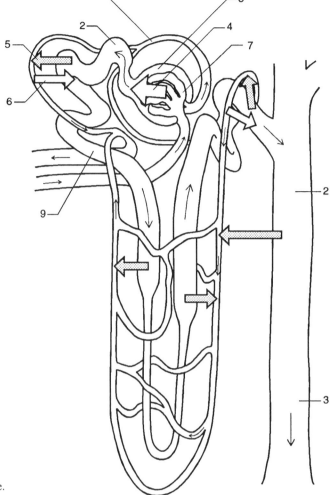

Figure 17.6. Formation and processing of filtrate.

Exercise 17.6:

_____ 1. Filtration occurs at the _____ (blood vessel).

_____ 2. During filtration, materials move from the _____ (blood, nephron) to the _____ (blood, nephron).

_____ 3. Reabsorption is the movement of materials from the _____ (blood, nephron) to the _____ (blood, nephron).

_____ 4. Secretion is the movement of materials from the _____ (blood, nephron) to the _____ (blood, nephron).

_____ 5. Reabsorption and secretion occur in _____ (nephron part).

_____ 6. The end product of filtrate processing is the fluid _____ .

Answers to Exercise 17.6: 1. glomerulus; 2. blood, nephron; 3. nephron, blood; 4. blood, nephron; 5. tubule; 6. urine.

G. Filtration

Color and label:
1. Bowman's capsule
   a. ○ parietal layer
   b. ○ visceral layer (podocytes)
   c. ○ capsular space (yellow)
2. ○ proximal convoluted tubule (cells)
3. ○ filtrate (yellow)

Color the blood vessels red. Label:
4. afferent arteriole
5. glomerulus
6. efferent arteriole

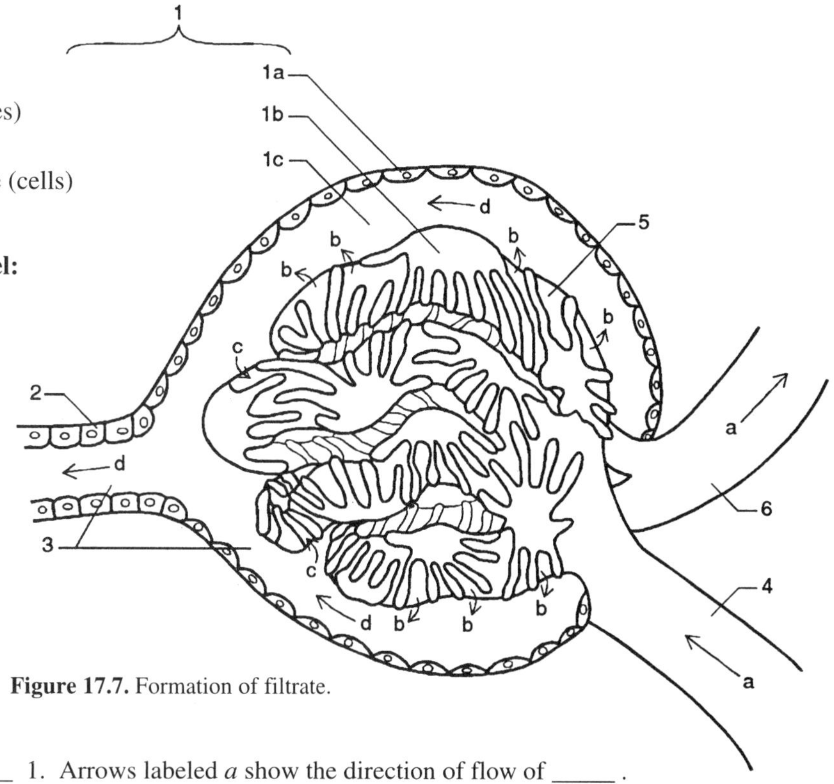

Figure 17.7. Formation of filtrate.

Exercise 17.7:

1. Arrows labeled *a* show the direction of flow of _____ .

   a. This fluid moves from the afferent arteriole to the efferent arteriole through the _____ .

   b. Fluids move from areas of _____ (high, low) pressure to areas of _____ (high, low) pressure. (see p. 188)

   c. The pressure in the afferent arteriole is caused by the _____ . (see p. 188)

   d. The force that causes fluid to move from the blood (glomerulus) to the capsular space is _____ .

   e. This process of filtration is shown by arrows labeled _____ .

2. Since the glomerular wall only allows small molecules to pass into the capsular space, will the following be filtered?

   a. cells
   b. plasma proteins
   c. ions (Na^+, K^+, Cl^-, HCO_3^-)
   d. nutrients (glucose, amino acids)
   e. urea
   f. water

3. The larger molecules (solute) in the plasma create a concentration gradient.

   a. Which larger molecules are in the plasma, but *not* in the filtrate?

   b. Which way is water expected to move because of this gradient? (see osmotic pressure, p. 30)

   c. Which arrows show osmotic pressure?

Filtration

G. Filtration continued

_____ 4. These forces (glomerular blood pressure and osmotic pressure) work in _____ (the same, opposite) directions.

_____ a. Which force is greater, glomerular blood pressure or the blood osmotic pressure?

_____ b. Therefore, glomerular filtration rate depends primarily upon _____ .

_____ 5. The end product of glomerular filtration is the formation of _____ (fluid).

_____ a. Filtrate is first found in _____ .

_____ b. What arrows label the flow of filtrate?

_____ 6. Normally 180 liters/day (125 ml/min) are filtered through the kidneys (glomerular filtration rate or GFR). Since only 1–2 liters of urine are produced, how much water must be reabsorbed?

Answers to Exercise 17.7: 1. blood; 1a. glomerulus; 1b. high, low; 1c. heart; 1d. blood pressure; 1e. b; 2a. no; 2b. only small proteins; 2c. yes; 2d. yes; 2e. yes; 2f. yes; 3a. plasma proteins; 3b. into the glomerulus; 3c. c; 4. opposite; 4a. glomerular blood pressure; 4b. blood pressure; 5. filtrate; 5a. Bowman's capsule; 5b. d; 6. 178–179 liters/day.

H. Regulation of Glomerular Filtration Rate

Color and label:
1. ○ afferent arteriole (red)
2. ○ glomerulus (red)
3. ○ efferent arteriole (red)
4. ○ capsular space (yellow)
5. ○ proximal convoluted tubule (yellow)
6. ○ blood pressure (arrow)

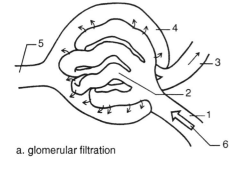

a. glomerular filtration

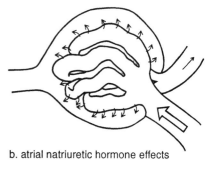

b. atrial natriuretic hormone effects

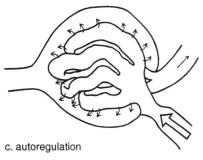

c. autoregulation

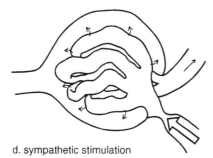

d. sympathetic stimulation

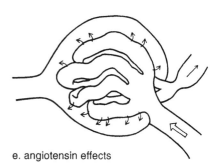

e. angiotensin effects

Figure 17.8. Effects of blood pressure on filtration.

Exercise 17.8:

1. Glomerular filtration rate (GFR) depends primarily upon _____ . (figure 17.8a)

2. An increase in blood pressure triggers the release of ANH (atrial natriuretic hormone) from the heart. (figure 17.8b)

   a. ANH causes the afferent arteriole to _____ (dilate, constrict).

   b. This means the GFR _____ (increases, decreases).

   c. An increased GFR serves to _____ (increase, decrease) blood volume, and therefore, _____ (increase, decrease) blood pressure.

3. When blood pressure increases, the afferent arteriole _____ (dilates, constricts). Does this serve to moderate the increase in GFR? (figure 17.8c)

4. The sympathetic nervous system raises blood pressure and severely constricts the _____ . (figure 17.8d)

   a. In this case, GFR _____ .

   b. Therefore, in emergencies, fluid retention _____ (increases, remains the same, decreases).

5. When blood pressure decreases, we might expect the GFR to _____ (increase, remain the same, decrease). (figure 17.8e)

   a. This means that kidney function would _____ (increase, remain the same, decrease).

   b. Angiotensin II, which is made in response to low blood pressure, causes constriction of the _____ .

   c. This constriction causes the GFR to be _____ (greater than, the same as, less than) expected.

   d. Therefore, when blood pressure is low, does angiotensin II serve to maintain kidney function?

Answers to Exercise 17.8: 1. blood pressure; 2a. dilate; 2b. increases; 2c. decrease, decrease; 3. constricts, yes; 4. afferent arteriole; 4a. decreases; 4b. increases; 5. decrease; 5a. decrease; 5b. efferent arteriole; 5c. greater than; 5d. yes.

Regulation of Glomerular Filtration Rate

I. Reabsorption

○ **Color red. Label:**
1. glomerulus
2. peritubular capillary
3. vasa recta

○ **Color (trace) the cells. Label:**
4. Bowman's capsule
5. proximal convoluted tubule
   a. microvilli
6. descending limb of the loop of Henle
7. ascending limb of the loop of Henle
8. distal convoluted tubule
   a. microvilli
9. collecting tubule

○ **Color yellow. Label:**
10. filtrate
11. urine

Color and label:
a. ○ amino acids
b. ○ glucose
c. ○ Na^+
d. ○ cations (K^+, Ca^{++})
e. ○ Cl^-
f. ○ anions (HCO_3^-, PO_4^{-3}, SO_4^{-2})
g. ○ urea
h. ○ water

Exercise 17.9:

_____ 1. Reabsorption is movement of materials from the _____ to the _____ .

_____ a. Materials reabsorbed from the convoluted tubules enter the _____ (capillaries).

_____ b. Materials reabsorbed from the loop of Henle enter the _____ (capillaries).

2. Which provides greater surface area for reabsorption?

_____ a. convoluted or straight tubules

_____ b. proximal or distal convoluted tubules

3. What materials are reabsorbed at each site?

_____ a. proximal convoluted tubule

_____ b. descending limb of the loop of Henle

_____ c. ascending limb of the loop of Henle

_____ d. distal convoluted tubule

_____ e. collecting tubule

_____ 4. The primary site for nutrient reabsorption is the _____ .

_____ 5. Water is not reabsorbed from the _____ .

_____ 6. The permeability of the distal convoluted tubule and collecting tubule is under hormonal control. Therefore, the amount of reabsorption at these sites would be _____ (constant, variable).

Answers to Exercise 17.9: 1. filtrate (nephron tubule), blood (capillaries); 1a. peritubular capillaries; 1b. vasa recta; 2a. convoluted; 2b. proximal (more microvilli); 3a. amino acids, glucose, Na^+, cations (K^+, Ca^{++}), Cl^-, anions (HCO_3^-, PO_4^{-3}, SO_4^{-2}), urea, water; 3b. water; 3c. Na^+, Cl^-; 3d. Na^+, water; 3e. Na^+, urea, water; 4. proximal convoluted tubule; 5. ascending limb of the loop of Henle; 6. variable.

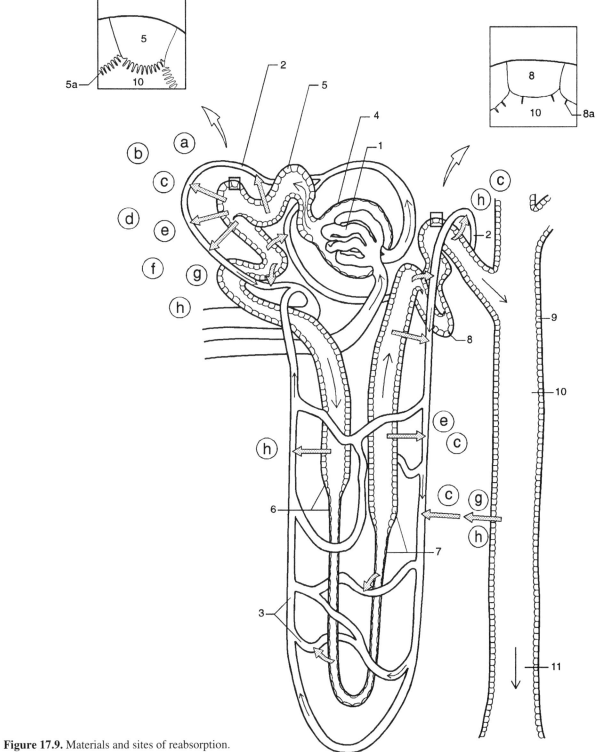

Figure 17.9. Materials and sites of reabsorption.

Reabsorption 279

J. Mechanisms of Reabsorption

Color and label:
1. solute
   a. ○ Na⁺
   b. ○ cotransport molecules (glucose, amino acids, phosphates, lactate, chloride)
   c. ○ glucose
   d. ○ urea
2. ○ blood
3. ○ interstitial fluid
4. ○ tubule cell
   a. ○ Na⁺K⁺ pump
   b. ○ basal membrane
   c. ○ apical membrane
   d. ○ carrier
5. ○ filtrate (yellow)
6. ○ water (arrow)

Exercise 17.10:

_____ 1. During reabsorption, filtrate materials cross _____ and _____ membranes, enter the _____ fluid, and cross into the _____.

_____ 2. The Na⁺ concentration is higher in the _____ (tubule cells, filtrate). (see figure 17.10a, b, and e)

_____ a. Low Na⁺ in tubule cells means that filtrate Na⁺ can enter across the apical membrane by _____ (transport mechanism).

_____ b. This low Na⁺ occurs because Na⁺ leaves via the basal membrane by _____.

_____ c. Na⁺ in interstitial fluid moves into the blood by _____.

_____ 3. Na⁺ reabsorption occurs primarily by cotransport with other molecules such as _____. (see figure 17.10b)

_____ a. Cotransport occurs via _____ embedded in the _____ membrane.

_____ b. Can this cotransport occur even if the cotransported molecule is moving against its concentration gradient?

_____ c. The cotransport molecules leave the tubule cell and move into the blood by _____ (transport mechanism).

_____ d. Does movement of cotransport molecules require energy?

_____ 4. For normal blood levels of nutrients, such as glucose, do the carriers become saturated? (see figure 17.10c)

_____ a. Should these nutrients appear in the urine?

_____ b. Therefore, does this carrier mediated mechanism normally provide for complete reabsorption of nutrients?

_____ 5. If blood glucose is high, the amount of glucose in the filtrate will be _____ (high, low). (see figure 17.10d)

_____ a. When filtrate glucose is high, can the carriers transport all the glucose out of the filtrate? Explain.

_____ b. Under these conditions, does glucose appear in the urine?

_____ 6. Urea, a nitrogenous waste, is reabsorbed by _____ (transport mechanism). Is some urea found in urine? (see figure 17.10e)

_____ 7. When solutes are reabsorbed, the concentration of filtrate water _____ (increases, remains the same, decreases). (see figure 17.10f)

_____ a. Therefore, water leaves the filtrate by _____.

_____ b. Is energy expended to move water?

_____ c. Most solute reabsorption occurs at the _____. (see page 278)

_____ d. Therefore, most water reabsorption occurs at the _____.

Answers to Exercise 17.10: 1. apical, basal, interstitial, blood; 2. filtrate; 2a. diffusion; 2b. active transport; 2c. diffusion; 3. glucose, amino acids, phosphates, lactate, chloride; 3a. carriers, apical; 3b. yes; 3c. diffusion; 3d. yes (indirectly dependent upon active transport at basal membrane); 4. no; 4a. no; 4b. yes; 5. high; 5a. no, carriers become saturated (transport maximum); 5b. yes; 6. diffusion, yes; 7. increases; 7a. diffusion (osmosis); 7b. no; 7c. proximal convoluted tubule; 7d. proximal convoluted tubule.

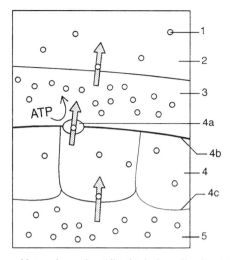
a. Na⁺ reabsorption (distal tubule, collecting tubule).

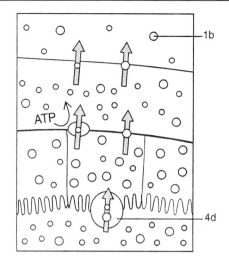
b. Na⁺ cotransport (proximal tubule, Henle's loop).

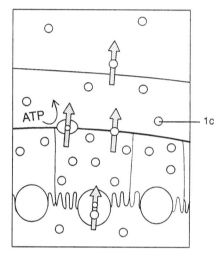
c. Glucose transport (normal blood glucose).

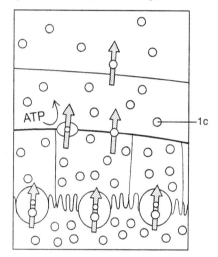
d. Glucose transport (high blood glucose).

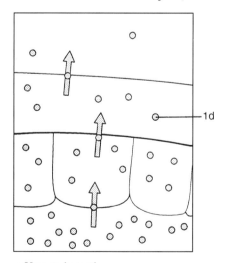
e. Urea reabsorption.

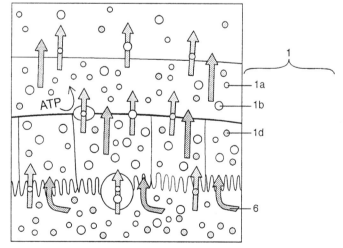
f. Water reabsorption.

Figure 17.10. Na⁺, glucose, urea, and water reabsorption.

Mechanisms of Reabsorption

K. Hormonal Regulation of Reabsorption

Color and label:
1. ○ filtrate (yellow)
2. ○ blood (red)
3. ○ Na$^+$
4. ○ urea
5. ○ water
6. ○ ADH (antidiuretic hormone)
7. ○ aldosterone

Label:
8. ○ interstitial fluid

Exercise 17.11:

_____ 1. While reabsorption at the proximal convoluted tubule and descending limb is primarily _____ (regulated, unregulated), reabsorption at the distal convoluted and collecting tubules is _____ (regulated, unregulated).

_____ 2. Two hormones that regulate reabsorption are _____ and _____ .

_____ 3. The solutes that leave the ascending limb, distal convoluted tubule, and collecting tubule are _____ and _____ .

_____ a. Is the ascending limb permeable to water?

_____ b. Therefore, the interstitial fluid becomes _____ (hypertonic, isotonic, hypotonic) and the ascending limb filtrate becomes _____ (hypertonic, isotonic, hypotonic).

_____ c. This hypotonic filtrate then flows through the _____ and _____ tubules.

_____ d. Does some water diffuse out of these tubules?

_____ 4. ADH increases membrane permeability to _____ .

_____ a. Therefore, the volume of urine _____ (increases, remains the same, decreases), and

_____ b. urine becomes more _____ (dilute, concentrated).

_____ 5. Aldosterone increases the reabsorption of _____ ions.

_____ a. Therefore, the amount of Na$^+$ in urine _____ (increases, remains the same, decreases).

_____ b. If Na$^+$ is reabsorbed, will water follow?

_____ c. Will the amount of water reabsorption always be the same?

_____ d. Therefore, the volume of urine _____ (increases, remains the same, decreases) and the blood volume _____ (increases, remains the same, decreases).

_____ e. Atrial natriuretic hormone inhibits ADH and aldosterone secretion, and therefore, causes Na$^+$ and water excretion to _____ (increase, remain the same, decrease).

Answers to Exercise 17.11: 1. unregulated, regulated; 2. ADH, aldosterone; 3. Na$^+$, urea; 3a. no; 3b. hypertonic, hypotonic; 3c. distal convoluted tubule, collecting tubule; 3d. yes; 4. water; 4a. decreases; 4b. concentrated; 5. Na$^+$; 5a. decreases; 5b. yes; 5c. no (this depends upon the membrane permeability to water, which depends upon the ADH levels); 5d. decreases, increases (plays minor role in regulation of blood volume); 5e. increase.

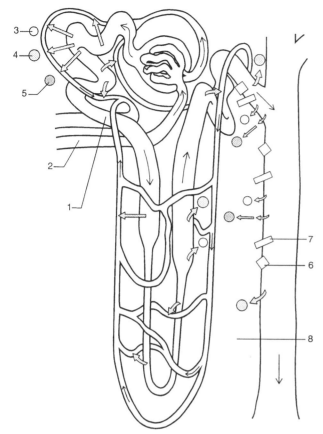

a. Solute and water reabsorption.

Figure 17.11. Hormonal effects on reabsorption.

b. Effects of ADH.

c. Effects of aldosterone.

Hormonal Regulation of Reabsorption

L. Secretion

Color and label:
1. ○ filtrate (yellow)
2. ○ blood (red)
3. ○ tubule cell
   a. ○ Na⁺K⁺ pump
   b. ○ carrier
4. ○ interstitial fluid
5. ○ K⁺
6. ○ H⁺
7. ○ drugs
8. ○ organic cations and anions
9. ○ Na⁺

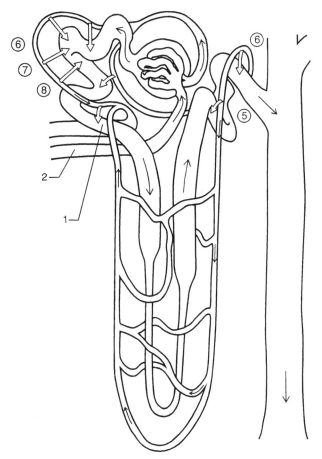

Figure 17.12a. Materials secreted.

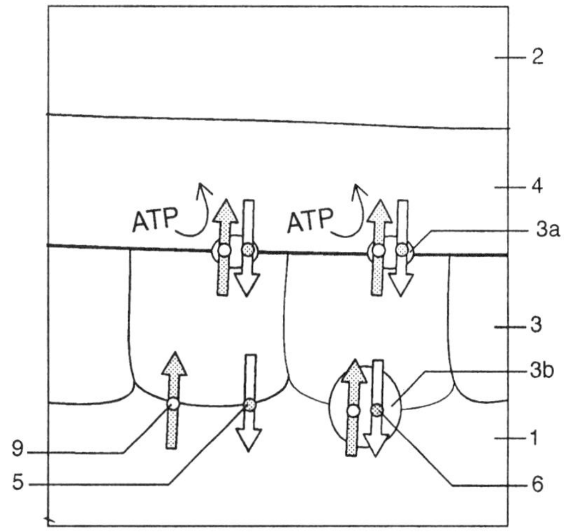

Figure 17.12b. Mechanisms of K⁺ and H⁺ secretion.

Exercise 17.12:

_____ 1. Secretion is the movement of materials from _____ to _____ .

_____ 2. Secretion occurs at the _____ and _____ (nephron parts).

_____ 3. Secreted materials include _____ , _____ , _____ , and _____ .

_____   a. Is K⁺ secretion energy dependent?

_____   b. H⁺ ion secretion depends upon an exchange with _____ .

_____   c. Is H⁺ secretion energy dependent?

_____ 4. Do secreted materials end up in urine?

Answers to Exercise 17.12: 1. blood, filtrate; 2. proximal convoluted tubule, distal convoluted tubule; 3. K⁺, H⁺, drugs, organic cations and anions; 3a. yes (depends upon Na⁺K⁺ pump); 3b. Na⁺; 3c. indirectly depends upon Na⁺K⁺ pump; 4. yes.

M. Juxtaglomerular Apparatus

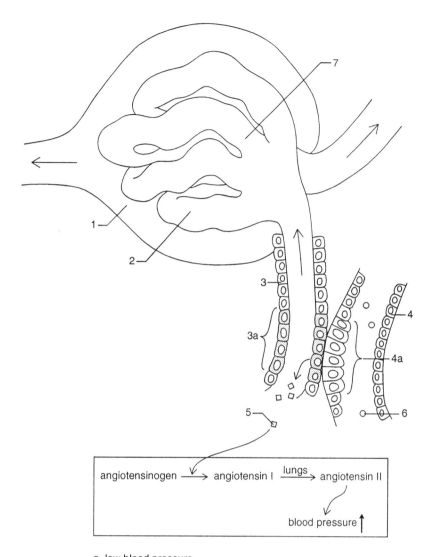

Color and label:
1. ○ filtrate (yellow)
2. ○ blood (red)
3. ○ afferent arteriole (cells)
   a. ○ juxtaglomerular cells
4. ○ distal convoluted tubule (cells)
   a. ○ macula densa
5. ○ renin
6. ○ Na$^+$ ions

Label:
7. glomerulus

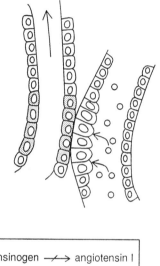

a. low blood pressure

b. high blood pressure

Figure 17.13. Juxtaglomerular regulation of blood pressure.

Exercise 17.13:

_____ 1. Juxtaglomerular cells are located in the wall of the _____ .

_____ a. When blood pressure is low, these cells make the enzyme _____ .

_____ b. Renin converts angiotensinogen in the plasma to _____ .

_____ c. Angiotensin II causes blood pressure to _____ (increase, remain the same, decrease).

_____ 2. Macula densa cells are located in the wall of the _____ .

_____ a. High Na$^+$ concentrations in blood plasma cause filtrate Na$^+$ concentrations to _____ (increase, decrease).

_____ b. High Na$^+$ retention means greater water retention, which, in turn, _____ (increases, decreases) blood pressure.

_____ c. Macula densa cells detect high Na$^+$ in filtrate and inhibit the production of _____ by the juxtaglomerular cells.

_____ d. This causes the blood pressure to _____ (increase, decrease).

Answers to Exercise 17.13: 1. afferent arteriole; 1a. renin; 1b. angiotensin I; 1c. increase; 2. distal convoluted tubule; 2a. increase; 2b. increases; 2c. renin; 2d. decrease.

Chapter 18: Fluids, Electrolytes, and Acid-Base Balance

A. Fluid Compartments

Color and label the major fluid compartments:
1. ○ intracellular fluid
2. ○ interstitial fluid
3. ○ blood plasma

Color and label:
4. ○ lymph

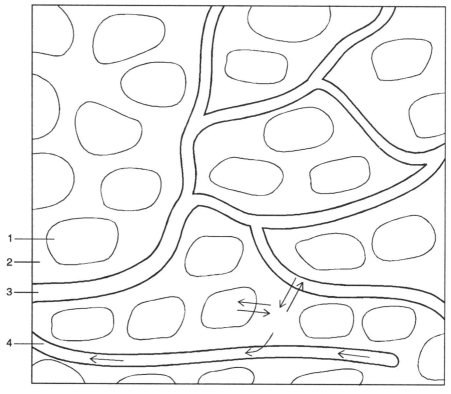

Figure 18.1. Fluid compartments and fluid exchange.

Exercise 18.1:

_____ 1. The major fluid compartments are _____ , _____ , and _____ .

_____ a. The major extracellular compartments are _____ and _____ .

_____ b. Lymph, cerebrospinal fluid, and synovial fluid belong to the _____ (extracellular, intracellular) compartment.

_____ 2. In order for fluid to move between the blood plasma and intracellular compartment, it must travel through the _____ .

_____ 3. Fluid enters lymphatic vessels from the _____ and drains into the _____ . (see lymph flow, p. 192)

Answers to Exercise 18.1: 1. intracellular fluid, interstitial fluid, blood plasma; 1a. blood plasma, interstitial fluid; 1b. extracellular; 2. interstitial fluid; 3. interstitial compartment, blood (at the right lymphatic duct and thoracic duct).

B. Electrolytes

Color each ion differently.

Label the compartments:
a. blood plasma
b. interstitial fluid
c. intracellular fluid

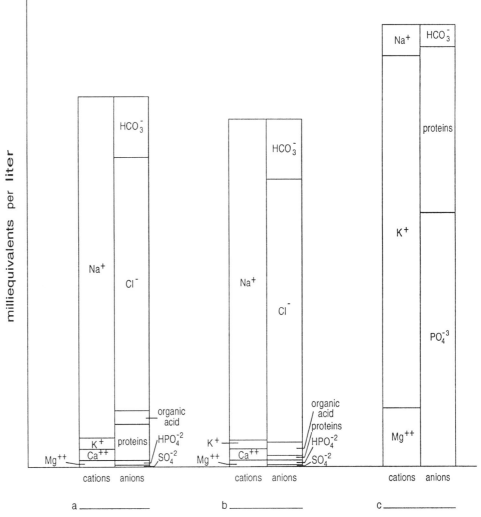

Figure 18.2. Concentrations of electrolytes in each compartment.

Exercise 18.2:

1. Do the concentrations of cations equal the concentration of anions in each compartment?

2. Which compartment shows the greatest variation in electrolyte concentrations?

   a. These differences in concentration are maintained because of the cell _____ (structure).

   b. The Na^+ ion concentration is greater in the _____ (extracellular, intracellular) fluid.

   c. The K^+ ion concentration is greater in the _____ (extracellular, intracellular) fluid.

   d. This unequal distribution of Na^+ and K^+ is important because _____ .

3. The protein concentration is greater in the _____ (plasma, interstitial fluid).

   a. Therefore, which compartment tends to retain water?

   b. If plasma proteins decrease or capillaries leak proteins, tissues swell (edema). Which compartment is retaining water now?

Answers to Exercise 18.2: 1. yes; 2. intracellular; 2a. membrane; 2b. extracellular; 2c. intracellular; 2d. it is the basis of the action potential; 3. plasma; 3a. plasma; 3b. interstitial fluid.

C. Fluid and Electrolyte Movement

Color and label:
1. ◯ intestines
2. ◯ skin
3. ◯ lungs
4. ◯ kidney
5. ◯ stomach

Label:
6. body tissues

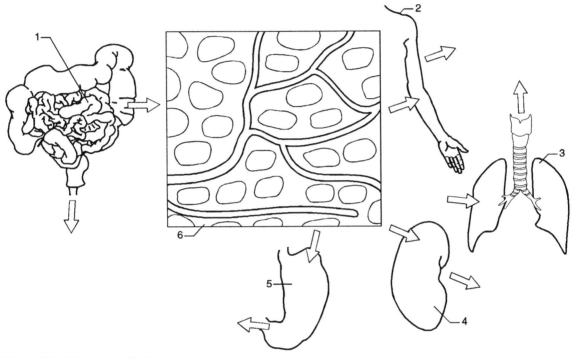

Figure 18.3. Movement of fluids.

Exercise 18.3:

_____ 1. Fluids enter the body via the _____ .

_____ 2. Fluids leave the body via the _____ , _____ , _____ , _____ , and _____ .

_____ 3. Electrolytes enter the body via the _____ .

_____ 4. What effect will gastrointestinal disorders have on salt absorption by the digestive tract?

Answers to Exercise 18.3: 1. intestines; 2. intestines, stomach, kidneys, lungs, skin; 3. digestive system; 4. decrease absorption.

D. Water Balance

Trace the arrows associated with each:

○ antidiuretic hormone (ADH)
○ aldosterone
○ atrial natriuretic hormone (ANH)
○ angiotensin II

Color and label:

1. ○ nephron
   a. juxtaglomerular apparatus
   b. glomerular filtration rate
2. hypothalamus
   a. ○ osmoreceptors
   b. ○ thirst center
3. ○ adrenal gland
4. ○ heart

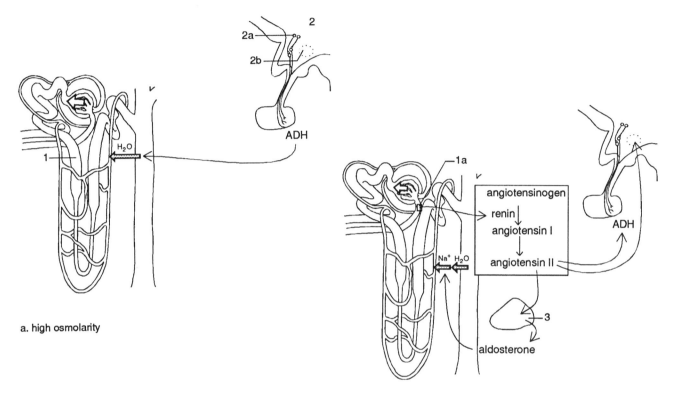

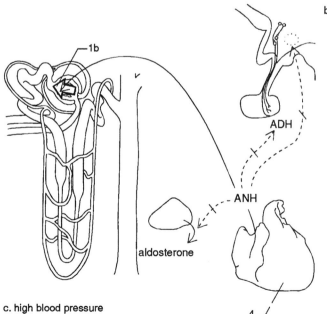

Figure 18.4. Regulation of water balance.

Exercise 18.4:

_____ 1. Water intake is controlled by the _____ center of the _____ .

_____ 2. Water loss is controlled by the _____ .

_____ 3. Dehydration _____ (increases, decreases) the osmolarity (solute concentration) of fluid compartments.

_____   a. Osmolarity is measured by _____ in the _____ .

_____   b. The osmoreceptors stimulate the production of _____ .

_____   c. ADH stimulates the nephron to _____ .

_____ 4. Dehydration _____ (increases, decreases) blood volume _____ , and therefore, _____ (increases, decreases) blood pressure .

_____   a. A decrease in blood pressure is measured by the _____ ,

_____   b. which releases _____ , which stimulate _____ production.

_____   c. Then, angiotensin II stimulates _____ , _____ , and _____ .

_____   d. Aldosterone causes reabsorption of _____ .

_____   e. Does this Na^+ reabsorption increase the likelihood of water reabsorption?

_____ 5. Water retention _____ (increases, decreases) blood volume, and therefore, _____ (increases, decreases) blood pressure.

_____   a. Increased blood pressure _____ (decreases, increases) the glomerular filtration rate.

_____   b. This means urine output _____ (decreases, increases).

_____   c. Therefore, the blood volume _____ (decreases, increases).

_____ 6. High blood pressure causes the heart to release _____ .

_____   a. ANH _____ (decreases, increases) thirst, which means water intake _____ (decreases, increases).

_____   b. ANH also inhibits _____ and _____ , which means water reabsorption _____ (decreases, increases).

_____   c. ANH causes the glomerular filtration rate to _____ (decrease, increase). (see regulation of GFR, p. 276)

_____   d. This means urine output _____ (decreases, increases).

_____   e. Therefore, the blood volume _____ (decreases, increases).

_____ 7. In summary, hormonal regulation of water balance depends upon measuring _____ and _____ .

Answers to Exercise 18.4: 1. thirst, hypothalamus; 2. nephron (kidney); 3. increases; 3a. osmoreceptors, hypothalamus; 3b. ADH; 3c. reabsorb water; 4. decreases, decreases; 4a. juxtaglomerular apparatus; 4b. renin, angiotensin II; 4c. thirst center, ADH production, aldosterone production; 4d. Na^+; 4e. yes (the exact amount of water reabsorbed depends upon the membrane permeability to water that is ADH dependent); 5. increases, increases; 5a. increases; 5b. increases; 5c. decreases; 6. ANH; 6a. decreases, decreases; 6b. ADH, aldosterone, decreases; 6c. increase; 6d. increases; 6e. decreases; 7. osmolarity, blood pressure.

E. Acid Regulation

Color and label:
- ○ HCO_3^- bicarbonate ion
- ○ H_2CO_3 carbonic acid

Color:
- ○ H^+
- ○ protein
- ○ Na^+
- ○ CO_2

Label:
1. gut
2. cells
3. circulatory system
4. lung
5. kidney

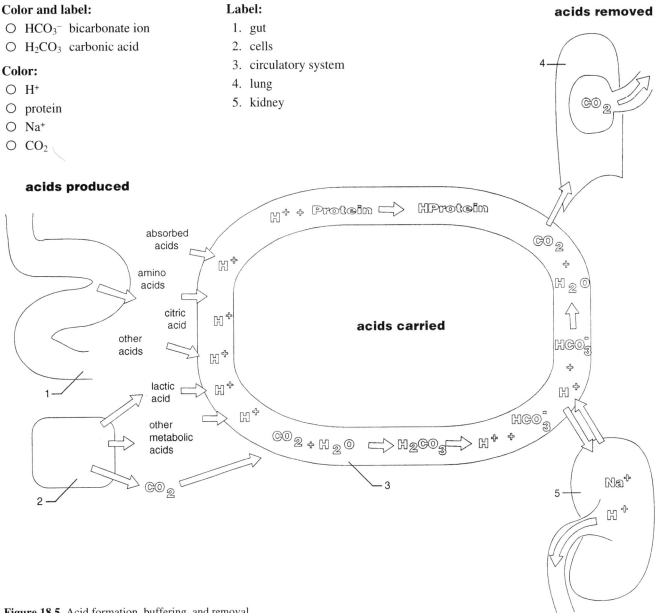

Figure 18.5. Acid formation, buffering, and removal.

Exercise 18.5:

_____ 1. Acids are produced by the metabolic activities of _____ .

_____ 2. Acids are taken into the body by the _____ .

_____ 3. When acids enter the circulatory system, they are buffered by _____ and _____ .

_____ 4. CO_2 reacts with water to produce _____ .

_____ 5. Carbonic acid dissociates (breaks down) to _____ and _____ .

_____ 6. Conditions in the lung favor the conversion of H^+ and HCO_3^- to _____ and _____ .

_____ 7. The kidneys exchange H^+ for _____ .

Answers to Exercise 18.5: 1. cells; 2. gut; 3. protein, bicarbonate; 4. carbonic acid; 5. carbon dioxide, water; 6. carbon dioxide, water; 7. sodium.

Chapter 19: Reproductive System

A. Reproductive Function

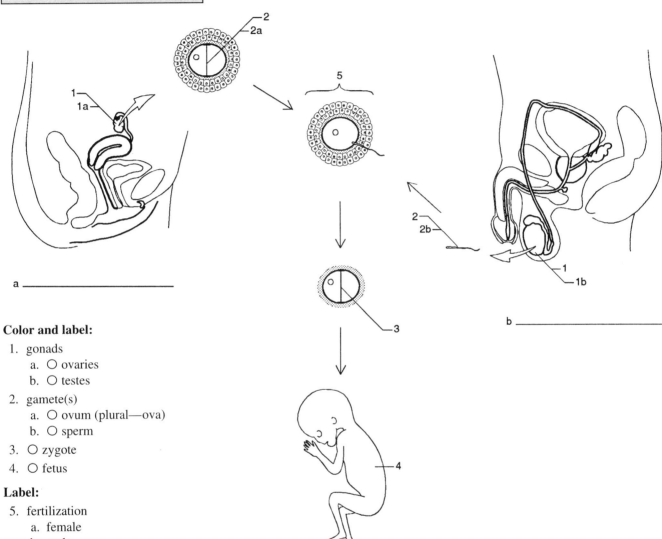

Figure 19.1. Zygote formation.

Color and label:
1. gonads
   a. ○ ovaries
   b. ○ testes
2. gamete(s)
   a. ○ ovum (plural—ova)
   b. ○ sperm
3. ○ zygote
4. ○ fetus

Label:
5. fertilization
   a. female
   b. male

Exercise 19.1:

_____ 1. Ova are made in the _____ , while sperm are made in the _____ .

_____ 2. The fertilization of an ovum by a sperm produces a _____ , which develops into a _____ .

_____ 3. The normal zygote has 23 pairs of chromosomes. Therefore, the ovum contributes _____ chromosomes and the sperm contributes _____ chromosomes.

Answers to Exercise 19.1: 1. ovaries, testes; 2. zygote, fetus; 3. 23, 23.

B. Development

Color:
- ○ Y chromosome
- ○ MIS (müllerian inhibiting substance)
- ○ testosterone
- ○ DHT (dihydrotestosterone)

Color and label:
1. ○ primitive gonads
   a. testes
   b. ovaries
2. ○ wolffian duct
   a. male duct system
3. ○ müllerian duct
   a. female duct system
4. ○ gubernaculum
   a. ligament
5. ○ primitive phallus (genital tubercle)
   a. glans (head) of the penis
   b. clitoris
6. ○ urogenital slit
   a. urethra
   b. vagina
7. ○ urogenital folds
   a. raphe
   b. labia minor
8. ○ labioscrotal folds
   a. scrotum
   b. labia major

Label:
9. anus
10. prepuce
11. shaft of penis
12. inguinal canal
13. vestibule

Exercise 19.2:

_____ 1. When the Y chromosome is present, the primitive gonads develop into _____ .

_____ a. The testes produce a hormone that acts on the wolffian ducts, causing them to develop into _____ (male, female) reproductive structures. The hormone is _____ .

_____ b. The testes also make _____ , which causes the breakdown of the _____ .

_____ c. Testosterone is converted to _____ , which, in turn, causes the development of _____ (male, female) external genitalia.

_____ 2. When the Y chromosome is absent, the primitive gonads develop into _____ .

_____ a. In the absence of testosterone, the wolffian ducts _____ .

_____ b. In the absence of MIS, the müllerian ducts become _____ .

_____ c. In the absence of dihydrotestosterone, the external genitalia become _____ .

3. Complete the table with the names of the external structures that form.

Embryonic Structure	Male	Female
a. primitive phallus		
b. urogenital slit		
c. urogenital folds		
d. labioscrotal folds		

_____ 4. The vestibule contains two openings, the _____ and _____ .

_____ 5. The fusion of the urethral folds in the male creates the tube within the penis called the _____ .

_____ 6. Usually before birth, the testes move from the abdominal cavity into the _____ , through the _____ .

Answers to Exercise 19.2: 1. testes; 1a. male, testosterone; 1b. MIS, müllerian ducts; 1c. dihydrotestosterone, male; 2. ovaries; 2a. break down; 2b. female ducts; 2c. female external genitalia; 3a. head of penis, clitoris; 3b. no structure (slit closes), vestibule; 3c. raphe (ventral surface of penis and scrotum) and urethra, labia minor; 3d. scrotum, labia major; 4. urethra, vagina; 5. urethra; 6. scrotum, inguinal canal.

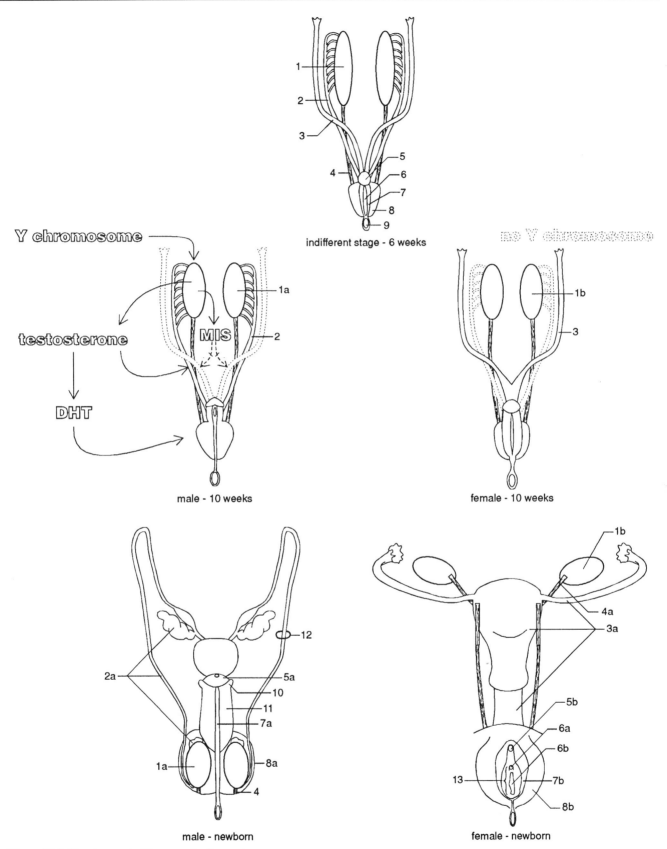

Figure 19.2. Development of the reproductive system.

Development 295

C. Male Anatomy

Color and label:
1. ○ testes
2. ○ epididymis
3. ○ vas deferens
    a. ampulla
4. ○ seminal vesicles
5. ○ ejaculatory duct
6. ○ prostate
7. ○ Cowper's gland
8. ○ urethra
9. penis
    a. ○ corpus cavernosum
    b. ○ corpus spongiosum
    c. ○ foreskin (prepuce)

Label:
10. scrotum
11. urinary bladder
12. inguinal canal
13. pubic bone
14. rectum

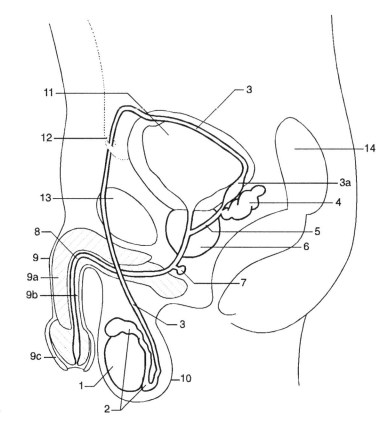

Figure 19.3. Lateral view of male anatomy.

Exercise 19.3:

1. Sperm made in the testes move into the _____ where they mature.

2. Sperm leave the scrotum via the _____ and are stored in the _____ .

3. During ejaculation, sperm leave the ampulla and enter the _____ , mixing with secretions released from the _____ .

4. This semen (sperm-secretion mixture) then enters the _____ , where more secretions are added from the _____ .

5. At or before ejaculation, a fluid may be released from _____ , which lubricates and neutralizes the urethra.

6. Given the location of the prostate, its enlargement can cause problems with _____ .

7. The cremaster muscles that line the scrotum contract in a cold environment. Therefore, the testes move _____ (closer to, away from) the body and are _____ (warmed, cooled).

8. The cremaster muscles relax in a warm environment. Therefore, the testes move _____ (closer to, away from) the body and are _____ (warmed, cooled).

Answers to Exercise 19.3: 1. epididymis; 2. vas deferens, ampulla; 3. ejaculatory ducts, seminal vesicles; 4. urethra, prostate; 5. Cowper's gland; 6. urination; 7. closer to, warmed; 8. away from, cooled.

D. Testes

Color and label:
1. ○ seminiferous tubules
2. ○ spermatogonia
3. ○ spermatocytes
4. ○ sperm
5. ○ sustentacular cells (Sertoli's cells)
6. ○ Leydig's cells (interstitial cells)
7. ○ testosterone

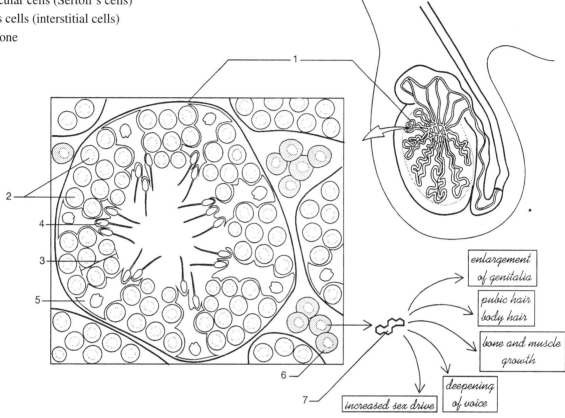

Figure 19.4. Midsagittal section of testis (right) and cross section of seminiferous tubule (left).

Exercise 19.4:

_____ 1. The testes contain _____ tubules.

_____ 2. The outer cells of the seminiferous tubules, the _____, divide to form _____.

_____ 3. Spermatocytes then develop into _____.

_____ 4. The cells within the seminiferous tubules that support sperm development are the _____.

_____ 5. The cells between the seminiferous tubules, _____ cells, make _____.

_____ 6. One effect of testosterone is _____.

_____ 7. To summarize, the functions of the testes are production of _____ and _____.

Answers to Exercise 19.4: 1. seminiferous; 2. spermatogonia, spermatocytes; 3. sperm; 4. sustentacular cells; 5. Leydig's, testosterone; 6. enlargement of genitalia, growth of pubic hair and body hair, bone and muscle growth, deepening of voice, increase in sex drive; 7. sperm, testosterone.

Testes

E. Male Hormonal Regulation

Color and label:

1. ○ Sertoli's cells
2. ○ inhibin
3. ○ Leydig's (interstitial) cells
4. ○ testosterone
   ○ GnRH (gonadotropic releasing hormone)
   ○ FSH (follicle stimulating hormone)
   ○ LH (ICSH-interstitial cell stimulating hormone)

Label:

5. seminiferous tubules
6. hypothalamus
7. pituitary

Exercise 19.5:

1. Testicular function is regulated by _____ and _____ .
   a. These hormones are made by the _____ .
   b. Their production is controlled by _____ , made by the part of the brain called the _____ . (see p. 156)

2. Sperm formation is initiated by the hormone _____ through its action on the _____ cells.
   a. Sertoli's cells make _____ .
   b. Inhibin inhibits _____ production.
   c. Therefore, the rate of sperm production is regulated by _____ (positive, negative) feedback. (see p. 158)

3. Testosterone is made by _____ cells when stimulated by _____ .
   a. Does testosterone support sperm production?
   b. Testosterone inhibits the production of _____ and _____ .
   c. Therefore, testosterone production is regulated by _____ (positive, negative) feedback.

4. Taking testosterone as a drug would cause sperm production to _____ (increase, decrease).

Answers to Exercise 19.5: 1. FSH, LH; 1a. anterior pituitary; 1b. GnRH, hypothalamus; 2. FSH, Sertoli's; 2a. inhibin; 2b. FSH; 2c. negative; 3. Leydig's (interstitial), LH (ICSH); 3a. yes; 3b. GnRH, LH; 3c. negative; 4. decrease (testosterone would inhibit GnRH, which would cause FSH levels to decrease).

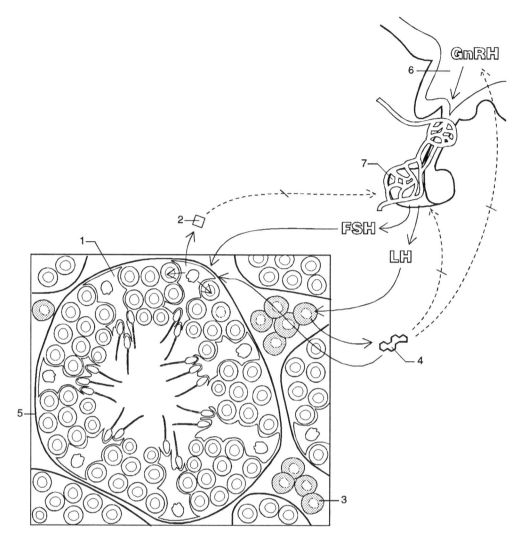

Figure 19.5. Hormonal control of testicular function.

F. Female Anatomy

Color (trace) and label:
1. ○ ovaries
2. ○ fallopian tubes (uterine tubes)
   a. fimbriae
3. uterus
   a. ○ perimetrium
   b. ○ myometrium
   c. ○ endometrium
   d. cervix
4. ○ vagina
5. ○ immature ovum (oocyte)
6. ○ sperm
7. ○ embryo
8. ○ ligaments
9. ○ clitoris

Label:
10. pubic bone
11. urinary bladder
12. urethra
13. labia minor
14. labia major
15. rectum
    a. ovulation
    b. fertilization
    c. implantation

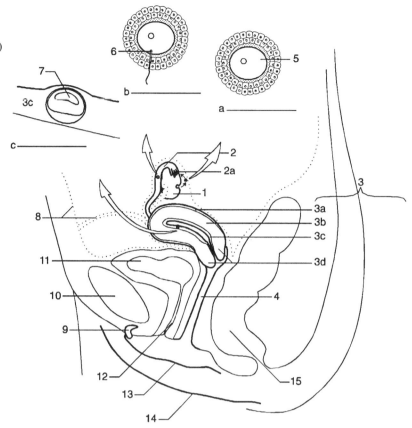

Figure 19.6. Lateral view of female anatomy.

Exercise 19.6:

1. The formation of ova begins in the _____ .

2. The immature ovum (oocyte) is released from the ovary in a process called _____ .

3. The ovum is drawn into the _____ by the fingerlike _____ .

4. During intercourse, sperm are deposited at the base of the _____ near the part of the uterus called the _____ .

5. Sperm move into the uterus through the _____ and then up the _____ .

6. Penetration of the ovum by a sperm (called _____) triggers ovum maturation. This normally occurs in the _____ .

7. The fertilized ovum, the zygote, undergoes mitosis as it moves down the fallopian tube, becoming a(n) _____ .

8. The embryo implants and develops in the _____ (layer) of the uterus.

9. The ovaries, fallopian tubes, and uterus are held in place by _____ .

Answers to Exercise 19.6: 1. ovaries; 2. ovulation; 3. fallopian tubes, fimbriae; 4. vagina, cervix; 5. cervix, fallopian tubes; 6. fertilization, fallopian tubes; 7. embryo; 8. endometrium; 9. ligaments.

G. Ovarian Cycle

Color:
- ○ estrogen
- ○ progesterone

Outline and label:
1. ○ ovary

○ Color the immature ova the same and label:
2. oogonium
3. primary oocyte
4. secondary oocyte

Color and label:
5. ○ follicular cells
6. ○ thecal cells

Label:
7. primordial follicle
8. primary follicle
9. Graafian follicle
   a. antrum
10. corpus luteum
11. corpus albicans
    (degenerating corpus luteum)

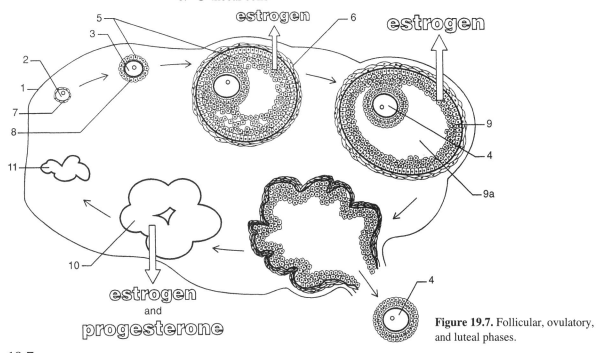

Figure 19.7. Follicular, ovulatory, and luteal phases.

Exercise 19.7:

_____ 1. The ovaries of a newborn female contain only primary follicles with their _____ oocytes.

2. When sexual maturity is reached, follicles mature during the follicular phase of the ovarian cycle. Follicle maturation means

_____   a. the chromosome number in the oocyte decreases to _____ (number).

_____   b. the number of follicular cells _____ (increases, remains the same, decreases).

_____   c. the growing follicle makes the hormone _____ .

_____ 3. The mature follicle is called the _____ .

_____   a. It ruptures, releasing the _____ in the process called _____ .

_____   b. The remaining follicular cells become the _____ .

_____ 4. During the luteal phase, the corpus luteum makes _____ and _____ .

_____ 5. If there is no pregnancy, the corpus luteum dies 11 days after ovulation, and the levels of estrogen and progesterone _____ (increase, decrease, remain the same).

Answers to Exercise 19.7: 1. primary; 2a. 23; 2b. increases; 2c. estrogen; 3. Graafian follicle; 3a. secondary oocyte, ovulation; 3b. corpus luteum; 4. estrogen, progesterone; 5. decrease.

H. Female Hormonal Regulation

Color:
- ○ GnRH (gonadotropic releasing hormone)
- ○ FSH (follicle stimulating hormone)
- ○ LH (luteinizing hormone)
- ○ estrogen
- ○ progesterone

Outline and label:
1. ○ ovary

Color and label:
2. ○ developing follicles
3. ○ corpus luteum

Label:
4. hypothalamus
5. anterior pituitary
    a. follicular phase
    b. ovulatory phase
    c. luteal phase

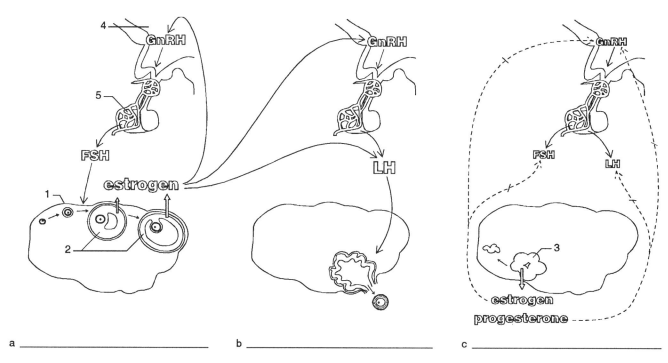

Figure 19.8. Hormonal regulation of ovarian function.

Exercise 19.8:

1. The hormones that control ovarian function are _____ and _____ .
   a. These hormones are made in the _____ .
   b. Their production is controlled by _____ made by the _____ .

2. FSH causes _____ .
   a. The follicular cells make _____ .
   b. Estrogen stimulates production of _____ and _____ .
   c. Increased GnRH means _____ (more, less) FSH, which, in turn, means _____ (more, less) estrogen.
   d. Thus, estrogen from developing follicles has a _____ (positive, negative) feedback effect on FSH and LH.

3. An LH surge causes _____ and the formation of the _____ .

4. The corpus luteum makes _____ and _____ .
   a. Estrogen made during the luteal phase _____ (inhibits, stimulates) GnRH and FSH by _____ (positive, negative) feedback.
   b. Progesterone _____ (inhibits, stimulates) GnRH and LH by _____ (positive, negative) feedback.
   c. Lower levels of GnRH, FSH, and LH mean new follicles _____ (will, will not) develop.

5. If there is no pregnancy, estrogen and progesterone levels _____ (increase, remain the same, decrease).
   a. This means the inhibition of GnRH, FSH, and LH _____ (is reversed, continues).
   b. Therefore, new follicles _____ (will, will not) develop.

Answers to Exercise 19.8: 1. FSH, LH; 1a. anterior pituitary; 1b. GnRH, hypothalamus; 2. maturation of follicles; 2a. estrogen; 2b. GnRH, LH; 2c. more, more; 2d. positive; 3. ovulation, corpus luteum; 4. estrogen, progesterone; 4a. inhibits, negative; 4b. inhibits, negative; 4c. will not; 5. decrease; 5a. is reversed; 5b. will.

Female Hormonal Regulation

I. Uterine Cycle

Color:
- ○ FSH (follicle stimulating hormone)
- ○ LH (luteinizing hormone)
- ○ estrogen
- ○ progesterone

Color and label:
1. ○ ovary
   a. ○ developing follicles
   b. ○ corpus luteum
2. endometrium (uterus)
   a. ○ stratum basale
   b. ○ stratum functionale
   c. ○ uterine debris and blood (red)
3. ○ arteries (red)
4. ○ veins (blue)
5. glands
   a. ○ glycogen

Label:
a. menstrual phase
b. proliferative phase
c. secretory phase

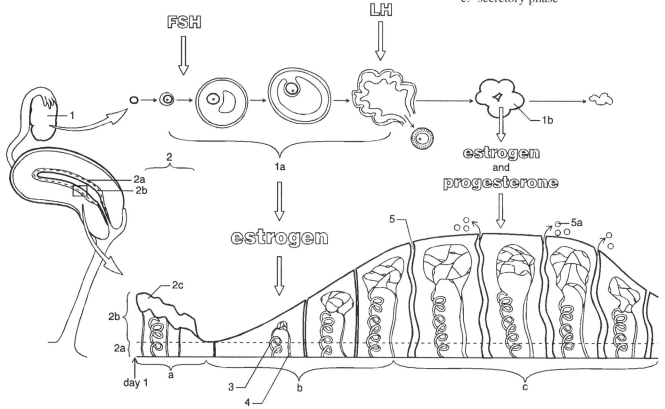

Figure 19.9. Uterine changes and hormonal control.

Exercise 19.9:

_____ 1. During the menstrual cycle, the stratum _____ remains the same, while the stratum _____ changes.

_____ 2. During the proliferative phase of the menstrual cycle, the ovary is in its _____ phase.
_____ a. The developing follicles make the hormone _____ .
_____ b. This hormone stimulates the endometrium to _____ .

_____ 3. During the secretory phase of the menstrual cycle, the ovary is in its _____ phase.
_____ a. The corpus luteum makes _____ and _____ .
_____ b. Progesterone _____ (dampens, enhances) the effects of estrogen on uterine growth.
_____ c. Progesterone also stimulates growth of _____ .

_____ 4. During the secretory phase, the uterine glands make _____ .
_____ a. Uterine glycogen provides _____ .
_____ b. If there is a pregnancy, is the uterus prepared for implantation?

304 Chapter 19 Reproductive System

5. If there is no pregnancy,
   a. the corpus luteum _____ (dies, lives), and
   b. estrogen and progesterone levels _____ (increase, remain the same, decrease).
   c. Therefore, the uterine lining _____ (dies, grows).

6. The sloughing off of this uterine debris is called _____ .
   a. This begins on day _____ (number) of the menstrual cycle.
   b. The menstrual phase corresponds to the beginning of the _____ (follicular, ovulatory, luteal) phase in the ovary.

7. During menopause, the ovaries become less responsive to FSH and LH.
   a. Will follicles develop?
   b. Therefore, the levels of estrogen _____ (increase, remain the same, decrease).
   c. Estrogen dependent effects, such as uterine proliferation, _____ (continue, stop).
   d. What happens to ovulation and menstruation?

Answers to Exercise 19.9: 1. basale, functionale; 2. follicular; 2a. estrogen; 2b. grow; 3. luteal; 3a. estrogen, progesterone; 3b. dampens; 3c. glands; 4. glycogen; 4a. nutrients for the embryo; 4b. yes; 5a. dies; 5b. decrease; 5c. dies; 6. menstruation; 6a. 1; 6b. follicular; 7a. no; 7b. decrease; 7c. stop; 7d. they stop.

J. Menstrual Cycle Timing

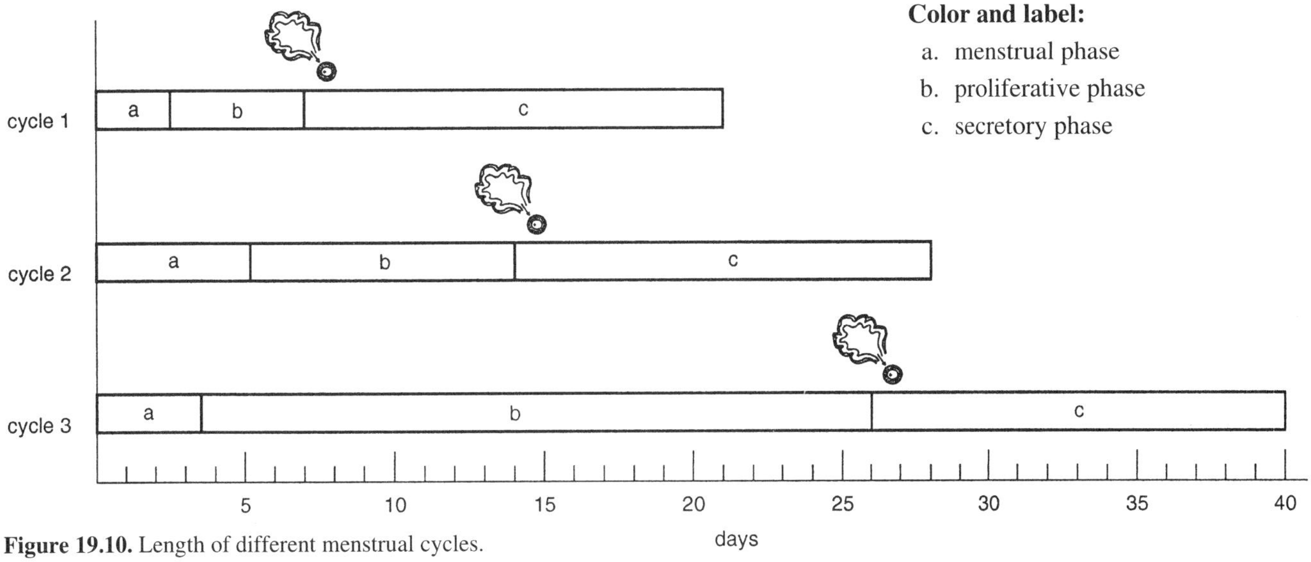

Color and label:
a. menstrual phase
b. proliferative phase
c. secretory phase

Figure 19.10. Length of different menstrual cycles.

Exercise 19.10:

1. Of the menstrual cycles shown, the shortest is _____ days and the longest is _____ days.
   a. The phases that vary in length are the _____ and _____ .
   b. The phase that is the same length in each cycle is the _____ .
   c. This phase is about _____ days long.

2. The secretory phase begins when _____ occurs.

3. Therefore, ovulation occurs _____ days before the next menstrual cycle.

4. Does ovulation ever occur at the midpoint in a cycle?

Answers to Exercise 19.10: 1. 21, 40; 1a. menstrual phase, proliferative phase; 1b. secretory phase; 1c. 14; 2. ovulation; 3. 14; 4. yes (in a 28 day cycle).

K. Pregnancy

Color:
- ○ HCG (human chorionic gonadotropic hormone)
- ○ estrogen
- ○ progesterone

Color and label:
1. ovary
   a. ○ Graafian follicle
   b. ○ corpus luteum
2. ○ fallopian tube
3. ○ ovum
4. ○ embryo
   a. ○ chorion
5. uterus
   a. ○ endometrium

Label:
6. sperm

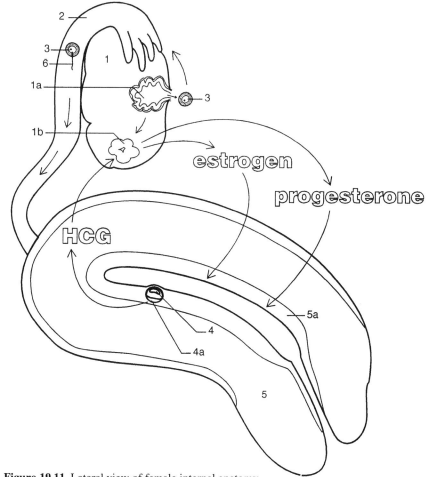

Figure 19.11. Lateral view of female internal anatomy.

Exercise 19.11:

_____ 1. In the absence of pregnancy, the lifespan of the corpus luteum is _____ days. (see menstrual cycle, p. 305)

_____ 2. If an ovum is fertilized, the embryo _____ in the uterus about 7 days after ovulation.

_____ 3. The chorion, the membrane that surrounds the embryo, makes _____ .

_____ 4. HCG stimulates the corpus luteum to continue making _____ and _____ .

_____ 5. Therefore, during pregnancy the uterine lining _____ (sloughs off, is maintained).

Answers to Exercise 19.11: 1. 11; 2. implants; 3. HCG; 4. estrogen, progesterone; 5. is maintained.

L. Mammary Glands

Color:
- ○ estrogen
- ○ progesterone
- ○ prolactin
- ○ PRH (prolactin releasing hormone)
- ○ PIH (prolactin inhibiting hormone)
- ○ oxytocin

Color (trace) and label:
1. ○ gland
   a. ducts
   b. ampulla
2. ○ suspensory ligaments
3. ○ adipose tissue
4. ○ muscle
5. ○ ribs

Label:
6. nipple
7. areola
8. hypothalamus
9. anterior pituitary
10. posterior pituitary

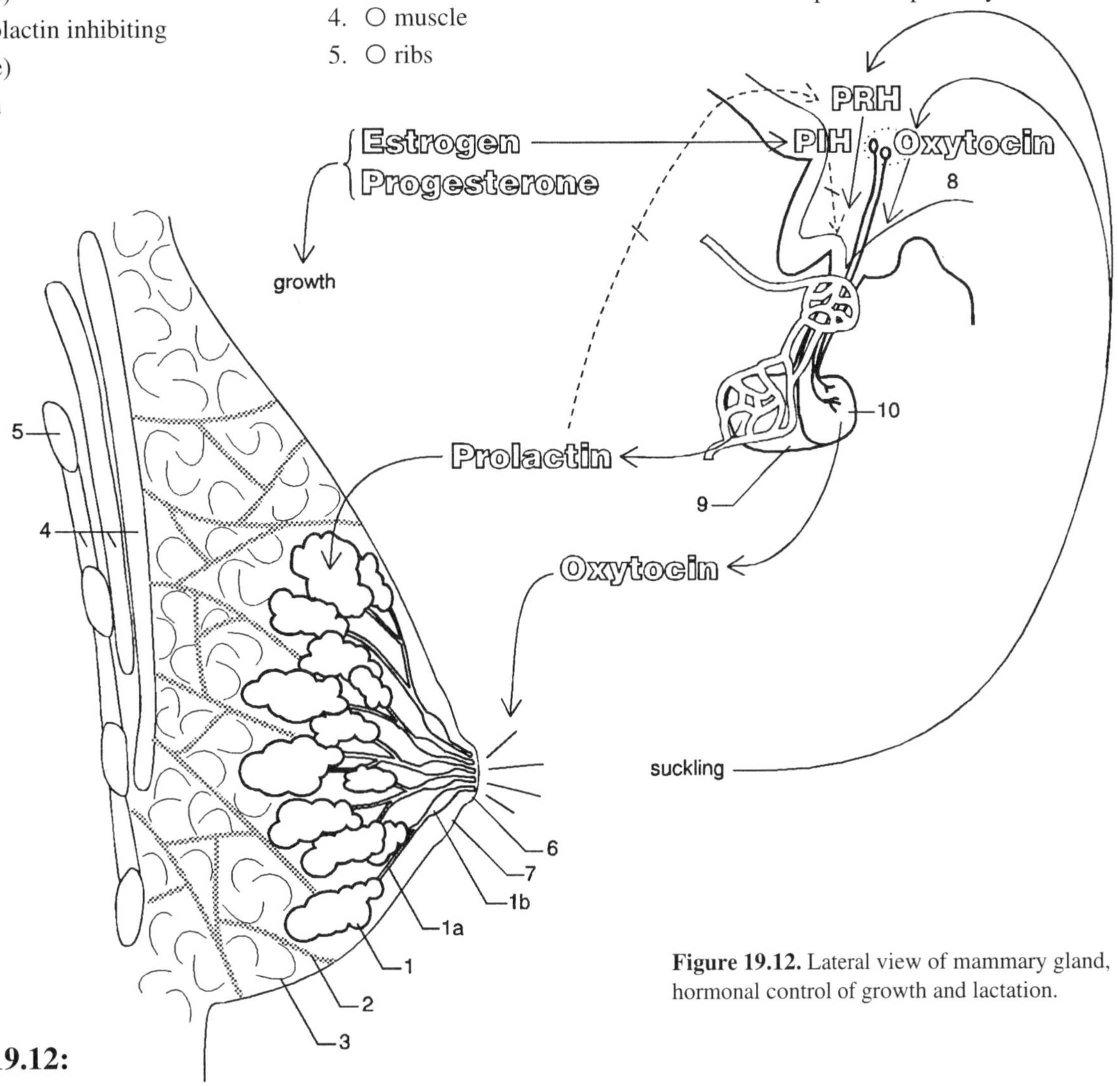

Figure 19.12. Lateral view of mammary gland, hormonal control of growth and lactation.

Exercise 19.12:

_____ 1. The mammary glands grow and develop when stimulated by _____ and _____ .

_____ 2. Estrogen _____ (stimulates, inhibits) PIH production.

_____ a. Therefore, when estrogen is high (i.e., pregnancy) milk production _____ (will, will not) occur.

_____ b. After delivery, estrogen levels decrease, PIH _____ (increases, decreases), and milk production _____ (can, cannot) occur.

_____ 3. Suckling stimulates production of _____ and _____ .

_____ a. Milk production occurs when _____ stimulates the glands.

_____ b. Milk "let down" (release) occurs with secretion of _____ .

Answers to Exercise 19.12: 1. estrogen, progesterone; 2. stimulates; 2a. will not; 2b. decreases, can; 3. prolactin, oxytocin; 3a. prolactin; 3b. oxytocin.

M. Human Sexual Response

○ Color red. Label:
1. vagina
2. clitoris (erectile tissue)
3. penis (erectile tissue)

Color and label:
4. ○ urethra
5. ○ sex program centers (hypothalamus)
6. ○ pleasure centers (hypothalamus)

○ Color:
S2–S4 (parasympathetic innervation)
vasodilation arrows

○ Color:
L2 (sympathetic innervation)
muscle contractions arrow

Label:
7. emission
8. expulsion

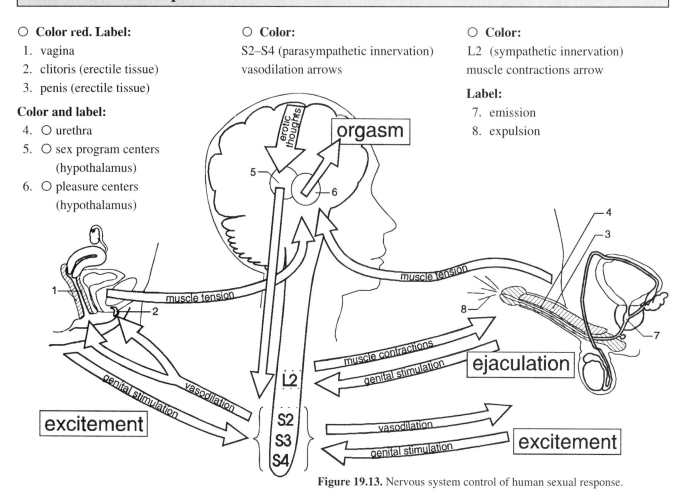

Figure 19.13. Nervous system control of human sexual response.

Exercise 19.13:

1. During excitation, stimulation of the genitalia and/or erotic thoughts increases _____ (parasympathetic, sympathetic) activity.

    a. This causes a(n) _____ (increase, decrease) in blood flow to the genitalia.

    b. The penis and clitoris become _____.

    c. Greater blood flow increases plasma seepage out of the capillaries in the vaginal wall. This causes vaginal lubrication to _____ (increase, decrease).

2. Ejaculation is controlled by the _____ nervous system.

    a. Sperm and semen move into the prostatic urethra during _____.

    b. Contraction of pelvic muscles sends semen out of the urethra during _____.

3. Muscle tension in males and females sends signals up the spinal cord to the _____ of the brain.

4. Stimulation of the pleasure centers causes _____.

5. Name one reflex action in human sexual response.

Answers to Exercise 19.13: 1. parasympathetic; 1a. increase; 1b. erect; 1c. increase; 2. sympathetic; 2a. emission; 2b. expulsion; 3. pleasure centers; 4. orgasm; 5. erection, vaginal lubrication, ejaculation.

Chapter 20: Development and Genetics

A. Gametogenesis

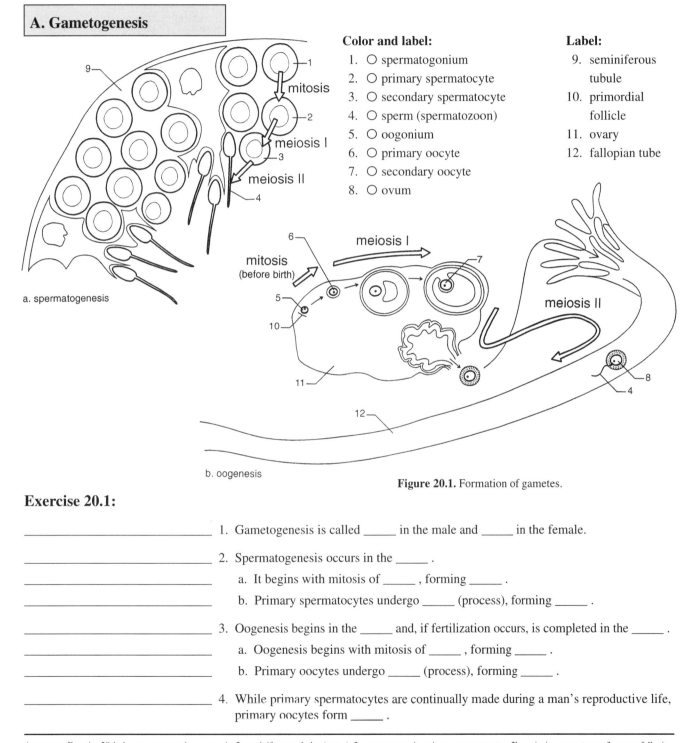

Color and label:
1. ○ spermatogonium
2. ○ primary spermatocyte
3. ○ secondary spermatocyte
4. ○ sperm (spermatozoon)
5. ○ oogonium
6. ○ primary oocyte
7. ○ secondary oocyte
8. ○ ovum

Label:
9. seminiferous tubule
10. primordial follicle
11. ovary
12. fallopian tube

a. spermatogenesis

b. oogenesis

Figure 20.1. Formation of gametes.

Exercise 20.1:

_____ 1. Gametogenesis is called _____ in the male and _____ in the female.

_____ 2. Spermatogenesis occurs in the _____ .

_____   a. It begins with mitosis of _____ , forming _____ .

_____   b. Primary spermatocytes undergo _____ (process), forming _____ .

_____ 3. Oogenesis begins in the _____ and, if fertilization occurs, is completed in the _____ .

_____   a. Oogenesis begins with mitosis of _____ , forming _____ .

_____   b. Primary oocytes undergo _____ (process), forming _____ .

_____ 4. While primary spermatocytes are continually made during a man's reproductive life, primary oocytes form _____ .

Answers to Exercise 20.1: 1. spermatogenesis, oogenesis; 2. seminiferous tubules (testes); 2a. spermatogonia, primary spermatocytes; 2b. meiosis, spermatozoa; 3. ovary, fallopian tube; 3a. oogonia, primary oocytes; 3b. meiosis, ova; 4. before birth.

B. Meiosis

Color and label:
1. ◯ short chromosome
2. ◯ long chromosome

Label:
3. chromatid
4. homologous chromosomes

5. tetrad
6. primary spermatocyte
7. secondary spermatocyte
8. spermatids
9. spermatozoa (sperm)
10. primary oocyte

11. secondary oocyte
12. ovum
13. polar body (bodies)
   a. prophase
   b. metaphase
   c. anaphase
   d. telophase

Exercise 20.2:

_____ 1. Both mitosis and meiosis are preceded by _____ .

_____ a. Mitosis yields cells with _____ (half the, the same) chromosome number as the original cell.

_____ b. Meiosis yields cells with _____ (half the, the same) chromosome number as the original cell.

_____ c. Therefore, meiosis depends upon _____ (1, 2) cell divisions to complete the chromosome reduction.

_____ 2. When chromosomes become visible, each contains two _____ .

_____ a. The chromatids within one chromosome are _____ (identical, similar, different). (see chromosomes and DNA replication, p. 35)

b. Chromosomes that are *similar* (but not identical) are called _____ chromosomes.

_____ 3. In mitosis, homologous chromosomes are arranged _____ (in pairs, randomly), while in meiosis they are arranged _____ .

4. In meiosis, each set of paired homologous chromosomes contains _____ chromatids and is, therefore, called a _____ .

5. Using the terms *chromatids, chromosomes,* and *homologous chromosomes,* complete the following:

_____ a. During mitosis, _____ separate to become _____ .

_____ b. During meiosis I, _____ separate to become _____ .

_____ c. During meiosis II, _____ separate to become _____ .

6. The number of sets of chromosomes can be described as *diploid* (or 2n) for two sets or *haploid* (or n) for one set. Use these terms to describe each of the following:

_____ a. original cell in mitosis

_____ b. end products of mitosis

_____ c. primary spermatocyte, primary oocyte

_____ d. secondary spermatocyte, secondary oocyte

_____ e. spermatid, ovum

_____ 7. Each meiosis yields _____ (number) sperm and _____ (number) ovum.

_____ a. Therefore, _____ (spermatogenesis, oogenesis) maximizes the numbers of sperm cells produced.

_____ b. Unequal cytokinesis in oogenesis provides the ovum with _____ .

c. Do the polar bodies serve to dispose of extra chromosome sets?

_____ 8. Meiosis begins before birth in _____ (males, females).

_____ a. Oogenesis is at _____ (stage) at birth.

_____ b. Oogenesis is at _____ (stage) at ovulation and fertilization.

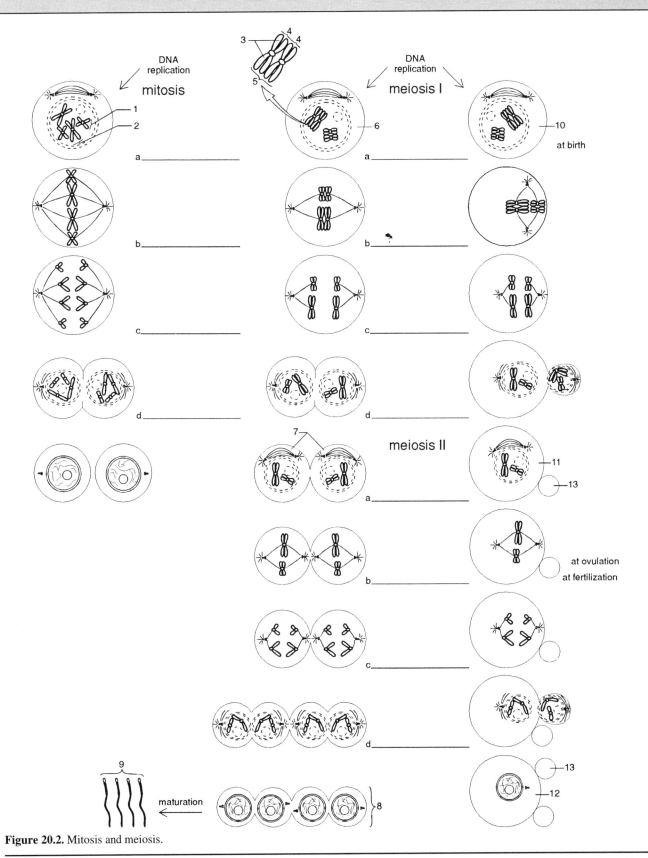

Figure 20.2. Mitosis and meiosis.

Answers to Exercise 20.2: 1. DNA replication; 1a. the same; 1b. half the; 1c. 2; 2. chromatids; 2a. identical; 2b. homologous; 3. randomly, in pairs; 4. 4, tetrad; 5a. chromatids, chromosomes; 5b. homologous chromosomes, chromosomes; 5c. chromatids, chromosomes; 6a. diploid (2n); 6b. diploid (2n); 6c. diploid (2n); 6d. haploid (n); 6e. haploid (n); 7. 4, 1; 7a. spermatogenesis; 7b. nutrients; 7c. yes; 8. females; 8a. prophase I; 8b. metaphase II.

Meiosis 311

C. Genes

Color the genes for the ABO blood group:
- ○ I^A (A antigen) allele
- ○ I^B (B antigen) allele
- ○ i (no A or B antigen) allele

Color the trait:
- ○ Type A red blood cell
- ○ Type B red blood cell
- ○ Type O red blood cell
- ○ Type AB red blood cell

Label:
1. chromosome
2. genotype
   a. homozygous
   b. heterozygous
3. phenotype

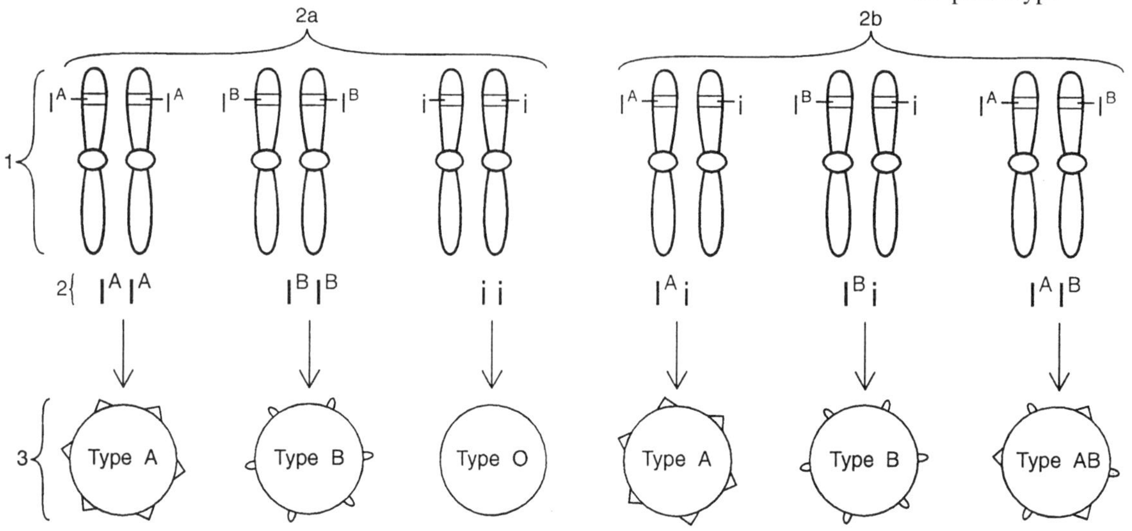

Figure 20.3. Genetics of blood types in six people.

Exercise 20.3:

1. Genes are located on _____ and code for a particular _____ .

2. Different forms of the same gene are called _____ .
   a. For the ABO blood group, the alleles are _____ , _____ , and _____ .
   b. Since homologous chromosomes exist in pairs, a person generally has _____ (number) copies of the same gene.

3. The genes present in an individual are called the _____ .
   a. When the same alleles are present, that person is said to be _____ . Examples of this are _____ , _____ , and _____ .
   b. When different alleles are present, that person is said to be _____ . Examples of this are _____ , _____ , and _____ .

4. The characteristics determined by the genes in an individual are called the _____ . For the blood group shown they are _____ , _____ , _____ , and _____ .

5. When an individual is heterozygous for a particular gene, only one of the genes may be expressed.
   a. The gene that is expressed is called the dominant gene. The dominant genes for blood type are _____ and _____ .
   b. The gene that is present but not expressed is called the recessive gene. The recessive gene for blood type is _____ .

6. When an individual is heterozygous for a particular gene and both genes are expressed, this is called codominance. The blood type that is an example of codominance is _____ .

Answers to Exercise 20.3: 1. chromosomes, trait; 2. alleles; 2a. I^A (A antigen), I^B (B antigen), i (no A or B antigen); 2b. 2; 3. genotype; 3a. homozygous, I^AI^A, I^BI^B, ii; 3b. heterozygous, I^Ai, I^Bi, I^AI^B; 4. phenotype, type A, type B, type O, type AB; 5a. I^A, I^B; 5b. i; 6. type AB.

D. Autosomal Inheritance

Color the genes:
- ○ H (Huntington's disease)
- ○ h (normal)
- ○ A (normal pigmentation)
- ○ a (albinism—lack of pigment)
- ○ P (normal for PKU)
- ○ p (PKU)

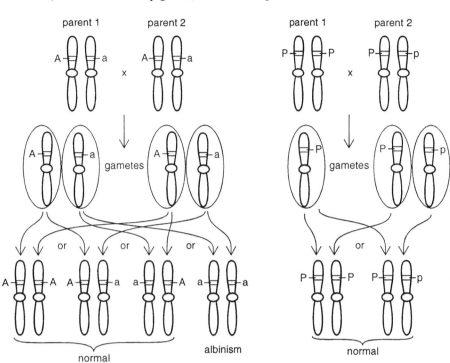

Figure 20.4. Variations in offspring from different matings.

Exercise 20.4:

1. Somatic (body) cells have _____ (number) pair(s) of chromosomes, 22 pairs of autosomes, and 1 pair of sex chromosomes.
    a. Do the autosomes exist as homologous pairs? (see figure)
    b. To produce offspring that have diploid cells, _____ (haploid, diploid) gametes must form.
    c. The chance that one of the homologous chromosomes will end up in a particular gamete is _____ .

2. How many different gametes form when
    a. someone is homozygous for a particular gene?
    b. someone is heterozygous for a particular gene?

3. Use the matings shown to complete the chart:

Mating	Genotypes (and Ratio) of Offspring	Phenotypes (and Ratio) of Offspring
Hh × hh	a	b
Aa × Aa	c	d
PP × Pp	e	f

4. These are examples of _____ (complete dominance, codominance).
    a. The genes that are dominant are _____ , _____ , and _____ .
    b. Is the most common gene in the population necessarily dominant?
    c. Are harmful genes always recessive?

Answers to Exercise 20.4: 1. 23; 1a. yes; 1b. haploid; 1c. 1/2; 2a. 1; 2b. 2; 3a. 1/2 Hh, 1/2 hh; 3b. 1/2 Huntington's disease, 1/2 normal; 3c. 1/4 AA, 1/2 Aa, 1/4 aa; 3d. 3/4 normal pigment; 1/4 albino; 3e. 1/2 PP, 1/2 Pp; 3f. all normal; 4. complete dominance; 4a. H (Huntington's disease), A (normal pigmentation), P (normal for PKU); 4b. no; 4c. no.

E. Sex-Linked Inheritance

Color the genes:
- ○ H (normal clotting)
- ○ h (hemophiliac)
- ○ SRY (sex determining region)

Color and label:
1. ○ X chromosome
2. ○ Y chromosome

Label:
3. ova
4. sperm

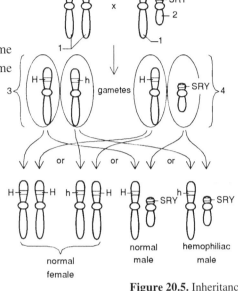
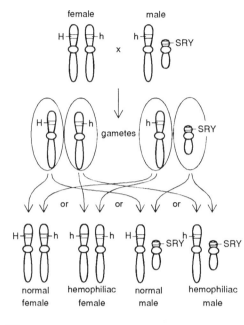

Figure 20.5. Inheritance of hemophilia.

Exercise 20.5:

1. The sex chromosomes are _____ and _____ .

2. Are the sex chromosomes homologous in females? males?

3. Females produce _____ (X, Y, X or Y) bearing ova, while males produce _____ (X, Y, X or Y) bearing sperm.

   a. Therefore, the sex of the offspring is determined by the _____ (male, female).

   b. The gene on the Y chromosome that triggers testis development is the _____ gene.

4. The gene for hemophilia is _____ (dominant, recessive).

   a. In the mating shown on the left, the proportion of offspring that are hemophiliacs is _____ .

   b. The albino offspring shown in figure 20.4 include both males and females. Is this so for this hemophiliac mating?

   c. If the male has one copy of an X-linked recessive gene, is that gene expressed? Why or why not?

   d. Males inherit their gene for hemophilia from their _____ (mother, father).

   e. Can homozygous recessive females for X-linked traits occur?

5. Hemophiliacs can be given clotting factor. Can we say that the environment can alter phenotypic expression?

6. Which gene shown is present only in males? Why?

Answer to Exercise 20.5: 1. X, Y; 2. yes, no; 3. X, X or Y; 3a. male; 3b. SRY; 4. recessive; 4a. 1/4; 4b. no (only male hemophiliacs in this mating); 4c. yes, no homologous chromosome containing dominant gene; 4d. mother; 4e. yes (see mating shown on right); 5. yes; 6. SRY gene, located on Y chromosome.

F. Genetic Variation

Color:
○ A (adenine)
○ T (thymine)
○ G (guanine)
○ C (cytosine)
○ gene A
○ gene a
○ gene B
○ gene b

Label:
1. original DNA
2. mutation
3. amino acid sequence
4. homologous chromosomes
5. tetrad
6. crossing over
7. gametes

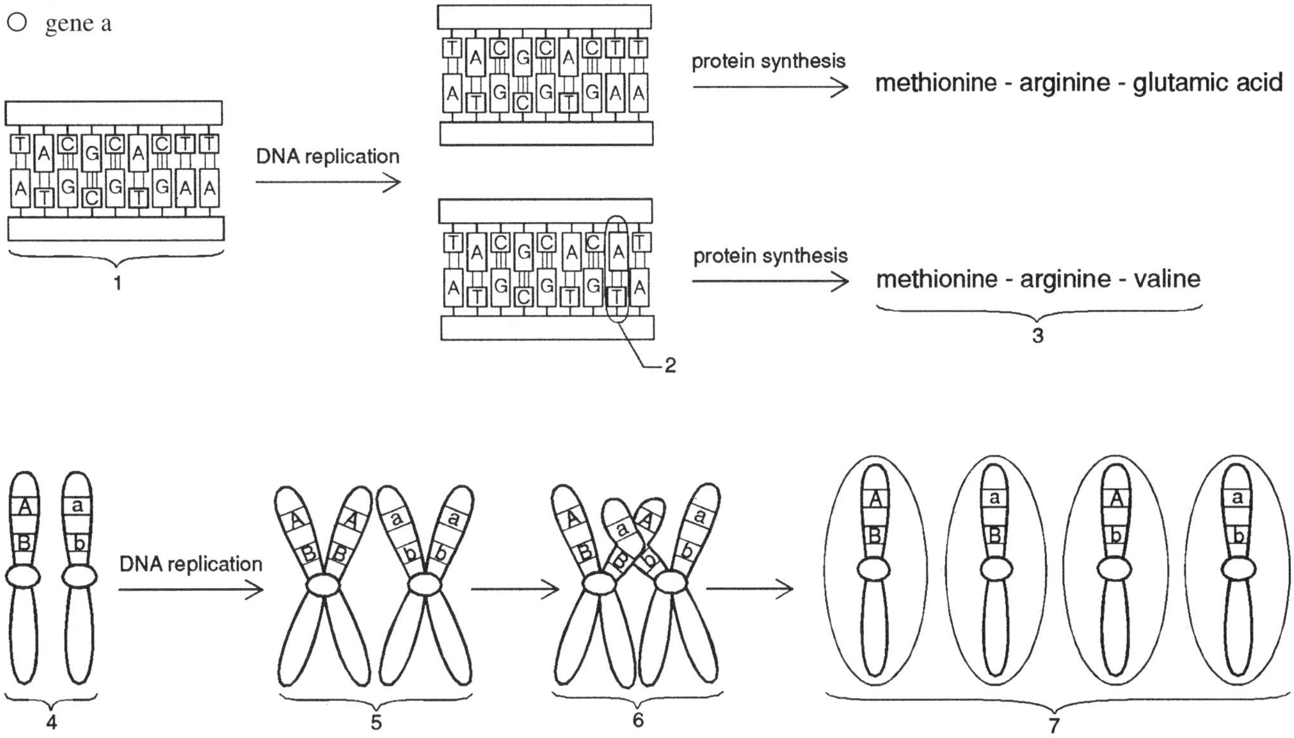

Figure 20.6. Possible sources of genetic variation.

Exercise 20.6:

1. A change in the sequence of DNA bases is called a _____ .

   a. The sequence of DNA bases determines the sequence of _____ in a protein. (see protein synthesis, p. 39)

   b. Could a mutation change the amino acid sequence?

   c. Depending upon the amino acid location, could protein function be altered?

   d. Could an accumulation of genetic changes account for variations in a person's proteins that might trigger immune responses during organ transplantation?

2. Homologous chromosomes may undergo _____ .

   a. Therefore, are genes on the same chromosome necessarily inherited together?

   b. Could this change the phenotypic expression of the offspring?

3. Could genetic variation account for variations in physiological functions, such as reactions to medication, susceptibility to disease, etc.?

4. Would you expect family members to be physiologically more similar than the general population?

Answers to Exercise 20.6. 1. mutation; 1a. amino acids; 1b. yes; 1c. yes; 1d. yes; 2. crossing over; 2a. no; 2b. yes; 3. yes; 4. yes.

Genetic Variation

G. Fertilization

Color and label the parts of the sperm:
1. head
   a. ○ acrosome (digestive enzymes)
   b. ○ sperm nucleus
2. midpiece
   a. ○ mitochondria
3. ○ tail (flagellum)

Color and label:
4. ○ corona radiata (follicular cells)
5. ○ zona pellucida
6. ○ ovum nucleus
7. ○ polar body

Label:
a. secondary oocyte
b. ovum
c. zygote

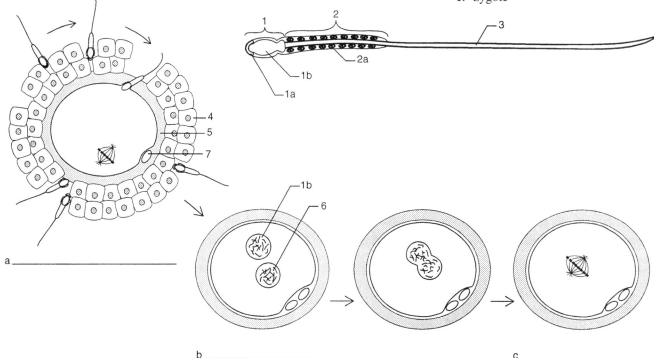

Figure 20.7. Steps in fertilization.

Exercise 20.7:

1. Fertilization occurs in the _____ . (see female anatomy, p. 300)
2. The sperm must penetrate the cellular _____ , the gelatinous _____ , and the cell membrane of the _____ .
   a. The sperm head is capped by a(n) _____ , which contains _____ .
   b. These enzymes digest a path to the membrane when present in high concentrations, i.e., when the number of sperm is _____ .
3. The membranes of the sperm and secondary oocyte fuse.
   a. The plasma membrane and zona pellucida change (zona reaction), so that no more _____ can enter the ovum.
   b. Sperm penetration triggers completion of _____ , forming a(n) _____ and a(n) _____ .
4. The sperm and newly formed ovum nuclei are each _____ (haploid, diploid).
5. Fusion of these nuclei completes fertilization. This forms the _____ , which is _____ (haploid, diploid).
6. If more than one ovum is released during ovulation and both are fertilized, _____ form. Are they genetically identical?

Answers to Exercise 20.7: 1. fallopian tubes; 2. corona radiata, zona pellucida, secondary oocyte; 2a. acrosome, digestive enzymes; 2b. high; 3a. sperm; 3b. meiosis, ovum, polar body; 4. haploid; 5. zygote, diploid; 6. twins, no.

H. Cleavage and Implantation

Color and label:
1. ○ cells (of embryo)
2. ○ zona pellucida
3. ○ trophoblast
4. ○ inner cell mass (embryoblast)
5. ○ endometrium
   a. uterine glands

Label:
   a. secondary oocyte
   b. mature ovum

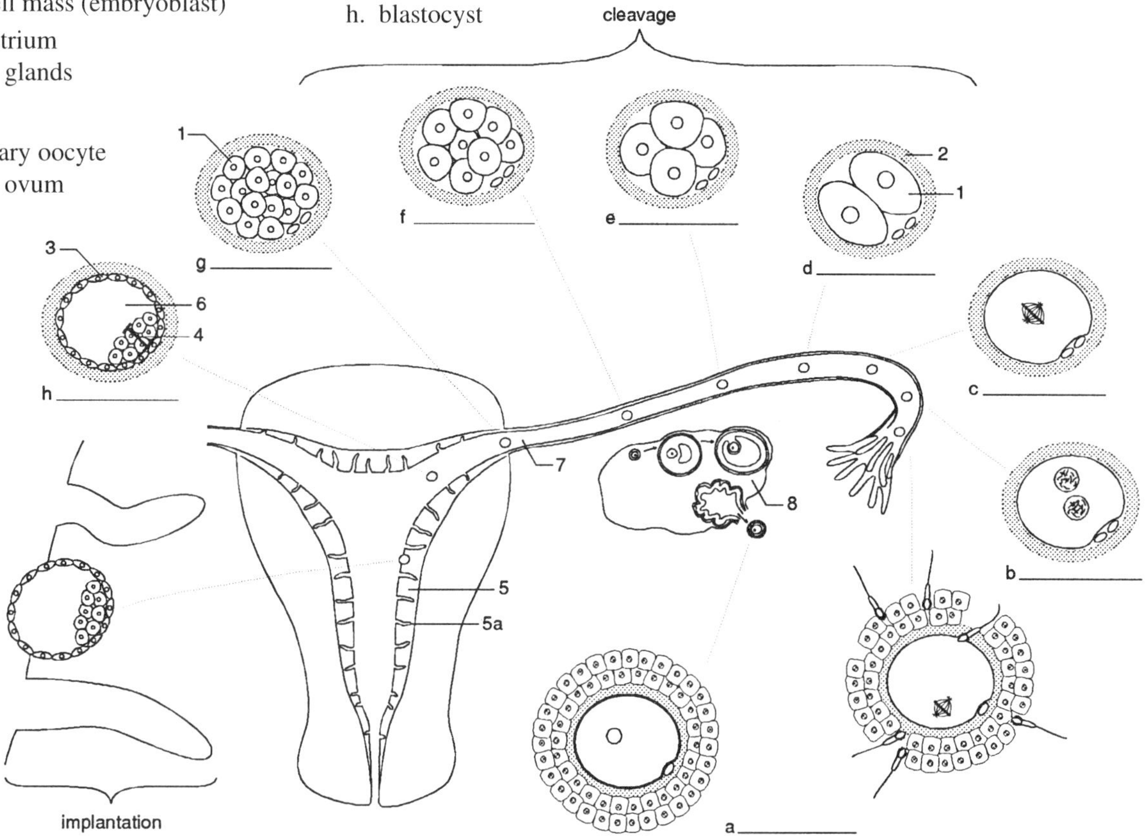

   c. zygote
   d. 2-cell stage
   e. 4-cell stage
   f. 8-cell stage
   g. morula
   h. blastocyst

6. blastocoele
7. fallopian tube
8. ovary

Figure 20.8. Early stages in embryonic development.

Exercise 20.8:

_____ 1. After fertilization, the cells divide by _____ (mitosis, meiosis).

_____   a. These cell divisions, known as cleavage, occur in the _____ .

_____   b. When the embryo reaches the uterus, it is a _____ .

_____   c. After 3–4 days, the blastocyst implants in the _____ of the uterus.

_____ 2. During cleavage, the size of the embryonic cells _____ (increases, remains the same, decreases).

_____   a. Nutrients and energy for development are supplied by the _____ .

_____   b. Upon implantation, nutrients are supplied by _____ .

_____ 3. If the embryonic cells separate into two masses, _____ form.

_____   a. These twins are _____ (monozygotic, dizygotic).

_____   b. Would these twins share membranes?

_____   c. If the separation is not complete, they are _____ .

Answers to Exercise 20.8: 1. mitosis; 1a. fallopian tubes; 1b. blastocyst; 1c. endometrium; 2. decreases; 2a. cytoplasm of ovum; 2b. uterine glands; 3. twins; 3a. monozygotic; 3b. no (if they form before blastocyst stage), they may if they form after implantation; 3c. Siamese twins.

Cleavage and Implantation   317

I. Germ Layers, Extraembryonic Membranes, Placentation

Color and label:
1. embryonic disc
   a. ○ ectoderm
   b. ○ mesoderm
   c. ○ endoderm

○ **Color the same. Label:**
2. trophoblast
3. chorion
   a. chorionic villi

Color (trace) and label:
4. ○ amnion
   a. amniotic cavity
5. ○ yolk sac

○ **Color the same. Label:**
6. allantois (body stalk)
7. umbilical cord

○ **Color red and label:**
8. umbilical vein
9. maternal artery
10. intervillous sinuses

○ **Color blue and label:**
11. umbilical arteries
12. maternal vein

Label:
13. inner cell mass
14. blastocoele
15. endometrium (decidua)
16. placenta

Exercise 20.9:

_____ 1. The inner cell mass at implantation becomes the _____ and _____ (germ layers).

_____ 2. A third embryonic germ layer that develops is the _____ .

3. Name the embryonic germ layer from which each of the following develops.

_____ a. nervous system

_____ b. muscle

_____ c. lining of gastrointestinal tract

_____ d. lining of respiratory system

_____ e. connective tissue

_____ f. epidermis

_____ 4. Organ formation occurs during the first trimester. Therefore, the trimester that is most sensitive to environmental agents (such as drugs, pollutants) is the _____ .

_____ 5. The trophoblast develops into the outermost membrane, called the _____ .

_____ a. The chorion grows fingerlike projections called _____ .

_____ b. These chorionic villi interdigitate with the maternal _____ .

_____ 6. Cells from the trophoblast attach to the edges of the ectoderm to form the _____ .

_____ a. This creates the _____ cavity.

_____ b. Amniotic fluid serves to _____ .

_____ 7. Cells arise from endoderm to form another cavity, the _____ .

_____ a. The yolk sac produces _____ in early development.

_____ b. It is the source of _____ that migrate into the gonads during development.

_____ 8. The outgrowth of the yolk sac is the vascular _____ .

_____ a. It develops into the _____ .

_____ b. The blood vessels of the umbilical cord supply the _____ .

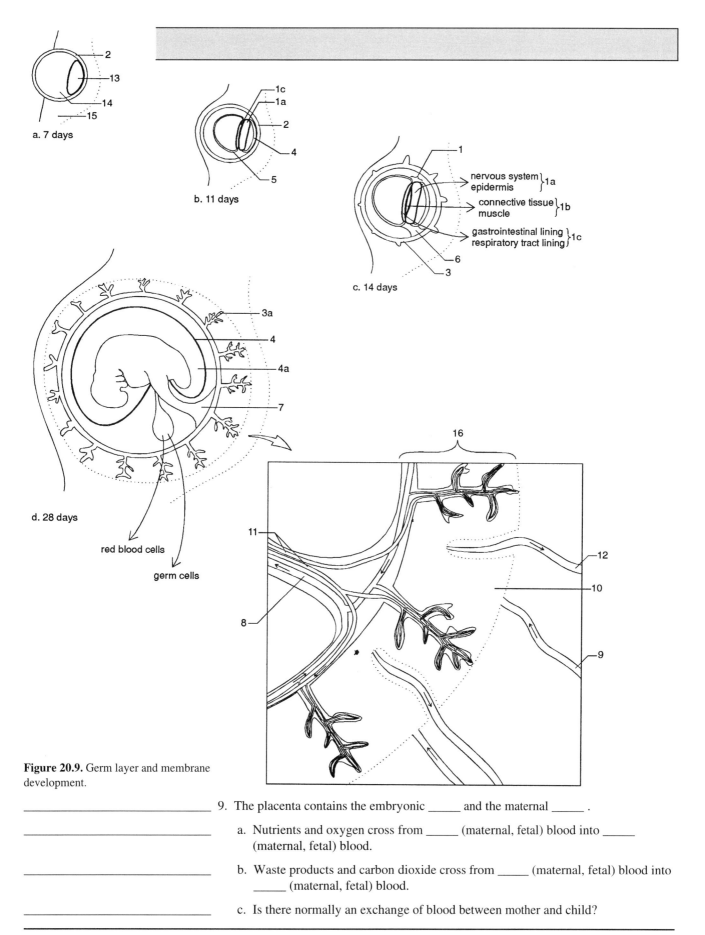

Figure 20.9. Germ layer and membrane development.

9. The placenta contains the embryonic _____ and the maternal _____ .

   a. Nutrients and oxygen cross from _____ (maternal, fetal) blood into _____ (maternal, fetal) blood.

   b. Waste products and carbon dioxide cross from _____ (maternal, fetal) blood into _____ (maternal, fetal) blood.

   c. Is there normally an exchange of blood between mother and child?

Answers to Exercise 20.9: 1. ectoderm, endoderm; 2. mesoderm; 3a. ectoderm; 3b. mesoderm; 3c. endoderm; 3d. endoderm; 3e. mesoderm; 3f. ectoderm; 4. first; 5. chorion; 5a. chorionic villi; 5b. intervillous sinuses; 6. amnion; 6a. amniotic; 6b. cushion the embryo (fetus); 7. yolk sac; 7a. red blood cells; 7b. germ cells; 8. allantois; 8a. umbilical cord; 8b. placenta; 9. chorionic villi, blood sinuses; 9a. maternal, fetal; 9b. fetal, maternal; 9c. no.

J. Hormonal Regulation of Pregnancy

Color:
- ○ HCG (human chorionic gonadotropic hormone)
- ○ estrogen
- ○ progesterone
- ○ prostaglandins
- ○ oxytocin
- ○ relaxin

Color and label:
1. ○ chorion
2. ○ corpus luteum
3. ○ placenta
4. ○ uterus (myometrium)
5. ○ pubic symphysis

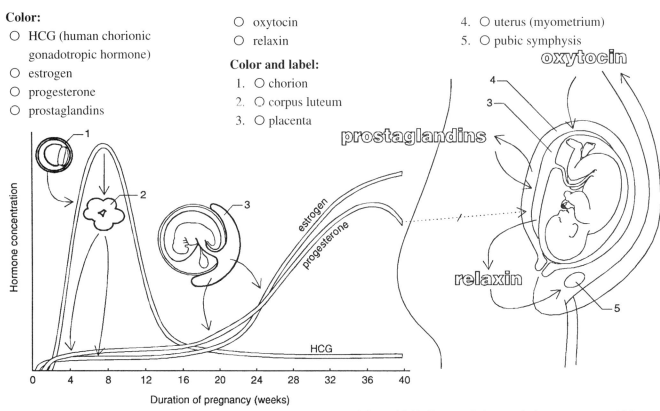

Figure 20.10. Hormonal changes during pregnancy and labor.

Exercise 20.10:

1. The chorion makes the hormone _____ .
   a. HCG stimulates the _____ to make _____ and _____ .
   b. The corpus luteum functions during the first _____ (number) months.

2. While the amount of estrogen and progesterone made by the corpus luteum decreases, the _____ takes over this function.

3. When estrogen and progesterone are high, FSH is _____ (high, low) and LH is _____ (high, low). (see female hormonal regulation, p. 302)

4. During pregnancy,
   a. new follicular growth does not occur because of _____ (low, high) FSH.
   b. new ovulations do not occur because of _____ (low, high) LH.
   c. menstruation does not occur because of _____ (low, high) estrogen and progesterone.

5. At the end of pregnancy,
   a. the levels of the hormone _____ decrease.
   b. the levels of the hormones _____ , _____ , and _____ increase.

6. Name the hormone(s) that do each of the following.
   a. increase uterine contractions (2 hormones)
   b. made in response to cervical pressure
   c. loosen ligaments in preparation for delivery
   d. prevents uterine contractions

Answers to Exercise 20.10: 1. HCG; 1a. corpus luteum, estrogen, progesterone; 1b. 3; 2. placenta; 3. low, low; 4a. low; 4b. low; 4c. high; 5a. progesterone; 5b. oxytocin, prostaglandins, relaxin; 6a. oxytocin, prostaglandins; 6b. oxytocin; 6c. relaxin; 6d. progesterone.

Index

abdominopelvic cavity, 2
abduction (movement), 97
ABO blood group, 176, 177, 312
accommodation, 139
acetabulum, 74, 75
acetyl CoA, 264
acetylcholine, 86, 87
    acetylcholine esterase, 86, 87
    acetylcholine receptor, 86, 87
acid, 15, 16
    acid-base balance, 287, 288, 292
acromion process, 70
acrosome, 316
actin, 81, 82
action potential, 108, 111, 112, 115
activation gate, 106, 108, 110
active site, 24, 26
active transport, 28
adduction (movement), 97
adenine, 22, 23, 38, 40, 315
adenohypophysis (anterior pituitary), 156–159
adenosine diphosphate (ADP), 23, 84, 85, 261
adenosine triphosphate (ATP), 23, 32, 83, 84, 261–263
adenylate cyclase, 153
adhering junction, 41
adipose tissue, 45
adrenal cortex, 162, 163
adrenal glands (suprarenal glands), 156, 157, 162, 163, 268, 269, 290, 291
adrenal medulla, 162–163
adrenocorticotropic hormone (ACTH), 156–159
afferent lymphatic vessels, 202, 203
afferent neuron, 117
agranulocytes, 172, 173
albinism, inheritance, 313
aldosterone, 162, 163, 282, 283, 290, 291
alimentary canal, 242, 243
allantois (body stalk), 318, 319
alveolus, 223, 240, 241
amino acid, 17–21
amino group, 17
aminopeptidases, 250, 251
amnion, 318, 319
amniotic cavity, 318, 319
ampulla (vas deferens), 296
ampulla (semicircular canals), 147
amylase, 248, 249
amylopectin, 247–249
amylose, 247
anaphase, 36, 310, 311
anatomical position, 97
angiotensin II, 276, 285, 290, 291
anterior chamber (eye), 138

anterior pituitary, 302, 303
anterior superior iliac spine, 74, 75
anti-A (A antibodies), 176, 177
anti-B (B antibodies), 176, 177
antibodies, 206, 207, 209, 211, 214–216
anticodon, 40
antidiuretic hormone (ADH), 156–159, 282, 283, 290, 291
antigen A (agglutinogen), 176, 177
antigen B (agglutinogen), 176, 177
antigen presenting cells (APC), 217, 220–222
antigens, 212, 214, 217
antrum, 301
anus, 294, 295
aorta, 195
    aortic chemoreceptors (CO_2, H^+, O_2), 236
    aortic semilunar valve, 180, 181
apneustic center, 237
apocrine (odoriferous) sweat gland, 49
aqueous chamber (anterior cavity), 138
aqueous humor, 138
arachidonic acid, 168
arachnoid, 119
arachnoid villus, 119
areola, 307
arrector pili, 49
arteries
    anterior tibial, 193
    aorta, 187, 198, 199, 246
    arcuate, 270, 272, 273
    axillary, 193
    brachial, 193
    brachiocephalic, 193
    celiac trunk, 196
    common carotid, 193, 195
    common iliac, 193, 196
    coronary, 193, 195
    descending aorta, 193
    dorsal aorta, 268, 269
    ductus arteriosus, 198, 199
    external carotid, 195
    external iliac, 193
    femoral, 193
    hepatic, 196, 246
    inferior mesenteric, 196
    interlobar, 270
    interlobular, 270, 272, 273
    internal carotid, 195
    internal iliac, 193
    left gastric, 196
    palmar arches, 193
    posterior tibial, 193
    pulmonary, 179, 198, 199
    pulmonary trunk, 198, 199

    radial, 193
    renal, 196, 268–270
    right coronary, 180, 181
    segmental, 270
    splenic, 196
    subclavian, 193, 195
    superior mesenteric, 196
    ulnar, 193
    umbilical, 198, 199, 318, 319
    vertebral, 195
arteriole, 188, 189
    afferent arteriole, 272, 273, 275, 276, 285
    efferent arteriole, 272, 273, 275, 276
articular capsule, 76
articular cartilage, 55, 62, 63, 76
articular discs, 76
articular facets for ribs, 70
ascending limb of loop of Henle, 278, 279
ascending tract, 130, 133
astrocytes, 104, 105
atlas, 70
atmospheric pressure, 230, 231
atom, 6, 9
    atomic mass, 5, 6
    atomic number, 5, 6
    atomic structure, 5
atrial natriuretic hormone (ANH), 276, 290, 291
atrial systole, 186
atrioventricular bundle, 184, 185
atrioventricular node, 184, 185
autonomic nervous system, 126, 127
    autonomic nerves, 256, 257
autosome, 313
axillary border, 70
axis, 70
axon, 48, 86–88, 103, 104, 114
    axon hillock, 114

bacteria, 206–209, 212
balance, part of brain controlling, 122, 123
basal body, 32
basal ganglia, 132
base, 15, 16
basement membrane, 42
basilar membrane, 144–146
basophil, 172, 173
behavior, control of, 122, 123, 128, 129
bicarbonate ion, 16, 292
bile acids, 252, 253, 258
bilirubin, 170, 171
binocular vision, 142
bipolar neurons, in retina, 140
bladder, 267–269

blastocoele, 317–319
blastocyst, 317
blister, 54
blood, 169
    blood plasma, 287, 288
    formed elements, 169
blood–brain barrier, 105
blood flow to the skin, control of, 128, 129
blood pressure, 4, 188
blood vessel histology, 187
blood vessels, 4, 56, 62, 63
B lymphocyte (B cell), 172, 173, 202, 203, 214, 215, 217
body wall, 2
bone, 45, 55, 62, 63
    bone development, 58, 59
    bone marrow, 64, 170, 171, 201, 215
    bone matrix, 64
bones
    acetabulum, 74, 75
    acromion process, 70
    anterior superior iliac spine, 74, 75
    articular facets for ribs, 70
    atlas, 70
    axillary border, 70
    axis, 70
    calcaneous, 74, 75
    capitulum (of humerus), 72
    carpals, 67, 72
    cervical vertebra(e), 70
    clavicle, 67, 70
    coccyx, 70
    condyloid process, 68
    coracoid process, 70
    coronoid fossa (of humerus), 72
    coronoid process, 68, 72
    coronoid process (of ulna), 72
    cribriform plate (of ethmoid bone), 68, 135
    deltoid tuberosity, 72
    ethmoid, 68
    external auditory meatus, 68
    femur, 67, 74, 75
    fibula, 67, 74, 75
    foramen magnum, 68
    frontal, 68
    frontal sinus, 68
    glenoid cavity, 70
    greater sciatic notch, 74, 75
    greater trochanter, 74, 75
    greater tubercle (of humerus), 72
    hip (coxal), 67
    humerus, 67, 72
    hyoid, 66
    iliac crest, 74, 75
    ilium, 74, 75
    incus (anvil), 143
    inferior ramus (of pubis), 74, 75
    intercondylar fossa (of femur), 74, 75
    intertubercular groove (of humerus), 72
    intervertebral discs, 70
    intervertebral foramen, 70
    ischial tuberosity, 74, 75
    ischium, 74, 75
    lacrimal, 68
    lamina, 70
    lateral condyle (of femur), 74, 75
    lateral epicondyle (of femur), 74, 75
    lateral epicondyle (of humerus), 72
    lateral malleolus, 74, 75
    lesser trochanter, 74, 75
    lesser tubercle (of humerus), 72
    linea aspera, 74, 75,
    lumbar vertebra(e), 70
    malleus (hammer), 143
    mandible, 68
    manubrium, 66
    mastoid process, 68
    maxilla, 68
    medial condyle (of femur), 74, **75**
    medial epicondyle (of femur), 74, 75
    medial epicondyle (of humerus), 72
    medial malleolus, 74, 75
    metacarpals, 67, 72
    metatarsals, 67, 74, 75
    nasal, 68
    neck of femur, 74, 75
    obturator foramen, 74, 75
    occipital, 68
    occipital condyles, 68
    odontoid process (dens), 70
    olecranon fossa, 72
    olecranon process, 72
    palatine, 68
    parietal, 68
    patella, 67, 74, 75
    pectoral girdle, 72
    pedicle, 70
    pelvic girdle, 74, 75
    phalanges, 67, 72, 74, 75
    posterior superior iliac spine, 74, 75
    pubic tubercle, 74, 75
    pubis, 74, 75
    radial tuberosity, 72
    radius, 67, 72
    ramus (of ischium), 74, 75
    ribs, 66, 71
    sacral foramina, 70
    sacrum, 70, 74, 75
    scapula, 67, 70
    sella turcica, 68
    sphenoid, 68
    spine (of ischium), 74, 75
    spine (of scapula), 70
    spinous process, 70
    stapes (stirrup), 143, 146
    sternum, 66, 71, 226
    styloid process, 68, 72
    superior ramus (of pubis), 74, 75
    symphysis pubis, 74, 75
    talus, 74, 75
    tarsals, 67, 74, 75
    temporal, 68
    thoracic vertebrae, 70
    tibia, 67, 74, 75
    tibial tuberosity, 74, 75
    transverse process, 70
    trochlea (of humerus), 72
    true pelvis, 75
    tubercle, 71
    ulna, 67, 72
    vertebral arch, 70
    vertebral border, 70
    vertebral foramen, 70
    vomer, 68
    xiphoid process, 66
    zygomatic, 68
    zygomatic process, 68
Bowman's capsule, 271–276, 278, 279
brain, 1, 2, 4, 118, 126
    brain stem, 122
    functions, 122
bronchiole, 223
brush border, 242, 243, 254
buffer, 16, 292
Bulb of Krause, 116

calcitonin, 160
calcium, 8
calcium binding factor, 254
calcium chloride ($CaCl_2$), 8
callus (in bone), 65
calmodulin, 153
canaliculus, 45, 56, 60, 61
canal of Schlemm, 138
capillaries, 179
capsular ligaments, 76
carbohydrate, 17–19, 27, 239, 247
carbon, 6
carbon dioxide transport, 234, 235
carbonic acid, 292
carboxyl group, 17
carboxypeptidase, 250, 251
cardiac conduction system, 184, 185
cardiac cycle, 186
cardiac muscle cells, 180, 181
cardiovascular center, 4
carotid chemoreceptors (CO_2, H^+, O_2), 236
carotid sinus, 124, 125
carrier, 27
cartilage, 59, 65
    elastic, 45
    fibrocartilage, 45
    hyaline, 45
cartilaginous joints, 76
cartilaginous matrix, 62, 63
catecholamines (epinephrine, norepinephrine), 151, 162, 163
C cells (parafollicular cells), 160
CD4 T cells, 172, 173, 218, 219, 221, 222
CD8 T cells, 172, 173, 218–222
cell, 1
    body, 48
cellular immunity, 214
cellulose, 248, 249
cementum, 240, 241
central canal (spinal cord), 119, 120
central chemoreceptors (CO_2, H^+), 236
central nervous system, 101, 102
central sulcus, 118
central tendon, diaphragm, 226, 227
centriole, 32, 36
centromere, 34, 36
centrosome, 32
cerebellum, 118, 122, 133
cerebral cortex, 118, 122, 133, 236
cerebrospinal fluid, 118, 119, 287, 288
cerebrum, 118
cervix, 300
chemical equilibrium, 13
chemical reactions, 12
chemoreceptors, 135
chemotaxis, 209
chief cell, 242, 243, 250
chlorine, 8

cholecystokinin (CCK), 258
cholesterol, 27, 265
choline, 86, 87
chondroblasts, 65
chondrocytes, 45, 62, 63
chordae tendineae, 180, 181
chorion, 306, 318–320
chorionic villi, 318, 319
choroid, 136, 140
    choroid plexus, 104, 105, 119
chromatid, 34, 310, 311
chromatin, 32, 36
chromosome, 34, 36, 38, 293, 310, 311
chylomicron, 252, 253
chyme, 248–250, 258
chymotrypsin, 250, 251
cilia, 32, 41, 42
ciliary body, 124–125, 136, 139
ciliary muscle, 136
ciliary process, 136, 138
ciliated columnar epithelium, 224, 225
circular muscle, 256–257
circulatory system, 179
cisterna chyli, 192
cleavage, 317
cleavage furrow, 36
clitoris, 209, 294–295, 300, 308
clotting, 174, 175
cochlea, 124, 125, 143–145
cochlear duct (scala media), 144–146
cochlear nerve, 143–145
codon, 40
coenzyme A, 264
collagen fibers, 44–46
collateral (axon), 87, 103
collecting tubule, 272, 273, 278, 279, 282, 283
colloid (thyroid), 160
colon, 255
common bile duct, 246
compact bone, 55, 56, 64, 65
competitive inhibition, 26
complement proteins, 206–208, 211, 212
complete tetanus, 89
compound, 5
conchae, 223
cones, in retina, 140
connective tissue, 44
    dense irregular, 44
    dense regular, 44
    elastic, 44
    loose (areolar), 44
    reticular, 44
connexons, 41
cornea, 136, 138, 139
corona radiata (follicular cells), 316
corpus albicans (degenerating corpus luteum), 301
corpus callosum, 118
corpus cavernosum, 296
corpus luteum, 301, 304–306, 320
corpus spongiosum, 296
cortical sinuses, 202, 203
corticotropin-releasing hormone (CRH), 158, 159
cortisol, 162, 163
costal cartilage, 66, 71
covalent bond, 7
Cowper's gland, 296
cranial cavity, 2
cranial nerves, 101, 124, 125

creatine, 84, 85
creatine phosphate, 84, 85
crista (ampullaris), 147
cross-bridge (muscle), 84, 85, 89
crossing over, 315
crown (tooth), 240, 241
crypt of Lieberkühn, 242, 243
cuboidal epithelium, 1
cupula, 147
cutaneous receptors, 116
cyclic AMP, 153, 168
cyclic GMP, 153
cystic duct, 246
cytokinesis, 310, 311
cytoplasm, 32, 39, 45
cytosine, 22, 38, 40, 315
cytotoxic (killer) T cell, 220–222

daughter cells, 36
dehydration, 290, 291
delayed type hypersensitivity cell, 220, 221
dendrites, 48, 103, 104
dendritic cell, 217
dental pulp, 240, 241
dentin, 240, 241
deoxyribonucleic acid (DNA), 22, 32, 34, 36, 38, 39, 315
deoxyribose, 22
depression (movement), 97
dermal papilla, 50
dermis, 49, 50, 116
    papillary layer, 50
    reticular layer, 50
descending limb of loop of Henle, 278, 279
descending tract, 131, 133
desmosomes, 41, 47
developing follicles, 304, 305
development, 294, 295
diacylglycerol (DAG), 153
diapedesis, 209
diaphysis, 55, 62, 63
diarrhea, 255
diastole, 186
diencephalon, 122
diffusion, 28
digestion, 239
    of carbohydrates, 248, 249
    of lipids, 252, 253
    of proteins, 250, 251
digestive system, 239
    movements, 244, 245
dihydrotestosterone (DHT), 294, 295
diploe, 55
disaccharide, 18, 19
distal convoluted tubule, 271–273, 278, 279, 282–285
dizygotic twins, 317
dorsal horn (spinal cord), 120
dorsal root (spinal cord), 120
dorsal root ganglion (spinal cord), 120
dorsiflexion, 97
down regulation, 154, 155
duodenum, 246, 248, 249, 258
dura mater, 119, 124, 125
dynamic equilibrium, 147

ear, 143–149
ear ossicles, 143–145

eating, drinking, part of brain controlling, 122, 123
eccrine sweat gland, 49
ectoderm, 318, 319
efferent lymphatic vessels, 202, 203
ejaculation, 296, 308
ejaculatory duct, 296
elastase, 250, 251
elastic cartilage, 46
elastic connective tissue, 224, 225
elastic fibers, 44–46
electrolytes, 287, 288
electron, 5–9
electron shell, 5
eleidin, 51
element, 5, 6
elevation (movement), 97
embryo, 59, 300, 306
embryonic development, 317
embryonic disc, 318, 319
emotions, part of brain controlling, 122, 123
enamel, 240, 241
endochondral bone, 59
    endochondral bone formation, 62–65
endocrine glands, 151
endoderm, 318, 319
endolymph (semicircular canals), 144, 145, 147–149
endometrium (decidua), 300, 304–306, 317–319
endoplasmic reticulum, 32
endosteum, 55, 65
energy of activation, 12, 25
enteric nervous system, 256, 257
enzyme, 24, 26, 27
    inhibitor, 26
eosinophil, 172, 173
ependymal cell, 1, 104, 105
epidermal pegs, 50
epidermis, 49, 50, 116
    layers, 50
epididymis, 296
epiglottis, 223, 244, 245
epinephrine, 151
epiphyseal cartilage, 59, 62, 63, 76
epiphyseal line, 55, 59, 62, 63
epiphysis, 55, 62, 63
epithelial cell, 242, 243
epithelial tissue, 42
    glandular, 42
    pseudostratified, 42
    simple columnar, 42
    simple cuboidal, 42
    simple squamous, 42
    stratified columnar, 42
    stratified cuboidal, 42
    stratified squamous, 42
    transitional, 42
epitope (antigen), 217
eponychium (cuticle), 53
erythrocytes, 46
erythropoietin, 170, 171
esophagus, 223, 239, 244, 245
essential amino acids, 266
estrogen, 301–307, 320
eustachian tube, 143
eversion (movement), 97
excitatory postsynaptic potential (EPSP), 112, 114
exocrine cells, 164, 165
exocrine glands, 42

exocytosis, 28, 210
expiration, 227
expiratory center, 237
expiratory reserve volume, 228, 229
extension (movement), 97
external auditory meatus, 68, 143
exteroceptors, 116
extraembryonic membranes, 318, 319
eye, 136–142

facilitated diffusion, 28
fallopian tubes, 300, 306, 309, 317
fats, 10
fatty acid, 17–19, 27, 264
    free fatty acids, 265
fear, part of brain controlling, 122,123
fenestration (lymph capillary), 189
fertilization, 293, 300, 316
fetal circulation, 198, 199
fetus, 293
fever, 210
fibrin, 174, 175, 209
fibrinogen, 206–209
fibroblasts, 44, 46
fibrocartilage, 46, 76
fibrous joints, 76
fibula, 67, 74, 75
fimbriae, 300
flagellum (a), 32
flexion (movement), 97
flexor withdrawal reflex, 121
follicle stimulating hormone (FSH), 156, 157, 298, 299, 302–305
    regulation, 158, 159
follicular cells (thyroid), 160, 301
foramen magnum, 68
foramen ovale, 198, 199
foreskin (prepuce), 296
formed elements in blood, 169
fourth ventricle of the brain, 119
fovea centralis, 136
free nerve endings, 116, 117
frontal (coronal) plane, 97
frontal sinus, 68
fructose, 247–249
functional residual capacity, 228, 229

galactose, 247–249
gallbladder, 246, 252, 253, 258
gamete(s), 293, 315
gametogenesis, 309
ganglion, 126
ganglion neurons, in retina, 140
gap junction, 41, 47
gas pressures, 230, 231
gastric glands, 242, 243, 250, 254
gastric lipase, 252, 253
gastric pits, 242, 243
gastrocolic reflex, 256, 257
gene, 38, 312
generator potential, 115
genotype, 312
germ layers, 318, 319
gingiva (gum), 240, 241
glans (head) of the penis, 294, 295
glenoid cavity, 70
glomerular filtration rate (GFR), 276, 290, 291

glomerulus, 272–276, 278, 279
glottis, 223
glucagon, 164, 165, 260
glucocorticoids, 162, 163
glucose, 247–249, 260, 280, 281
    metabolism, 260
    regulation, 164, 165
glycerol, 18, 19, 264
glycocalyx, 27
glycogen, 32, 80, 84, 85, 260
glycolipid, 27
glycolysis, 84, 85, 261
glycoprotein, 27, 41
goblet cell, 42, 224, 225, 242, 243
Golgi apparatus, 32
Golgi tendon organ, 117
gomphosis, 76
gonadotropin releasing hormone (GnRH), 158, 159, 298, 299, 302, 303
gonads, 293–295
"goose bumps", 52, 159
Graafian follicle, 301, 306
granulocytes, 172, 173
gray matter, 120, 121
growth hormone (GH), 156, 157
    regulation, 158, 159
growth hormone-inhibiting hormone (GHIH), 158
growth hormone-releasing hormone (GHRH), 158, 159
guanine, 22, 38, 40, 315
guanylate cyclase, 153
gubernaculum, 294, 295
gustatory receptors, 135
gyres (convolutions), 118

hair, 49
hair cells, 144, 145, 147
hair follicle, 52
hair root, 52
hair shaft, 52
hard palate, 244, 245
Haversian canal, 45, 56, 64
Haversian system (osteon), 56, 64
hearing, 144, 145
heart, 2, 4, 290, 291
    structure, 179
helper T cell, 215, 217
heme, 170, 171
hemoglobin, 170, 171, 232–235
hemophilia, inheritance, 314
heparin, 206–208
hepatic ducts, 246
heterozygous, 312
high density lipoproteins (HDL), 265
hilus, kidney, 268, 269
histamines, 206, 207, 211
holocrine glands, 42
homeostasis, 4
homologous chromosomes, 310, 311, 315
homozygous, 312
hormone action, 151
hormones,
    mechanisms of action, 151, 152
    regulation of, 154, 155
human chorionic gonadotropic hormone (HCG), 306, 320
human leukocyte antigens, 217
human sexual response, 308

humoral immunity, 214
Huntington's disease, inheritance, 313
hyaline cartilage, 46, 62, 63
hydrogen, 7, 9, 10, 14
    ion, 14
    bond, 9, 20, 22, 34
hydrophilic attraction, 10
hydrophobic interactions, 10
hydroxide ion, 14
hyperextension (movement), 97
hypertonic solution, 30, 31
hypodermis, 49, 50, 116
hyponychium, 53
hypothalamus, 118, 122, 128, 129, 156–159, 166, 210, 290,
hypotonic solution, 30, 31

I^A (A antigen), 312
I^B (B antigen), 312
immature ovum, 301
immunity, 201, 209, 214
immunoglobulins (Ig), type of, 216
immunological surveillance, 212
implantation, 300, 317
inactivation gate, 106, 108, 110
inclusions, 32
incomplete tetanus, 89
inflammation, 206–209, 212
infundibulum, 156–159
inguinal canal, 294–296
inhibin, 298, 299
inhibitory postsynaptic potential (IPSP), 112, 114
inner cell mass (embryoblast), 317–319
inner ear, 143
inorganic chemical, 1
inositol triphosphate, 153
inspiration, 227
inspiratory center, 237
inspiratory reserve volume, 228, 229
insulin, 164, 165, 260
integral protein, 27
integument, 49
intellectual processes, part of brain controlling, 122, 123
intercalated discs, 47
interferon, 213
interleukins, 221, 222
    interleukin-1, 210
intermediate filaments, 32
interoceptors, 117
interosseous membrane, 76
interphase, 36
interstitial cell stimulating hormone (ICSH), 298, 299
interstitial fluid, 287, 288
    balance, 189
intervertebral discs, 70, 76
intervertebral foramen, 70
intervillous sinuses, 318, 319
intestines, 289
intracellular fluid, 287, 288
intramembranous bone, 59, 60, 64, 65
intrinsic factor, 254
inversion (movement), 97
ion, 8
ion channels, 112
ionic bond, 8
iris, 124, 125, 136, 138

irritant and stretch receptors (airways and alveoli), 236
Islets of Langerhans, 164, 165
isometric muscle contraction, 89
isotonic muscle contraction, 89
isotonic solution, 30, 31
isotope, 6

joint capsule, 117
joint kinesthetic receptor, 117
joint receptor, 133
juxtaglomerular apparatus, 285, 290, 291
juxtamedullary nephron, 271

keratin, 51
keratohyalin, 51
kidney, 267–285, 289
kinins, 206–208
Krebs cycle, 262, 263

labia major, 294, 295, 300
labia minor, 294, 295, 300
labioscrotal folds, 294, 295
lacrimal gland, 124, 125
lactase, 248, 249
lacteal, 242, 243, 252, 253
lactic acid, 84, 85, 261
lactose, 247–249
lacuna, 45, 56, 60, 61
lamellae, 45, 56
lamina, 70
lamina propria, 242, 243
large intestine, 239
larynx, 223, 244, 245
latent period, 89
lateral geniculate body (of thalamus), 142
lens, 136, 138, 139
leucocytes, 46
levers, 100
Leydig cells (interstitial cells), 297–299
limbic system, 122
linea aspera, 74, 75
lipase, 252, 253
lipid, 17–19, 32, 239
lipid transport, 265
lipoprotein, 265
liver, 246, 255, 258, 265
lobar bronchus, 223
longitudinal muscle, 256, 257
loop of Henle, 271–273
low density lipoproteins (LDL), 265
lungs, 2, 226, 289
lung volumes, 228, 229
lunula, 53
luteinizing hormone (LH), 156, 157, 302–305
    LH regulation, 158, 159
lymph, 202, 203, 287, 288
    capillary, 192
    flow, 192
    nodes, 192, 201
    nodule, 202, 203
    vessel, 189, 192, 287, 288
lymphatic system, 201
lymphatic tissue, 201
lymphocyte, 172, 173, 202, 203, 209
lymphokines, 221, 222
lysosome, 32, 210
lysozymes, 204, 205

macrophage cells, 44, 46, 172, 173, 202, 203, 209, 212
macula, 148, 149
macula densa, 285
macula lutea, 136
major calyces, 270
major histocompatibility complex (MHC), 214, 217–221
maltase, 248, 249
maltose, 247–249
mammary glands, 307
marrow (medullary) cavity, 55, 62, 63, 65
mast cells, 44, 46, 206–208, 211
matrix, 44, 45
    calcified, 56
    calcified cartilaginous, 62, 63
mature ovum, 317
medulla oblongata, 118, 122, 237
medullary sinuses, 202, 203
megakaryocyte, 173
meiosis, 310, 311
Meissner's corpuscle, 116
melanin, 32, 51
melanocytes, 156, 157
melanocyte stimulating hormone (MSH), 156, 157
membrane,
    cell (plasma), 27, 32, 41
    hyperpolarization, 110
    polarization, 109
    potentials, 106
    repolarization, 110
memory, part of brain controlling, 122, 123
memory cell, 221, 222
meninges, 119
Merkel's corpuscle, 116
merocrine glands, 42
mesenchymal cells, 44, 60, 61
mesenchyme (embryonic tissue), 44
mesentery, 256, 257
mesoderm, 318, 319
messenger RNA, 22, 38–40
metabolism, 259
    control of, 128, 129
metaphase, 36, 310, 311
micelle, 252, 253
microfilaments, 32, 41
microglia, 104, 105
microtubules, 32, 41
microvilli, 32, 41, 242, 243, 278, 279
midbrain, 118, 122, 142
middle ear, 143
milk "let down" (release), 307
mineralocorticoids, 162, 163
minor calyx (renal sinus), 270, 271
mitochondrion, 32, 80, 84, 85–87, 261–263
mitosis, 36, 310, 311
mitral valve, 179–181
molarity, 11
mole, 8
molecular weight, 8
monocyte, 172, 173, 209
monosaccharide, 18, 19
monozygotic twins, 317
morula, 317
motor (cerebral) cortex, 131, 132
motor effector, 121
motor neuron, 88, 103, 121
motor unit, 88
mouth, 239–241

mucosa, 242, 243, 256, 257
mucus, 240, 241
mucus cell, 242, 243
mucous membranes, 204, 205
müllerian duct, 294, 295
müllerian inhibiting substance (MIS), 294, 295
multiunit smooth muscle, 47
muscle,
    circular, 242–245
    cross-bridges, 82
    filaments, 81
    longitudinal, 242–245
    skeletal, 47
    skeletal muscle control, 131
    smooth, 47
    spindle, 117, 133
    twitch, 89
    types, 47
muscles
    adductor longus, 95
    adductor magnus, 95
    biceps brachi, 91
    biceps femoris (long head), 96
    biceps femoris (short head), 96
    brachialis, 91
    cardiac, 47
    circular, 242–245
    coracobrachialis, 91
    cremaster, 296
    diaphragm, 190, 226, 227, 237, 268, 269
    extensor carpi radialis brevis, 94
    extensor carpi radialis longus, 94
    extensor carpi ulnaris, 94
    extensor digitorum communis, 94
    extensor digitorum longus, 95
    extensor hallucis longus, 95
    external intercostal, 227, 237
    external oblique, 90
    flexor carpi radialis, 93
    flexor carpi ulnaris, 93
    flexor digitorum profundus, 93
    flexor digitorum superficialis, 93
    frontalis, 90
    gastrocnemius, 96
    gluteus maximus, 96
    gracilis, 95
    iliopsoas, 95
    iliotibial tract, 96
    infraspinatus, 92
    inguinal ligament, 95
    intercostal, 226
    internal intercostal, 227, 237
    internal oblique, 90
    latissimus dorsi, 92
    levator anguli oris, 90
    levator labii superioris, 90
    levator scapuli, 92
    masseter, 90
    orbicularis oculi, 90
    orbicularis oris, 90
    palmaris longus, 93
    pectineus, 95
    pectoralis major, 91
    pronator quadratus, 93
    pronator teres, 93
    rectus abdominis, 90
    rectus femoris, 95
    rhomboid major, 92
    rhomboid minor, 92

risorius, 90
sartorius, 95
semimembranosus, 96
semitendinosus, 96
soleus, 96
sternocleidomastoid, 90, 124, 125
supinator, 93
supraspinatus, 92
temporalis, 90
tensor fasciae latae, 95
teres major, 92
teres minor, 92
tibialis anterior, 95
transverse abdominis, 90
trapezius, 92, 124, 125
triceps brachii, 92
vastus lateralis, 95
vastus medialis, 95
muscularis mucosa, 242, 243
mutation, 315
myelin (sheath), 48, 103–105
    myelinated axon, 111
myenteric plexus, 256, 257
myocardium, 180, 181
myofibril, 80, 81
myometrium, 300
myoneural (neuromuscular) junction, 86, 87, 89
myosin (thick filaments), 81, 82, 84, 85

NAD, 261–263
nail, structure, 53
nasal bone, 68
nasal cavity, 223
nasopharynx, 135
natural killer (NK) cell, 172, 173, 212
negative feedback, 4, 154, 155
nephron, 270–272, 274, 290, 291
nerve cell, structure, 48
nervous system, 1
neuroglia, 103
neurohypophysis (posterior pituitary), 158, 159
neuromuscular junction, 86, 87, 89
neuron, 88, 103
neurotransmitter, 112
neutron, 5, 6
neutrophil, 172, 173, 209
nipple, 307
Nissl bodies, 103
nitrogen, 5
node of Ranvier, 48, 103
noncompetitive inhibition, 26
nonessential amino acids, 266
norepinephrine, 151
nuclear membrane, 32, 36, 39
nuclear pores, 32, 39
nucleic acids, 22
nucleolus, 32, 36
nucleotide, 34
nucleus (atomic), 5, 6
nucleus (cell), 32, 39, 45, 80
    of nerve cell, 48
nucleus in descending tract, 133
nutrient artery (in bone), 55

Oddi's sphincter, 246
olfactory nerve, 124, 135
    receptors, 124, 125

oligodendrocytes, 104, 105
oocyte, 300
oogenesis, 309
oogonium, 301, 309
opsonization, 209
optic chiasma, 118, 142
optic disk, 136
optic tract, 142
oral cavity, 223
organ, 1, 2
organelle, 1, 32
organic chemical, 1
organ of Corti, 144, 145
organs of Ruffini, 116
organ system, 1
osmoreceptors, 128, 129, 290, 291
osmosis, 28
ossification center, 59
osteoblast, 60–65
osteoclast, 62, 63, 64
osteocyte, 45, 56, 60–64
osteogenic (osteoprogenitor) cells, 60, 61
osteoid, 57
osteon, 56, 64
otolith crystals, 148, 149
otolith membrane, 148, 149
outer ear, 143
ova, 314
oval window, 144–146
ovarian cycle, 301
ovaries, 293–295, 300–306, 309, 317
ovulation, 300
ovum (pl. ova), 293, 306, 309–311, 316
oxidative phosphorylation, 262, 263
oxygen, 7, 9, 10, 14
oxygen dissociation curve, 232, 233
oxygen transport, 232, 233
oxyntic cells (parietal), 242–243, 250
oxytocin (OCT), 156–159, 307, 320

pacemaker cell, 182, 183
Pacinian corpuscles, 116
palate, 135, 223
pancreas, 164, 165, 246, 248–253, 255, 258
    alpha cells, 164, 165
    beta cells, 164, 165
pancreatic amylase, 248, 249
pancreatic duct, 246
pancreatic lipase, 252, 253
papillary duct, 271
papillary muscle, 180, 181, 184, 185
parasympathetic nervous system, 126, 256, 257
parathyroid glands, 161
parathyroid hormone (PTH, parathormone), 161
parietal cell (oxyntic), 242, 243, 250
parietal membrane, 2
parietal pericardium, 2, 180, 181
parietal peritoneum, 2, 268, 269
parietal pleura, 2, 226
parotid gland, 124, 125, 240, 241
partial pressures, 230, 231
pectoral girdle, 72
pelvic girdle, 74, 75
penis, 294–296, 308
    glans (head), 294, 295
pepsin, 250, 251
pepsinogen, 250, 251

peptide bond, 18–20, 40
perforin, 212, 221, 222
pericardial cavity, 2
perichondrium, 62, 63
perilymph, 144, 145
perimetrium, 300
periodontal membrane, 240, 241
periosteal bone collar, 62, 63
periosteal buds, 63
periosteum, 55, 56, 62, 63, 65
peripheral nervous system, 101
peripheral protein, 27
peristaltic reflex, 256, 257
peritoneal cavity, 268, 269
peritubular capillary, 272, 273, 278, 279
peroxisome, 32
Peyer's patches, 201–203
pH, 15, 16
phagocyte, 210, 211, 214
phagosome, 210
phallus (genital tubercle), 294, 295
pharynx, 223, 244, 245
phenotype, 312
phosphate, 23, 27
phospholipid, 27, 252, 253
phosphorus, 5
photoreceptors, 140, 141
pia mater, 119
pineal gland, 118, 166
pinna, 143
pinocytosis, 28
pituitary gland, 118, 156, 157, 298, 299, 307
    regulation, 158, 159
    stalk, 118
PKU, inheritance, 313
placenta, 198, 199, 318–320
    placentation, 318, 319
plantar flexion (movement), 97
plasma, 169
plasma cell, 214, 215
plasma membrane (cell membrane), 27, 32, 41
pleasure centers (hypothalamus), 308
pleurae, 223
    pleural cavity, 2
pneumotaxic center, 237
podocytes, 275, 276
polar body, 310, 311, 316
polypeptide chain, 20
polysaccharides, 247
pons, 118, 122, 237
portal veins (hypothalamo-hypophyseal), 158, 159
postcentral gyrus of cerebral cortex, 130
posterior chamber, 138
posterior pituitary (neurohypophysis), 156, 157
posterior superior iliac spine, 74, 75
postganglionic neurons, 126, 162, 163
postsynaptic neuron, 112, 114
potassium channels, 106, 110
    potassium gate, 106, 108, 110
phagocytosis, 208–210
preganglionic neurons, 126, 162, 163
pregnancy, 306
premotor cortex, 131
prepuce, 294, 295
pressure receptors, 4
presynaptic neuron, 114
primary bronchi, 223

primary follicle, 301
primary motor cortex (somatotopic map), 131
primary oocyte, 301, 309–311
primary ossification center, 62, 63
primary spermatocyte, 309–311
primary visual cortex (occipital lobe of cerebrum), 142
primordial follicle, 301, 309
progesterone, 301–307, 320
prohormone, 154, 155
prolactin (PRL), 156, 157, 307
    regulation, 158, 159
prolactin-inhibiting hormone (PIH), 158, 159, 307
prolactin-releasing hormone (PRH), 158, 159, 307
pronation (movement), 97
propagation of nerve impulse, 111
prophase, 36, 310, 311
proprioceptors, 117, 236
prostaglandins, 168, 320
prostate, 296
protein, 18–21, 239
    synthesis, 32, 39
proton, 5, 6
protraction (movement), 97
proximal convoluted tubule, 271–273, 275–279, 282–284
pseudostratified columnar epithelium, 224, 225
pubic symphysis, 76, 320
pulmonary artery, 195
pulmonary semilunar valve, 180, 181
pupil, 136
purines, 22
Purkinje fibers, 184, 185
pyrimidines, 22
pyrogens, 210
pyruvic acid, 84, 85, 261

rage, part of brain controlling, 122, 123
raphe, 294, 295
receptor(s), 28, 115–117, 236
receptor-mediated endocytosis, 28
receptor potential, 115
rectum, 268, 269, 296, 300
red blood cell, 169–171, 234, 235
red marrow, 55
regulation
    of heartbeat, 184, 185
    of respiration, 237
relaxin, 320
renal
    columns, 270
    cortex, 271
    filtrate, 278–282, 284
    filtration, 274–276, 278, 279
    pelvis, 270
    pyramid, 270, 271
    reabsorption, 274, 278–281
    secretion, 274, 284
    sinus, 270
renin, 285
reproduction, part of brain controlling, 122
reproductive system, 293
residual volume, 228, 229
respiratory
    centers, 236, 237
    membranes, 230, 231

    system, 223
    tract, 230, 231
reticular activating system, 122
reticular cells, 44
reticular fibers, 42, 44, 46
retina, 124, 125, 136, 138–140
retraction (movement), 97
Rh factor, 178
RhoGAM, 178
ribonucleic acid (RNA), 22
ribose, 22, 23
ribosomal RNA, 22, 32
ribosome, 32, 39, 40
rigor mortis, 83
RNA formation, 38
rods, in retina, 140
root canal, 240, 241
root hair plexuses, 116
rotation (movement), 97
rough endoplasmic reticulum, 32
round window, 144–146

saccule, 143, 148, 149
sacral nerves, 256, 257
sagittal plane, 97
salivary amylase, 240, 241, 248, 249
salivary gland, 240, 241, 248, 249, 255
sarcolemma, 47, 80, 87
sarcomere, 81
sarcoplasmic reticulum, 80, 82, 89
satellite cell, 103
saturated fatty acids, 18, 19
scala tympani, 144–146
scala vestibuli, 144–146
Schwann cell, 103
sclera, 136, 140
scrotum, 294–296
sebaceous gland, 49
secondary lymphatic tissue, 215, 220, 221
secondary oocyte, 301, 309–311, 316, 317
secondary ossification center, 62, 63
secondary spermatocyte, 309–311
second messenger, 153
secretin, 258
self antigen, 218, 219
self defense, part of brain controlling, 122, 123
semen, 296
semicircular canals, 124, 125, 143, 147
seminal vesicles, 296
seminiferous tubules, 297–299, 309
sensory (afferent) neuron, 103, 115, 116, 121
sensory receptor, 115–117, 121, 133
seromucus gland (cells), 224, 225
serosa, 242, 243, 256, 257
serous fluid, 2
serous membrane, 2
Sertoli cells, 298, 299
sex determining region (SRY), 314
sex linked inheritance, 314
sex program centers (hypothalamus), 308
sex steroids, 162, 163
sexual performance, part of brain controlling, 122, 123
Siamese twins, 317
single unit smooth muscle, 47
sinoatrial node, 184, 185
skin, 289
small intestine, 239, 242, 243, 255

smooth endoplasmic reticulum, 32
smooth muscle, 224, 225
sodium, 8
sodium channels, 106, 108, 110
sodium chloride, 8, 10
sodium-potassium pump, 108, 110
soft palate, 244, 245
solute, 11, 28, 30
solvent, 11, 28, 30
somatic nervous system, 126
    motor system, 102, 131, 132
    sensory system, 130
somatostatin, 158, 159
somatotopic map, 130
sound perception, 146
speech, part of brain controlling, 122, 123
sperm, 293, 296, 297, 300, 306, 309–311, 314, 316
spermatids, 310, 311
spermatocyte, 32, 297
spermatogenesis, 309
spermatogonium (a), 297, 309
sphere of hydration, 10
spinal
    cord, 2, 101, 118, 120, 121, 126, 130
    nerves, 101, 120
    reflex, 121
spindle fibers, 32, 36
spleen, 170, 171, 201–203
spongy (cancellous) bone, 55–56, 64, 65
squamous (type I) cell, 224, 225
static equilibrium, 148, 149
steroid, 17
    hormones, 152
sterol, 252, 253
stimulus, 108
stomach, 239, 242, 243, 250–253, 255, 258, 289
stratum
    basale, 304, 305
    corneum, 50
    functionale, 304, 305
    germinativum, 50
    granulosum, 50
    lucidum, 50
    spinosum (Malpighii), 50
subarachnoid space, 119
sublingual gland, 240, 241
submandibular gland, 240, 241
submucosa, 242, 243, 256, 257
submucosal plexus, 256, 257
substrate, 24, 26
subthreshold stimulus, 108, 114
sucrase, 248, 249
sucrose, 247–249
summation, 114
supination (movement), 97
supplementary motor cortex, 131
supporting cells, 135
suppressor cell, 220, 221
surface antigen, 209
suspensory ligaments (breasts), 136, 139, 307
sustentacular cells (Sertoli cells), 297
suture, 76
swallowing, 244, 245
sweat, 204, 205
sympathetic
    ganglion, 256, 257
    nerves, 256, 257
    nervous system, 126, 162, 163

symphysis, 76
    pubis, 74, 75
synapse, 103, 112
synaptic cleft, 86, 87, 112
synaptic vesicle, 86, 87, 112
synchondrosis, 76
syndesmosis, 76
synovial fluid, 76, 287, 288
synovial joints, 76
synovial membrane, 76
synthesis of triglycerides, 264

T (transverse) tubules, 80, 89
taste bud, 135
taste pore, 135
T_C (cytotoxic or killer cells), 172, 173
TCR, 220–222
T_d (mediate delayed type hypersensitivity cell), 221, 222
T_D (mediate delayed type hypersensitivity cell), 172, 173
tectorial membrane, 144, 145
telophase, 36, 310–311
tendon, 79, 117
terminal branches, 48
    terminal button of presynaptic neuron, 103, 112
testes, 293–297
testosterone, 294, 295, 297–299
tetrad, 310, 311, 315
thalamus, 118, 122, 130, 132, 133
thecal cells, 301
T_H helper cell, 220–222, 172, 173
thick filament, 82, 84, 85
thin filament, 82
third ventricle of the brain, 118, 119
thirst center, 290, 291
thoracic cavity, 226, 227
thoracic duct, 192
threshold potential, 182, 183
threshold stimulation, 108
thrombocytes, 46
thromboxane A_2 (TXA), 168
thymine, 22, 38, 315
thymocyte (thymus cell), 218, 219
thymus, 166, 172, 173, 201, 218, 219
thyroid cartilage of larynx, 160
thyroid gland, 156, 157, 160
    follicle, 160
thyroid hormones, 151, 152, 160
thyroid stimulating hormone (TS), 156, 159
thyrotropin-releasing hormone (TRH), 158, 159
tidal volume, 228, 229
tight junction, 41, 204, 205
timbre of sound, 146
tissue, 1
tissue macrophage, 208, 209
T lymphocytes, 166, 172, 173, 202, 203, 214, 218, 219
tongue, 135, 240, 241, 244, 245
tonicity, 30
tonofilaments, 41
tonsils, 201–203
tooth, 240, 241
trabecula, 56, 64

trachea, 223
transferrin, 254
transfer RNA, 22, 39, 40
transverse plane, 97
triacylglycerol, 18, 19
tricuspid valve, 179–181
triglyceride, 18, 19, 252, 253, 264, 265
trophoblast, 317–319
tropomyosin, 81, 82
troponin, 81–82, 89
true pelvis, 75
trypsin, 250, 251
T_S (suppressor cells), 172, 173, 221, 222
tumor specific antigen, 212
tunica adventitia, 187
tunica intima, 187
tunica media, 187
twins, 316
tympanic membrane, 143–145
type II cell, 224, 225

umbilical cord, 318, 319
unmyelinated axon, 111
unsaturated fatty acid, 18, 19
up regulation, 154, 155
uracil, 22, 38, 40
urea, 278–283
ureter, 267–270
urethra, 267–269, 294–296, 300, 308
urinary bladder, 296, 300
urine, 267, 274, 278, 279
urogenital folds, 294, 295
urogenital slit, 294, 295
uterine cycle, 304, 305
uterine glands, 317
uterus, 268, 269, 300, 304, 305, 320
    endometrium (decidua), 306, 317–319
    layers, 320
    phases, 304, 305
utricle, 143, 148, 149

vagina, 300, 308
vasa recta, 272, 273, 278, 279
vas deferens, 296
veins
    arcuate, 270, 272, 273
    basilic, 194
    brachial, 194
    carotid sinus, 195
    cephalic, 194
    communicating, 194
    ductus venosus, 198, 199
    external iliac, 194
    external jugular, 194
    femoral, 194
    great saphenous, 194
    hepatic, 197, 246
    hepatic portal, 198, 199, 246, 252–254
    hepatic portal system, 197
    inferior vena cava, 194, 195, 197–199, 246, 268, 269
    interlobular, 270, 272, 273
    intermediate antebrachial, 194
    internal iliac, 194

    internal jugular, 194, 195
    left subclavian, 192
    pulmonary, 179, 195, 198, 199
    renal, 268–270
    right subclavian, 192
    segmental, 270
    subclavian, 194, 195
    superior vena cava, 195
    tibial, 194
    umbilical, 198, 199, 318, 319
ventilation, 227
ventral horn (spinal cord), 120
ventral root (spinal cord), 120
ventricular septum, 180, 181
ventricular systole, 186
venule, 189
vertebrae, 226
vertebral arch, 70
vertebral border, 70
vertebral canal, 2
vertebral foramen, 70
very low density lipoproteins (VLDL), 265
vestibular appparatus, 133
vestibular membrane, 144, 145
vestibular nerve, 143, 147–149
vestibule, 143
vestibulocochlear nerve, 143
villus, 242, 243
virus, 213
visceral membrane, 2
visceral pericardium, 2, 180, 181
visceral peritoneum, 2
visceral pleura, 2, 226
visual fields, 142
visual pathway to brain, 142
vital capacity, 228, 229
vitamin B_{12}, 254
vitamin D, 254
vitreous chamber (posterior cavity), 138
vitreous humor, 138
Volkmann's canal, 56

water, 7, 9, 14, 18, 19
water balance, 290, 291
white blood cell, 169, 172, 173, 224, 225
    formation, 172, 173
white matter (spinal cord), 120
wolffian duct, 294, 295

X chromosome, 314

Y chromosome, 294, 295, 314
yellow marrow, 55
yolk sac, 318, 319

z line (muscle), 81, 82
zona fasciculata, 162, 163
zona glomerulosa, 162, 163
zona pellucida, 316, 317
zona reticularis, 162, 163
zygote, 293, 316, 317